Mathematics for Industry

Volume 20

Aims & Scope

The meaning of "Mathematics for Industry" (sometimes abbreviated as MI or MfI) is different from that of "Mathematics in Industry" (or of "Industrial Mathematics"). The latter is restrictive: it tends to be identified with the actual mathematics that specifically arises in the daily management and operation of manufacturing. The former, however, denotes a new research field in mathematics that may serve as a foundation for creating future technologies. This concept was born from the integration and reorganization of pure and applied mathematics in the present day into a fluid and versatile form capable of stimulating awareness of the importance of mathematics in industry, as well as responding to the needs of industrial technologies. The history of this integration and reorganization indicates that this basic idea will someday find increasing utility. Mathematics can be a key technology in modern society.

The series aims to promote this trend by (1) providing comprehensive content on applications of mathematics, especially to industry technologies via various types of scientific research, (2) introducing basic, useful, necessary and crucial knowledge for several applications through concrete subjects, and (3) introducing new research results and developments for applications of mathematics in the real world. These points may provide the basis for opening a new mathematics-oriented technological world and even new research fields of mathematics.

More information about this series at http://www.springer.com/series/13254

Yoshiharu Soeta · Yoichi Ando

Neurally Based Measurement and Evaluation of Environmental Noise

Springer

Yoshiharu Soeta
National Institute of Advanced Industrial Science and Technology
Osaka
Japan

Yoichi Ando
Kobe University
Kobe
Japan

ISSN 2198-350X ISSN 2198-3518 (electronic)
Mathematics for Industry
ISBN 978-4-431-55431-8 ISBN 978-4-431-55432-5 (eBook)
DOI 10.1007/978-4-431-55432-5

Library of Congress Control Number: 2015937752

Springer Tokyo Heidelberg New York Dordrecht London

Printed on acid-free paper

Springer Japan KK is part of Springer Science+Business Media (www.springer.com)

Preface

Environmental noise is a big problem for human beings. Initially, noise level was the biggest problem. It has caused many bad effects, such as headache, discomfort, hearing loss, sleep disturbance. Long-term environmental noises also have deep effects and are integrated in the brain and body without any conscious awareness of them, as is discussed in Chap. 9. Particularly, the effects of the developments of unborn babies and the specialization of the cerebral hemispheres in growing children are serious. Numerous new technologies, such as noise insulation and active noise cancelling, have been developed and have contributed to the reduction of noise levels. But people feel annoyed even when the noise level is quite low because of the qualitative aspects of noises. In addition, the method of measuring and evaluating noise is still premature because the model of the auditory-brain system has ended at the peripheral level. Therefore, the development of measuring and evaluating the quality of noise based on our brain function is becoming more important for human beings.

This book deals with the methods of measurement and evaluation of environmental noise based on an auditory neural and brain-oriented model. The model consists of the autocorrelation function (ACF) and the interaural cross-correlation function (IACF) mechanisms for signals arriving at the two ear entrances. This model was based on neural evidence. First, we focused on the human auditory system, which was investigated mainly through the human brain and psychological response, such as by electroencephalography (EEG), magnetoencephalography (MEG), loudness, and annoyance, because the features of the human auditory system have to be taken into account in evaluating the quality of noises. The results suggest that the human auditory system has ACF and IACF mechanisms and the factors extracted from those mechanisms are useful as cues for temporal and spatial sensation of sounds.

It is hoped that the survey presented here will encourage researchers, students, and engineers in a wide range of fields, such as the automotive industry, the aerospace industry, mechanical engineering, railways, electronics industries, soundscape, architecture, and acoustics.

This book largely serves as a record of the research carried out at the Ando Laboratory, Graduate School of Science and Technology, Kobe University, between 1969 and 2009, even after the authors' graduation or retirement; as well as the studies conducted at the Living Informatics Research Group, National Institute of Advanced Industrial Science and Technology (AIST) between 2002 and 2015. The authors thank Dr. Shin-ichi Sato, Dr. Hiroyuki Sakai, Dr. Ryota Shimokura, Dr. Kenji Fujii, Dr. Seiji Nakagawa, and Dr. Mitsuo Tonoike for their collaboration and help.

March 2015

Yoshiharu Soeta
Yoichi Ando

Contents

Abbreviations

ABR	Auditory brain stem response
AC	Auditory cortex
ACF	Autocorrelation function
AEF	Auditory evoked magnetic field
AEP	Auditory evoked potential
AGC	Automatic gain control
ANOVA	Analysis of variance
ASW	Apparent source width
BC	Bone chain
BIR	Binaural impulse response
BIRs	Binaural impulse responses
BM	Basilar membrane
BMM	Boring machine method
CCF	Cross-correlation function
CCM	Cut-and-cover method
CN	Cochlear nucleus
DL	Difference limen
EC	External canal
ECD	Equivalent current dipole
ED	Eardrum
EEG	Electroencephalography
EOR	Energy onset response
EPNL	Effective perceived noise level
FFR	Frequency following response
GME	Ear canal sound pressure to cochlear vestibule pressure gain
HC	Hair cell
HPL	Human placental lactogen
HRTF	Head-related transfer function
HVAC	Heating, ventilation, and air conditioning
IACC	Interaural cross-correlation coefficient
$IACC_{IR}$	IACC of impulse response

$IACC_{SR}$	IACC of other signal response through a hall
IACF	Interaural cross-correlation function
IC	Inferior colliculus
IRN	Iterated rippled noise
ITD	Interaural time delay
JND	Just-noticeable difference
LEV	Listener envelopment
LLN	Lateral lemniscus nucleus
LSO	Lateral superior olive
MEG	Magnetoencephalography
MGB	Medial denticulate body
MLS	Maximum length sequence
MNTB	Medial nucleus of the trapezoid body
MSO	Medial superior olive
NACF	Normalized autocorrelation function
NATM	New Austrian tunneling method
NC	Noise criterion
NCB	Balanced noise criterion
NI	Nonidentification
PLG	Electroplethysmography
PNC	Preferred noise criterion
PNL	Perceived noise level
PNLT	Tone-corrected perceived noise level
POR	Pitch onset response
PSDs	Platform screen doors
PSE	Point of subjective equality
RC	Room criteria
RMS	Root-mean-squares
RNC	Room noise criterion
SOC	Superior olivary complex
SV	Scale values
SVTF	Ear canal sound pressure to stapes footplate velocity transfer function
TNIS	Train noise in a station
WECPNL	Weighted equivalent continuous perceived noise level

Chapter 1
Introduction

Abstract First, background and issues of this book and related literatures are described. Second, our solution to the issues based on human auditory system is briefly explained.

Keywords Auditory system · Autocorrelation · Interaural cross-correlation

For the evaluation of environmental noise, a large number of noise indices have been proposed by many researchers. A-weighted equivalent continuous sound pressure level (SPL), L_{Aeq}, is probably the most widespread noise index. It accounts for the magnitude of a noise and the sensitivity at different frequencies. It is simple to measure and correlate well with many psychological response to noise (e.g., Namba and Kuwano 1984; Kuwano et al. 1989; Ayr et al. 2003).

Noise criterion (NC) curves were proposed for rating indoor noise, noise from air-conditioning equipment, and so on. The method consists of a set of criteria curves extending from 63 to 8000 Hz and a tangency rating procedure. The criteria curves define the limits of octave band spectra that must not be exceeded to meet occupant acceptance in certain spaces. NC curves apply to the total noise in a room, with normal office activities in progress and all mechanical systems operating (Beranek 1956). Preferred noise criterion (PNC) curves are modified versions of the original NC curves. They are lower in both the low and high frequencies than originally specified in the NC curves (Beranek 1971). Noise rating (NR) curves are based on similar assumptions. Balanced noise criterion (NCB) curves are a further improvement of NC and PNC curves (Beranek 1989). Room criteria (RC) curves were derived by Blazier (1981) for application to the acoustical design and rating of heating, ventilation, and air conditioning (HVAC) systems. The RC method has been revised to the RC Mark II method (Blazier 1995, 1997). This latter method is more complicated to use but provides more thorough information about the character of the noise. In order to reach a technical compromise between NCB and RC curves, a set of curves named room noise criterion (RNC) was proposed

Y. Soeta and Y. Ando, *Neurally Based Measurement and Evaluation of Environmental Noise*, Mathematics for Industry 20,
DOI 10.1007/978-4-431-55432-5_1

(Shomer 2000; Shomer and Bradley 2000). It can be used with a tangency method for determining an RNC rating of a room sound spectrum, but it allows to evaluate temporal variations in low frequency sound as well.

However, these indices are determined based on SPL, that is, quantitative aspects of a noise, and frequency characteristics. For evaluation of noise, qualitative aspects of a noise are important because people may feel annoyed due to the aspects of sound quality even when the SPL of a noise is only about 35 dBA (Kitamura et al. 2002). The aspects can be formulated by the factors extracted from the autocorrelation function (ACF) and the interaural cross-correlation function (IACF) of noises arriving at two ear entrances.

This book deals with indices extracted from the ACF and IACF for the evaluation of sound quality although basic psychoacoustic indices such as loudness, sharpness, roughness, and fluctuation strength have widely used as sound quality indices (Zwicker and Fastl 1999). One rationale is that the perception of most auditory quality is based on information embedded in the timing of spikes, that is, temporal correlation representations arise from spike timing patters in the auditory nerve (Yin et al. 1987; Ando et al. 1991; Cariani and Delgutte 1996a, b; Saberi et al. 1998). Another rationale is that ACF and IACF factors describe temporal primary sensations, such as pitch, loudness, and timbre, and spatial primary sensations, such as localization in the horizontal plane, apparent source width, and subjective diffuseness, respectively (Ando et al. 1999; Ando 2001; Ando and Cariani 2009).

In Chap. 2, main and basic idea of this book, that is, signal processing model based on human auditory system, is described. In Chap. 3, concept of correlation and definitions of temporal and spatial factors obtained from ACF and IACF are explained. Chapters 2 and 3 are fundamental frameworks of this book. In Chaps. 4 and 5, temporal and spatial sensations, such as loudness, pitch, and apparent source width, are formulated based on the ACF and IACF factors. In Chap. 6, some examples of noise measurements based on the ACF and IACF factors are explained. In Chap. 7, subjective annoyance to some noises is described by the ACF and IACF factors. In Chaps. 8 and 9, short-term and long-term effects of noise on human body are discussed. In Chap. 10, some applications of the model to sound design are exemplified.

References

Ando Y (2001) A theory of primary sensations and spatial sensations measuring environmental noise. J Sound Vib 241:3–18

Ando Y, Cariani P (2009) Auditory and visual sensations. Springer, New York

Ando Y, Yamamoto K, Nagamastu H, Kang SH (1991) Auditory brainstem response (ABR) in relation to the horizontal angle of sound incidence. Acoust Lett 15:57–64

Ando Y, Sato S, Sakai H (1999) Fundamental subjective attributes of sound fields based on the model of auditory brain system. In: Sendra JJ (ed) Computational acoustics in architecture. WIT Press, Southampton, pp 63–99

Ayr U, Cirillo E, Fato I, Martellotta F (2003) A new approach to assessing the performance of noise indices in buildings. Appl Acoust 64:129–145

Beranek LL (1956) Criteria for office quieting based on questionnaire rating studies. J Acoust Soc Am 28:833–852

Beranek LL (1971) Noise and vibration control McGraw-Hill, New York

Beranek LL (1989) Balanced noise-criterion (NCB) curves. J Acoust Soc Am 86:650–664

Blazier WE (1981) Revised noise criterion for application in the acoustical design and rating of HVAC systems. Noise Control Eng J 162:64–73

Blazier WE (1995) Sound quality consideration in rating noise from heating, ventilating and air-conditioning (HVAC) systems in buildings. Noise Control Eng J 43:53–63

Blazier WE (1997) RC Mark II: A refined procedure for rating the noise of heating, ventilating, and airconditioning (HVAC) systems in buildings. Noise Control Eng J 45:243–250

Cariani PA, Delgutte B (1996a) Neural correlates of the pitch of complex tones. I. Pitch and pitch salience. J Neurophysiol 76:1698–1716

Cariani PA, Delgutte B (1996b) Neural correlates of the pitch of complex tones. II. Pitch shift, pitch ambiguity, phase invariance, pitch circularity, rate pitch, and the dominance region for pitch. J Neurophysiol 76:1717–1734

Kitamura T, Sato S, Shimokura R, Ando Y (2002) Measurement of temporal and spatial factors of a flushing toilet noise in a downstairs bedroom. J Temporal Des Arch Environ 2:13–19

Kuwano S, Namba S, Hajime Miura H (1989) Advantages and disadvantages of A-weighted sound pressure level in relation to subjective impression of environmental noise. Noise Control Eng J 33:107–115

Namba S, Kuwano S (1984) Psychological study on Leq as a measure of loud-ness of various kinds of noises. J Acoust Soc Jpn (E) 5:135–148

Saberi K, Takahashi Y, Konishi M, Albeck Y, Arthur BJ, Farahbod H (1998) Effects of interaural decorrelation on neural and behavioral detection of spatial cues. Neuron 21:789–798

Shomer PD (2000) Proposed revisions to room noise criteria. Noise Control Eng J 48:85–96

Shomer PD, Bradley JS (2000) A test of proposed revisions to room noise criteria curves. Noise Control Eng J 48:124–129

Yin TC, Chan JCK, Carney LH (1987) Effects of interaural time delays of noise stimuli on low-frequency cells in the cat's inferior colliculus. III. Evidence for cross-correlation. J Neurophysiol 58:562–583

Zwicker E, Fastl H (1999) Psychoacoustics: facts and models. Springer, Berlin

Chapter 2
Signal Processing Model of Human Auditory System

Abstract To evaluate environmental noise, we need to use methods based on functioning of our auditory system. In this chapter, basics of human auditory system are described. First, the ear sensitivity of the human ear from a sound source to the auditory system consisting of the external canal, eardrum, bone chain with oval window, auditory nerve, cochlear nucleus, thalamus, brain stem, and cortex, and signal processing model for evaluation of environmental noise are described. Second, brain responses in relation to basic perception such as loudness and annoyance are described.

Keywords Auditory system · Autocorrelation model · Interaural cross-correlation model · Auditory brain stem response · Auditory evoked magnetic fields · Loudness · Annoyance

2.1 Human Hearing System

The main function of hearing system is to get information about the outside, which is carried by pressure variations in the air, that is, sound wave. Sound waves are generated by the movement or vibration of an object, that is, sound source. As the vibrating object moves out and in, the nearby air molecules create a slight increase and decrease in pressure, called condensation and rarefaction, respectively. From the pressure variations, we perceive what the sound source is and where it comes from.

We perceive a sound wave, which is a continual time series signal, by the ears. We also perceive three-dimensional acoustic space by the ears, mainly because the head-related transfer function (HRTF) between a point of a sound source and the two ear entrances has directional characteristics from the shapes of the head and the pinnae. The pinnae significantly modify the incoming sound, particularly at high frequencies, and this is important in our ability for sound localization (Gardner and Gardner 1973; Butler and Belundiuk 1977). HRTF includes the interaural time and interaural level differences.

Y. Soeta and Y. Ando, *Neurally Based Measurement and Evaluation of Environmental Noise*, Mathematics for Industry 20,
DOI 10.1007/978-4-431-55432-5_2

Figure 2.1 shows the example of amplitude of the HRTF, $H(\xi, \eta, \omega)$, as a parameter of angle of incidence ξ (Mehrgardt and Mellert 1977). The angles ξ and η indicates azimuth and elevation angle, respectively. The angle $\xi = 0°$ corresponds to the frontal direction and $\xi = 90°$ corresponds to the lateral direction toward the side of the ear being examined.

After a sound wave arrives nearby, it passes through the peripheral auditory system, the outer ear, middle ear, and inner ear. The outer ear is the external part of the auditory system, including the pinnae and the ear canal. Sound travels down the ear canal and causes the eardrum, or tympanic membrane, to vibrate. Because of the resonance of the outer ear, we are more sensitive to sound frequencies between 1000 and 6000 Hz. The transfer function of the ear canal, $E(\omega)$, is shown in Fig. 2.2 (Wiener and Ross 1946; Shaw 1974; Mehrgardt and Mellert 1977).

The middle ear is the air-filled space between the eardrum and the cochlea that contains the ossicles. The acoustic vibrations on the eardrum are transmitted through the middle ear by three small bones, malleus, incus, and stapes, to the oval window of the cochlea. The middle ear acts as an impedance-matching device or transformer that improves sound transmission and reduces the amount of reflected sound. This is accomplished mainly by the differences in effective areas of eardrum and the oval window and to a small extent by the lever action of the ossicles. Transmission of sound through the middle ear, $C(\omega)$, is most efficient at frequencies between 500 and 4000 Hz as shown in Fig. 2.3 (Puria et al. 1997; Aibara et al. 2001).

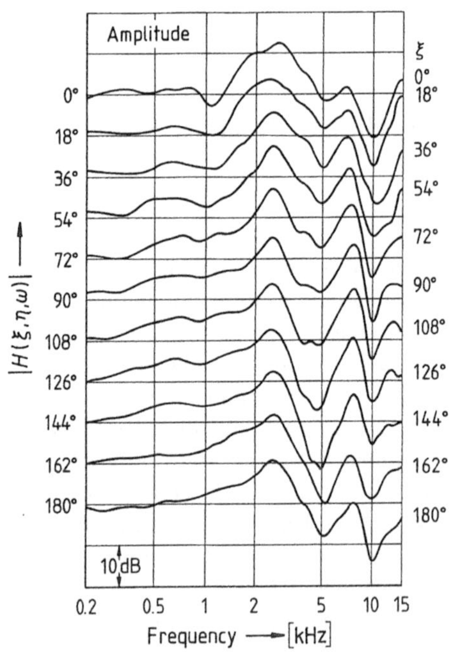

Fig. 2.1 Transfer functions (amplitude) from a free field to the ear canal entrance as a parameters of angle of incidence ξ (Mehrgardt and Mellert 1977)

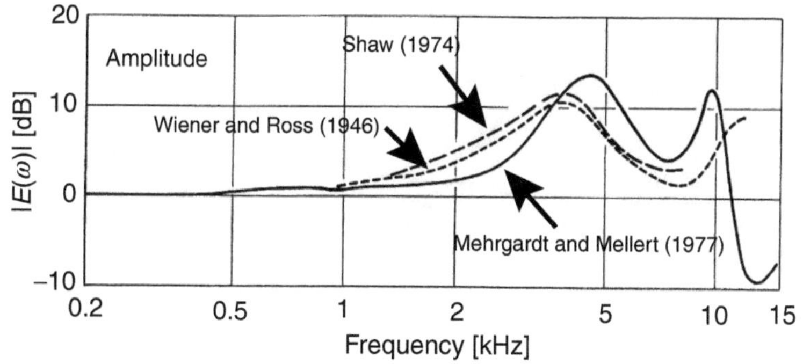

Fig. 2.2 Transfer functions (amplitude) of the ear canal (Mehrgardt and Mellert 1977)

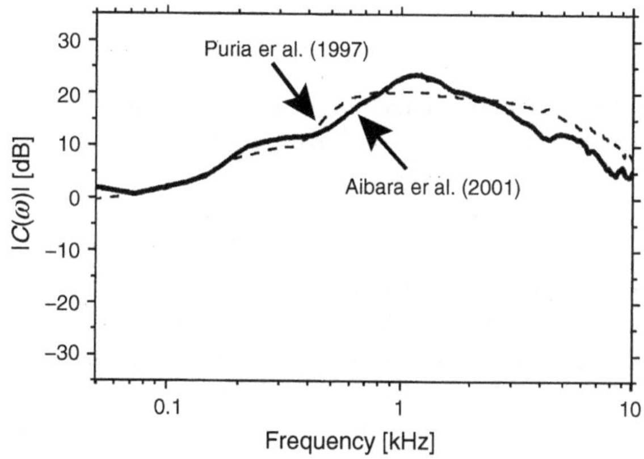

Fig. 2.3 Transfer functions (amplitude) of the middle ear (Aibara et al. 2001)

For the usual sound field, the transfer function between a sound source located in front of the listener and the cochlea, $S(\omega)$, may be represented by

$$S(\omega) = H(\xi, \eta, \omega)E(\omega)C(\omega). \tag{2.1}$$

The values are plotted in Fig. 2.4 (Ando 1998). The pattern of the transfer function agrees with the ear sensitivity for people with normal hearing estimated from equal-loudness-level contours at 40 phon (ISO 226:2003).

The inner ear is the part of the ear that is filled with fluid, including the cochlea and semicircular canals. Sound enters the cochlea through the oval window covered by a membrane. When the oval window moves due to the pressure from the stapes, Reissner's membrane and the basilar membrane are pushed down, and the round window moves out. It follows that vibration of the stapes leads to vibration of the basilar membrane. The basilar membrane separates out the frequency components of a sound. At the base of the cochlea, near the oval window, the basilar membrane is narrow and stiff and is sensitive to high frequencies. At the apex of the cochlea,

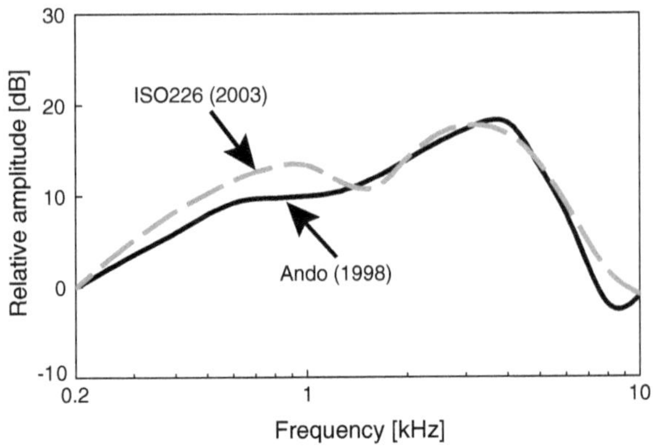

Fig. 2.4 Sensitivity of the human ear to a sound source in front of the listeners estimated from transformation characteristics between the sound source and the cochlea, $S(\omega) = H(\omega)E(\omega)C(\omega)$, (Ando 1998) and equal-loudness-level contours at 40 phon (ISO 226)

the other end of the membrane is wide and loose and is sensitive to low frequencies. The basilar membrane behaves as a band of overlapping band-pass filters, which is called auditory filters.

The mechanical vibrations of the basilar membrane are converted into electrical activity in the auditory nerve. This task is accomplished by the inner hair cells. Vibration of the basilar membrane causes a displacement of a stereochilia at the tips of the hair cells which lies within the organ of Corti on the basilar membrane, and this leads to action potentials (spikes) within the nerve fibers of the auditory nerve. Because each inner hair cell is attached to a specific place on the basilar membrane, each neuron in the auditory nerve carries information about the vibration of the basilar membrane at a single place in the cochlea. This means each neuron in the auditory nerve is sensitive to each characteristic frequency.

The auditory nerve carries the information about incoming sound from the cochlea to the cochlea nucleus. Cells of the cochlear nucleus project to higher nuclei through ventral and dorsal streams (Pickles 2008). Cells that project via a ventral stream primarily project to the superior olivery complex (SOC). The SOC is divided into three primary nuclei: the medial superior olive (MSO), lateral superior olive (LSO), and the medial nucleus of the trapezoid body (MNTB). The timing and intensities of the stimuli at the two ears are compared in the SOC, and the information is used for sound localization. Cells that project via a dorsal stream project mainly to the contralateral lateral lemniscus nuclei (LMN) and inferior colliculus (IC). They are involved in the complex analysis of a sound.

The ventral stream, mainly involved in sound localization, runs primarily to the SOC. The stream has two divisions. In the first division, the intensities of the stimuli at the two ears are compared in the LSO. In the second division, the timing of the stimuli in the two ears is compared in the MSO. The dorsal stream, mainly involved in sound identification, runs primarily to the IC of the opposite side, some fibers synapsing in the LMN on the way.

The IC is the main receiving station for the ascending pathways from lower stages of the brain stem. It forms the primary site of convergence of the sound identification and sound localization streams. It is suggested that this is a critical stage in transformation from responses dominated by the simple acoustic characteristics, to those which integrate acoustic properties in a way that begins to define an auditory object (Pickles 2008).

The medial geniculate body (MGB) is the specific thalamic auditory relay of the auditory system, receiving afferent from the IC, and projecting to the cerebral cortex. It also has heavy reciprocal connections back from the cortex, indicating that the cortex and MGB are grouped together as a functional unit.

The auditory cortex consists of core areas, surrounded by belt and parabelt areas. Previous studies suggest that the core area is necessary for the response to basic features of a sound, while the belt and parabelt areas are necessary for the response to complex features. It is suggested that the auditory cortex is necessary for the representation of auditory objects, that is, the assembly of information about all features of a sound (Pickles 2008).

To evaluate sound qualities, it is necessary to consider our auditory functioning, that is, how incoming sound is processed from the peripheral to central auditory system.

2.2 Neural Evidences of the Autocorrelation Model in the Auditory Pathways

A model for evaluations of environmental noise has been proposed based on human auditory system (Ando 2001). The model of the auditory-brain system includes the autocorrelation function (ACF) mechanism, which might exist in the auditory nerve, and the interaural cross-correlation function (IACF) mechanism, which might exist in the IC. Temporal and spatial sensations may be processed in the left and right hemisphere according to the temporal factors extracted from the ACF and the spatial factors extracted from the IACF, respectively, which is discussed in Chaps. 3 and 4. The overall subjective responses, for example, subjective preference and annoyance, may be processed in both hemispheres in relation to the temporal and spatial factors (Ando 2002).

The internal ACF may provide a representation for judging pitch salience. In temporal models of pitch perception, it is assumed that the pitch is extracted with autocorrelation (Licklider 1951; Bilsen and Ritsma 1969; Wightman 1973; Yost and Hill 1979; Meddis and Hewitt 1991; Patterson et al. 1995). Regarding the pitch salience or strength of pitch, psychophysical research has revealed that the strength of the pitch corresponds well to the peak amplitude of the ACF of the auditory signal, which represents the degree of temporal regularity of the sound (Wightman 1973; Yost et al. 1996; Yost 1996; Ando et al. 1999). One type of sound that allows for systematic manipulation of pitch salience is iterated rippled noise (IRN). IRN is produced by adding a delayed copy of a noise to the original noise and then repeating this delay-and-add process (Bilsen 1966; Yost 1996). A normalized ACF of IRN reveals a peak at the reciprocal of the delay, whose magnitude grows with increasing number of iterations reflecting the increasing periodicity.

Physiologically, recordings of responses to IRN stimuli from auditory nerve fibers (Fay et al. 1983; ten Kate and van Bekkum 1988) and cochlear nucleus neurons (Bilsen et al. 1975; Shofner 1991, 1999; Winter et al. 2001) show that the pitch of IRN is represented in the firing patterns of action potentials locked to either the temporal fine structure or envelope periodicity. That is, there is temporal regularity in the fine structure of the neural firing patterns, and it produces peaks in the autocorrelogram. These data suggest that the pitch of IRN stimuli is based on ACF mechanism. Indeed, the pooled interspike interval distributions of auditory nerve discharge patterns in response to complex sounds resemble the ACF of the stimulus waveform, and the magnitude of the ACF peak corresponds well with pitch salience (Cariani and Delgutte 1996a, b).

Electroencephalography (EEG), the measurement of electric potential differences on the scalp, is a widely applied method for investigating the functions of the human brain. Magnetoencephalography (MEG) is closely related to EEG. In both methods, the measured signals are generated by the same synchronized neuronal activity in the brain. The time resolution of EEG and MEG is in the millisecond range. Thus with EEG and MEG, it is possible to follow the rapid changes in cortical activity that reflect ongoing signal processing in the brain; the electrical events of single neurons typically last from one to several tens of milliseconds (Hämäläinen et al. 1993).

Several types of stimulus-evoked brain stem neural activity may be recorded using the EEG. Best known and most extensively studied EEG is the auditor brain stem response (ABR). Another type of brain stem neural activity is the frequency-following response (FFR). Unlike the ABR, the FFR reflects sustained neural activity (integrated over a population of neural elements) that is phase-locked to the individual cycles of the stimulus waveform and/or the envelope of periodic stimuli (Krishnan 2007).

The ABR and auditory evoked magnetic field (AEF) were recorded and analyzed to identify such ACF mechanism in human brain. FFRs were recorded from seven listeners in response to IRNs, which varied only in their degree of pitch salience (Krishnan et al. 2010). The FFR reflects sustained phase-locked activity in a population of neural elements within the rostral brain stem (e.g., Krishnan 2007; Chandrasekaran and Kraus 2010).

To create IRN stimuli with a dynamic fundamental frequency (F0) contour whose pitch varies as a function of time, a time-varying delay-and-add algorithm to a filtered Gaussian noise (10–3000 Hz) was applied (Denham 2005). The pitch increases in a curvilinear fashion from about 100 to 135 Hz over the 250-ms stimulus duration. By using a different number of iterations (n) in the IRN generating circuit, the F0 contour's pitch salience was varied. FFRs were recorded from each listener in response to monaural stimulation of the right ear at 80 dB sound pressure level (SPL) through a magnetically shielded insert earphone (Etymotic, ER-3A) (Krishnan et al. 2010). Neural responses were recorded differentially between a noninverting (+) electrode placed on the midline of the forehead at the hairline (Fz) and inverting electrodes (−) placed on the left (A1) and right (A2) mastoid, and the 7th cervical vertebra (C7). Another electrode placed on the mid-forehead (Fpz) served as the common ground. FFRs were recorded simultaneously from the three different electrode configurations and subsequently averaged for each stimulus condition to yield a response with a higher signal-to-noise ratio.

To analyze the robustness of encoding to stimuli differing in pitch salience, the neural pitch strength of each response waveform was quantified. From each FFR, the normalized ACF calculated over the entire duration of the response was computed in order to determine the dominant periodicities contained within the response. The height of the first peak in the ACF from time-delay zero was taken as the magnitude of neural pitch strength (Krishnan et al. 2005). In all cases, this peak fell at a delay of 10 ms, the fundamental pitch period of the input stimulus (Fig. 2.5). The growth in FFR pitch strength (derived from peak magnitude of the FFR ACF) with increasing iteration steps suggests an increase in the degree of neural phase-locking to the pitch relevant periodicity resulting from increased temporal regularity in the stimulus.

In addition, behavioral frequency difference limens (F0 DLs) were measured from each listener to obtain a perceptual estimate related to pitch salience. F0 DLs decreased with increasing stimulus periodicity revealing better pitch change detection for more salient stimuli. The strong correlation observed between the neural and behavioral measures supports the view that ACF-related pitch encoding at a subcortical and sensory level of processing plays an important role in shaping pitch perception.

MEG has been used to investigate how features of sound stimuli related to pitch are represented in the human auditory cortex. Focused on a spatial representation of pure tone in the auditory system according to their frequency, tonotopic organization of the human auditory cortex has been investigated (e.g., Elberling et al. 1982; Romani et al. 1982; Pantev et al. 1988, 1995). Focusing on the temporal structure of the sound, the periodicity pitch-related cortical response has been investigated (Pantev et al. 1989; Langner et al. 1997; Cansino et al. 2003; Fujioka et al. 2003; Seither-Preisler et al. 2003).

AEFs in relation to bandwidth variations of band-pass noise have been examined (Soeta et al. 2005a, 2006). The results indicate that the peak amplitude of N1m, which is found above the left and right temporal lobes around 100 ms after the stimulus onset, decreases with increasing bandwidth of the band-pass noise. The peak amplitude of the ACF increases with decreasing bandwidth of the auditory stimuli (Merthayasa et al. 1994; Sato et al. 2002; Soeta et al. 2004a). The pitch strength of band-pass noises was found to increase with decreasing bandwidth (Fastl and Stoll 1979). Therefore, it was assumed that sounds that have larger peak amplitude of the ACF or stronger pitch could lead to more cortical activity, which would cause an increase in the strength of the N1m response.

To evaluate responses related to pitch salience, which is characterized by the peak amplitude of the ACF of the sound, in auditory cortex, the AEFs elicited by

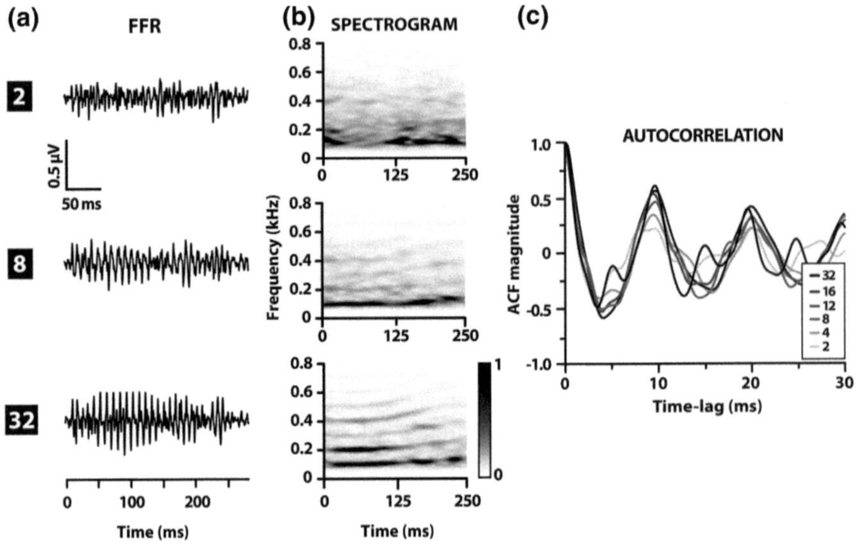

Fig. 2.5 FFR waveforms (a), spectrograms (b), and ACFs (c) as a function of iteration steps (n) computed from grand averaged brain stem responses. Spectrograms were computed by sliding a 50-ms analysis window by a 1 ms increment and computing the FFT in each time bin. Normalized magnitudes are indicated by the *gray-scale* gradient; *darker shades* indicate stronger encoding of pitch relevant harmonics (i.e., more pronounced phase-locked activity). Minimal periodicity is observed in FFRs at low iteration steps ($n = 2$; *top row*). By $n = 8$ iterations, FFR phase-locked activity captures periodicity related to the fundamental frequency (F0) and its harmonics (*middle row*). Robust encoding is even more pronounced at $n = 32$ iterations when the stimulus is maximally salient (*bottom row*). Temporal waveforms and ACFs derived from the FFRs reveal increasing periodicity in the neural response with increasing iteration steps, thus indicating more robust brain stem activity for salient pitch (Krishnan et al. 2010)

IRN with different iteration numbers were recorded (Soeta et al. 2005b). It was anticipated that the N1m amplitude would increase with an increase in the number of iterations of the IRN.

Ten normal hearing listeners (22–31 years, all right-handed) took part in the experiment. The IRN was produced by a delay-and-add algorithm applied to band-pass noise that was filtered using fourth-order Butterworth filters between 400 and 2200 Hz. The number of iterations of the delay-and-add process was set at 0, 1, 4, and 16. The delay was fixed at 1 ms, corresponding to a pitch of 1000 Hz. The stimulus duration used the experiment was 0.5 s, including rise and fall ramps of 10 ms. The auditory stimuli were delivered to the listeners through plastic tubes and inserted earpieces at a comfortable listening level adjusted separately for each listener. Figure 2.6 shows the temporal waveforms and the power spectra of some of the stimuli measured with an ear simulator that includes a microphone, a pre-amplifier, and an adaptor connected to the earpiece. Figure 2.7 shows the ACF of some of the stimuli measured with the ear simulator. The τ_1 of IRN corresponds to the delay. The ϕ_1 increases as the number of iterations increases.

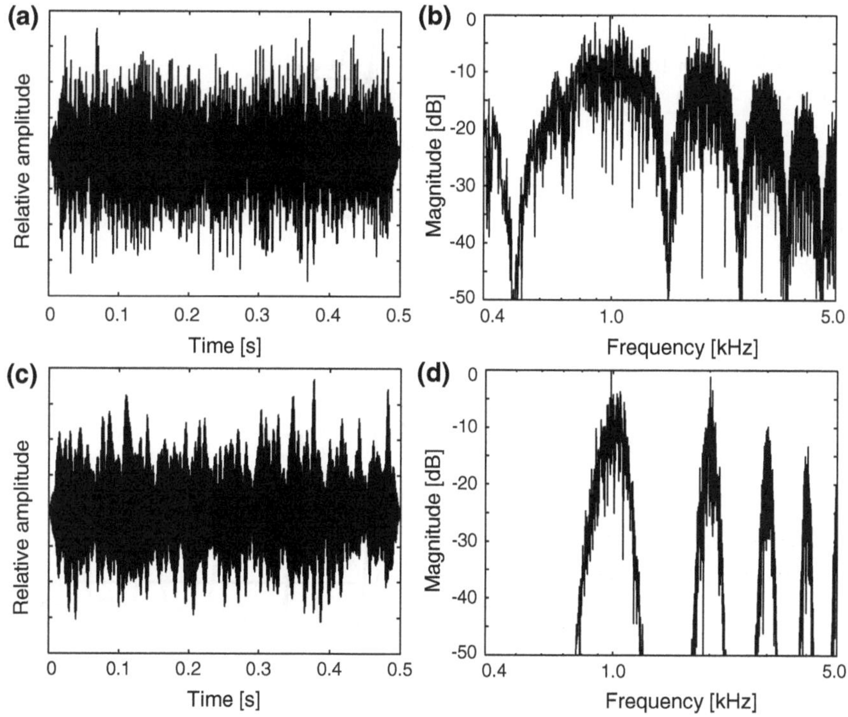

Fig. 2.6 Temporal waveforms and power spectra of the IRN with the number of the iterations (**a**, **b**) 1 and (**c**, **d**) 16 (Soeta et al. 2005b)

The AEFs were recorded using a 122 channel whole-head DC superconducting quantum interference device (DC-SQUID) magnetometer (Neuromag-122^{TM}; Neuromag Ltd., Helsinki, Finland) in a magnetically shielded room (Hämäläinen et al. 1993). The IRNs were presented in randomized order with a constant inter-stimulus interval of 1.5 s. Listeners were instructed to watch a self-selected silent movie and to ignore the stimuli. The magnetic data were sampled at 0.4 kHz after being band-pass filtered between 0.03 and 100 Hz and then averaged approximately 50 times. The averaged responses were digitally filtered between 1.0 and 30.0 Hz. The analysis time was 0.7 s from 0.2 s prior to the stimulus onset, with an average prestimulus period of 0.2 s serving as the baseline. The Neuromag-122^{TM} has two pick-up coils in each position that measure two tangential derivatives, $\partial Bz/\partial x$ and $\partial Bz/\partial y$, of the field component Bz. With such a coil configuration, the largest signal occurs just above a dipolar source where the gradient is steepest (Knuutila et al. 1993). To evaluate the amplitude and latency of the N1m peak, the root-mean-squares (RMS) of $\partial Bz/\partial x$ and $\partial Bz/\partial y$ were determined as the amplitude of the responses at each recording position. The N1m peak amplitude and latency were defined as the RMS peak and latency in the latency range from 70 to 130 ms over

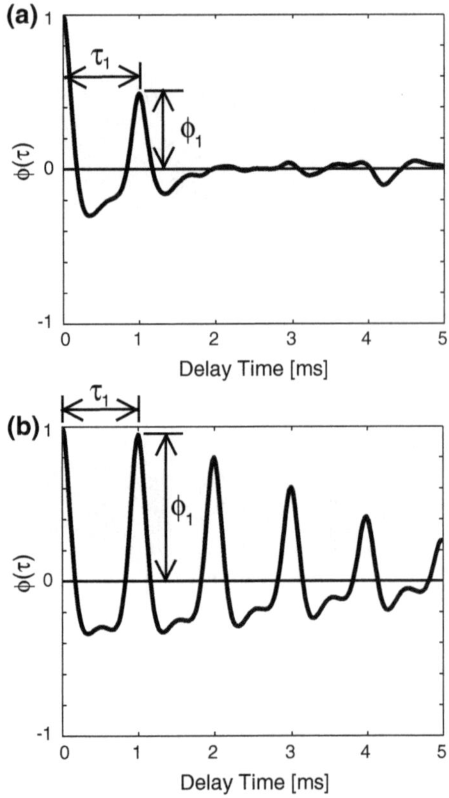

Fig. 2.7 ACFs of the IRN with the number of the iterations **a** 1 and **b** 16 (Soeta et al. 2005b)

the right and left hemispheres. In each listener, we employed the N1m peak latency and amplitude with a channel that showed the maximum amplitude placed at each hemisphere.

To estimate the location and strength of the underlying neural activity of the N1m wave, a single equivalent current dipole (ECD) was assumed as the source of the magnetic field of the N1m wave in a head-based coordinate system. The ECDs that best described the measured magnetic field at the N1m peak latencies were found by least squares fitting in a spherical volume conductor (Kaukoranta et al. 1986). A one-dipole model was used separately for the left and right hemispheres, with a subset of channels over each hemisphere. The origin of this coordinate system was set at the midpoint of the medio-lateral axis (*y*-axis) which joined the center points of the entrance to the acoustic meatuses of the left and right ears. The posterior–anterior axis (*x*-axis) was oriented from the origin to the nasion, while the inferior–superior axis (*z*-axis) was perpendicular to the *x*–*y* plane. All ECDs with goodness-of-fit values exceeding 80 % were used in further analyses.

Clear N1m responses were observed in both the right and left temporal regions in all listeners (Fig. 2.8). The N1m latencies were not systematically affected by the number of iterations of the IRN. Figure 2.9 depicts the mean N1m amplitude

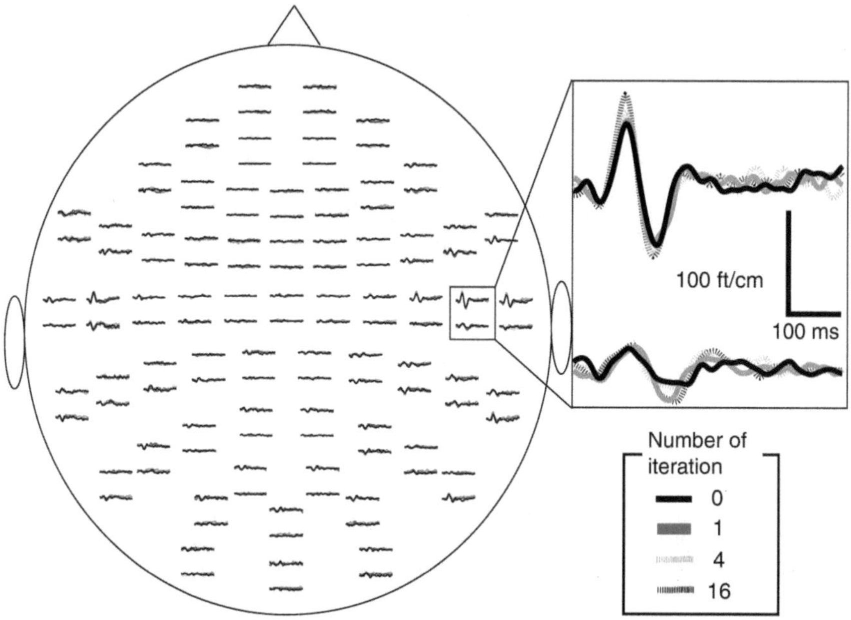

Fig. 2.8 Typical waveforms of auditory evoked magnetic fields from 122 channels in a listener (Soeta et al. 2005b)

(across ten listeners) as a function of the number of iterations. A greater number of iterations of the IRN produced a larger N1m amplitude, that is larger ϕ_1 of the stimulus produced a larger N1m response. A two-way analysis of variance (ANOVA) (number of iterations × hemisphere) revealed a significant main effect of the number of iterations on the N1m peak amplitude. This result is consistent with the previous study using band-pass noise with variation of bandwidth (Soeta et al. 2005a, 2006) and IRN with a delay of 2, 4, or 16 ms (Krumbholz et al. 2003; Soeta and Nakagawa 2008a). The amplitude of the AEF component in response to periodic stimuli was compared with simulated peripheral activity patterns of the auditory nerve (Seither-Preisler et al. 2003). The results showed that the amplitude of the N1m is correlated with the pitch strength estimated on the basis of auditory nerve activity. This result is consistent with the present results.

For the dipole strength, similar results to those for the N1m peak amplitude were obtained. A greater number of iterations of IRN produced a larger N1m ECD moment. A two-way ANOVA revealed a significant main effect of the number of iterations of IRN on the ECD moments. The ECD locations did not show any systematic variation across the listeners as a function of the number of iterations of IRN.

Figure 2.10 shows the relationship between ϕ_1 of the stimulus and ECD moment of the N1m response. To compare the previous result, the results of band-pass noise with center frequency of 1000 Hz were also included in Fig. 2.10 (Soeta et al. 2006). Psychophysical studies have indicated that first peak of the ACF, ϕ_1, could

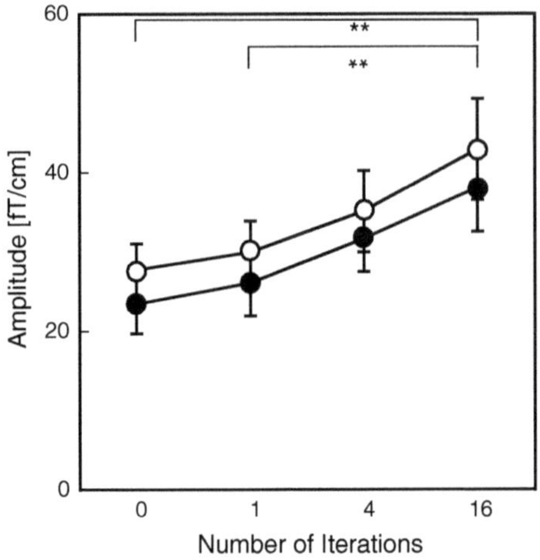

Fig. 2.9 Mean amplitude of the N1m (±SEMs) as a function of the number of iterations from the right (*filled circle*) and left (*open circle*) hemispheres. The *asterisks* indicate statistical significance ($*P < 0.05$, $**P < 0.01$; Post hoc Bonferroni test) (Soeta et al. 2005b)

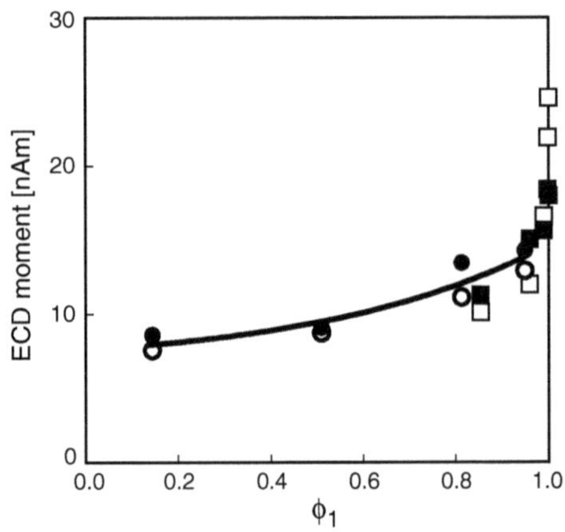

Fig. 2.10 Relationship between peak amplitude of the ACF, ϕ_1, and ECD moment of the N1m from the right (*filled circle*) and left (*open circle*) hemispheres. The curve is of the form $6.6 + 100.9*\phi_1$ (Soeta et al. 2005b). The previous ECD moment of the N1m in response to band-pass noise with center frequency of 1000 Hz from the right (*filled square*) and left (*open square*) hemispheres were also included for comparison (Soeta et al. 2006)

account for the pitch strength of the stimulus (Wightman 1973; Yost et al. 1996; Ando et al. 1999) and pitch strength of IRN was an exponential function of ϕ_1, namely of the form $a + 10b\phi_1$, where "a" and "b" are constants (Yost 1996). Note that, the ECD moment derived in the current study could also be described in the form $a + 10b\phi_1$, where $a = 6.6$ and b = 0.9, as shown in Fig. 2.10.

The pitch onset response (POR) was introduced to circumvent the energy onset response (EOR), such as N1m, and thereby isolate the response of those neural elements specifically in pitch processing (Krumbholz et al. 2003). The POR is a negative component of the AEF, which can be elicited by the transition from a noise to a tone even when there is no concurrent change in sound energy. The results showed that the amplitude of the POR is correlated with the pitch strength, that is to say, the number of iterations of the IRN, suggesting that the underlying generators are part of a pitch-processing network. This is consistent with N1m results (Soeta et al. 2005b; Soeta and Nakagawa 2008a). The results also show that the source of the POR lies somewhat anterior and inferior to that of the N1m. Then, whether the POR originates from the same generator as the EOR was examined (Seither-Preisler et al. 2004). The results suggest an interaction between the POR and the EOR, which may be based on common generators. Given this, the N1m responses derived in our study (Soeta et al. 2005b; Soeta and Nakagawa 2008a) are considered to include both the EOR and POR responses. However, the stimulus used in the study had the same SPL and the same stimulus onset. Consequently, the effect of the number of iterations of the IRN on N1m is contributed mainly by a pitch-processing network.

2.3 Neural Evidences of the Interaural Cross-Correlation Model in the Auditory Pathways

The internal binaural IACF may provide a representation for judging subjective preference and spatial sensations. The psychological responses to IACF factors have largely been obtained in humans, while the neurophysiological responses are confined to animal studies (Yin et al. 1987; Saberi et al. 1998; Palmer et al. 1999). The ABR and AEF were recorded and analyzed to identify such a mechanism in human brain.

To probe the neural correlates of horizontal sound direction (azimuth), the source signals $p(t)$ of trains of clicks (50-μs pulses) were presented every 100 ms for 200 s (2000 times) and the left and right ABRs were recorded through electrodes placed on the vertex and the left and right mastoids (Ando and Hosaka 1983; Ando 1985). Signals were supplied to loudspeakers positioned at various horizontal angles (0°–180°) with respect to the front of the listener, all on the listener's right-hand side. The distance between each loudspeaker and the center of the head was kept at 68 ± 1 cm. The speakers had a frequency response of ±3 dB for 100 Hz–10 kHz.

Typical examples of recorded ABR waveforms as a function of the horizontal angle of sound incidence are shown in Fig. 2.11. It can be appreciated that waves I–VI differ in peak amplitude and latency as the sound location changes its angle of incidence relative to the listener's head. Similar ABR waveforms were obtained from each of four listeners (males, 23 ± 2 years of age). Their ABRs were averaged together, and the mean amplitude of the ABR waveform peaks (waves I–VI) was computed as a function of the horizontal angle (Fig. 2.12a–f). Of particular interest is

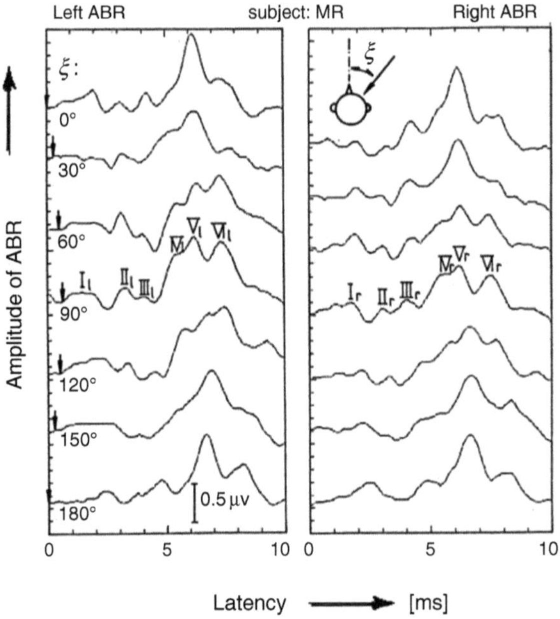

Fig. 2.11 Examples of ABR obtained between the vertex and left and right mastoids as a function of the latency less than 10 ms and a parameter of *horizontal* angle of sound incidence. The abscissa is the latency in the auditory pathways relative to the time when the single pulse arrives at the right ear entrance. *Arrows* indicate the time delay, which depends upon the sound source location on the right-hand side of the listener, and the null amplitude of ABR. The wave number is defined by the symbol from $I_{l,r}$ to $VI_{l,r}$, which is reflected by the activity at each nucleus (relay station) in the auditory systems. The suffix signifies the left and right auditory pathway (Ando et al. 1991)

that the average peak I amplitudes from the right electrode are greater than those from the left, $r > l$ for angles $\xi = 30°–120°$ ($p < 0.01$), which may reflect interaural differences in sound pressure (head shadowing) produced by the source location on the right-hand side. This tendency is reversed for wave II for two angles $\xi = 60°–90°$ ($l > r$, $p < 0.05$; Fig. 2.12b). The behavior of wave III (Fig. 2.12c) is similar to that of wave I ($r > l$, $p < 0.01$). This tendency again reverses for wave IV ($l > r$, $p < 0.05$; Fig. 2.12d) and is maintained further in wave VI ($l > r$, $p < 0.05$; Fig. 2.12f) even though absolute values are amplified.

From these patterns, it could be inferred that the flow of the left and right neural signals is interchanged 3 times at the cochlear nucleus, the SOC, and the LLN as shown in Fig. 2.13. The interaction at the IC in particular may be operative for the interaural signal processing as discussed later. In wave V as shown in Fig. 2.12e, such a reversal cannot be seen, and the relative behavior of amplitudes of the left and the right is parallel and similar. Thus, these two amplitudes were averaged and plotted in Fig. 2.14 (V symbols). For comparison, the amplitudes of wave IV [left (*l*) and right (*r*)] normalized to their respective ABR amplitudes at the frontal sound

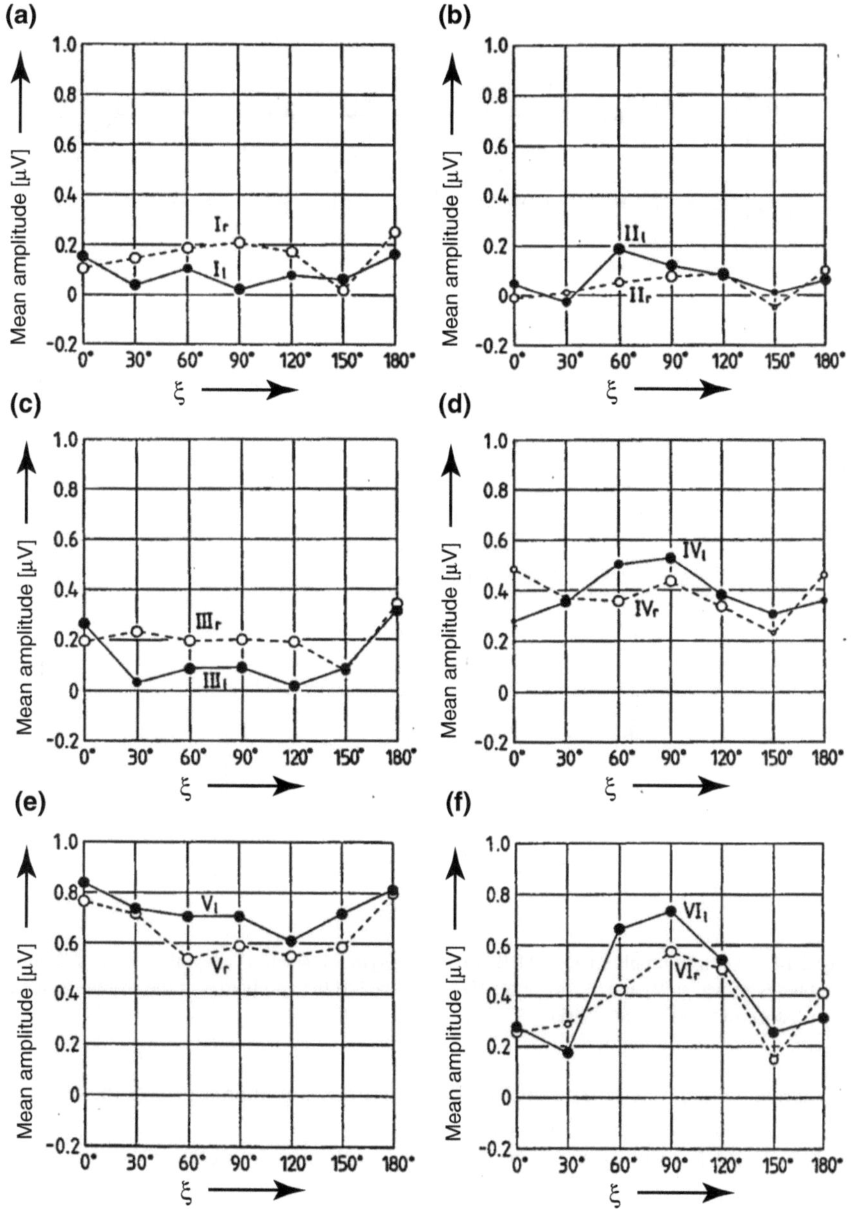

Fig. 2.12 Averaged amplitudes of ABR for each wave I–VI. The size of *circles* indicated the number of available data from four listeners. *Filled circles* left ABRs; *open circles* right ABRs. **a** Wave I. **b** Wave II. **c** Wave III. **d** Wave IV. **e** Wave V. **f** Wave VI (Ando et al. 1991)

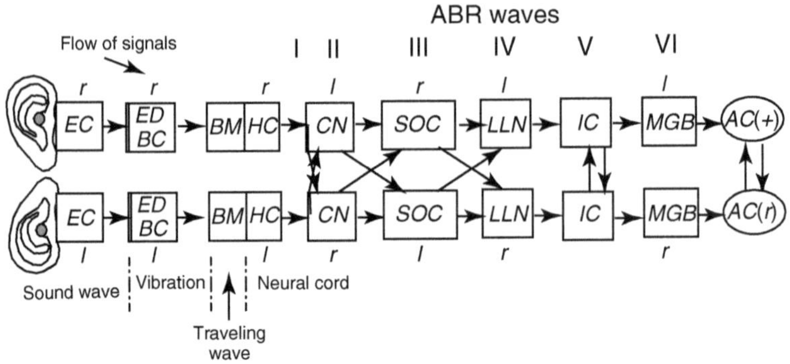

Fig. 2.13 Schematic illustration of the flow of neural signals in auditory pathways. *EC* external canal; *ED* and *BC* eardrum and bone chain; *BM* and *HC* basilar membrane and hair cell; *CN* cochlear nucleus; *SOC* superior olivary complex; *LLN* lateral lemniscus nucleus; *IC* inferior colliculus; *MGB* medial denticulate body; and *AC* auditory cortex of the right and left hemispheres

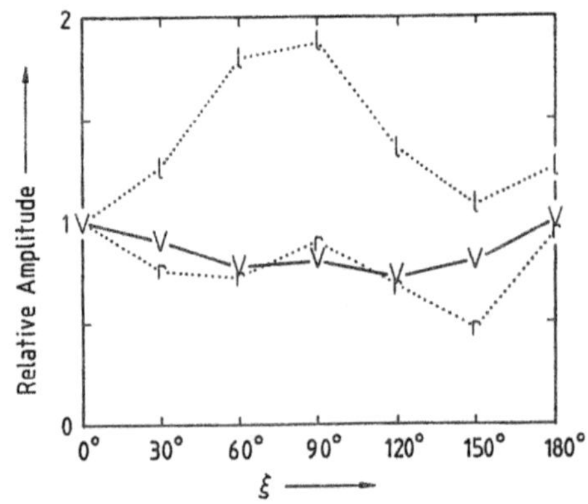

Fig. 2.14 Averaged amplitudes of ABR waves IV_l (symbol: *l*) and IV_r (symbol: *r*), and averaged amplitudes of waves V_l and V_r (symbol: V) normalized to the amplitudes at the frontal incidence (four listeners) (Ando et al. 1991)

incidence, which may correspond to the normalized sound pressures at the right and left ear entrances, respectively, are also plotted.

Neural ABR responses can be compared with cross-correlations derived from acoustical measurements at the two ears of a dummy head. A-weighted signals were presented and free-field sound pressure measurements were taken at the two ear entrances of a dummy head as a function of the horizontal angle of the sound source. Figure 2.15 depicts the signal power at the two ears (R, L) for different angles (the zero-lag term of the ACFs) and the maximum value of the IACF (ϕ), which are normalized only by the respective values at $\xi = 0°$. Received signal power is the greatest for the ear ipsilateral to the speaker R when it is situated 90° and least for the contralateral ear. These acoustic measures can be compared with

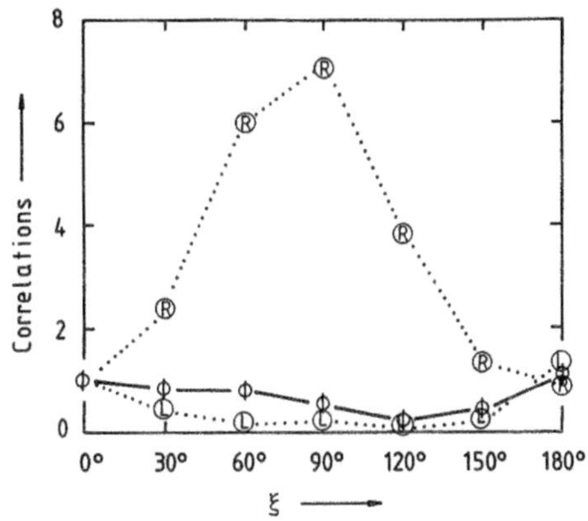

Fig. 2.15 Correlations of sound signals arriving at the left and right ear entrances of a dummy head, which are normalized by the respective values at $\xi = 0°$. L $\Phi_{ll}(0)$ measured at the left ear; R $\Phi_{rr}(0)$ measured at the right ear; Φ: maximum interaural cross-correlation, $|\Phi_{lr}(\tau)|_{max}$, $|\tau| < 1$ ms (Ando et al. 1991)

the neurally generated ABR potentials (see Fig. 2.14). Here, the neural correlate of the relative power of the received signals at the left and right ears is the average of the peak amplitudes of waves IV and V (left and right), normalized to those at the frontal incidence ($\xi = 0°$). Similar results are obtained when amplitudes are normalized to those at $\xi = 180°$. Although differences in units and scaling confound direct comparison between the results in Figs. 2.14 and 2.15, there are nevertheless qualitative similarities between these acoustic and physiologic responses. The relative behavior of wave IV (l) in Fig. 2.14 is similar to $\Phi_{rr}(0)$ in Fig. 2.15, which was measured at the right ear entrance r. Also, the relative behavior of wave IVr is similar to $\Phi_{ll}(0)$ at the left ear entrance l. In fact, amplitudes of wave IV (left and right) are proportional to $\Phi_{rr}(0)$ and $\Phi_{ll}(0)$, respectively, due to the interchange of signal flow. The behavior of wave V is similar to that of the maximum value, $|\phi_{lr}(\tau)|_{max}$, $|\tau| < 1$ ms. Because correlations have the dimensions of the power of the sound signals (i.e., the square of ABR amplitude), the interaural cross-correlation coefficient (IACC), which is defined by the maximum value of the IACF, may correspond to

$$P = \frac{A_V^2}{A_{IV,r}A_{IV,l}} \quad (2.2)$$

where A_V is the amplitude of the wave V, which may be reflected by the "maximum" synchronized neural activity ($\approx|\Phi_{lr}(\tau)|_{max}$) in the inputs to the IC (see Fig. 2.13). $A_{IV,r}$ and $A_{IV,l}$ are amplitudes of wave IV from the right and left, respectively. The results obtained by Eq. (2.2) are plotted in Fig. 2.16. It is clear that the behavior of the IACC and P are in good agreement ($r = 0.92$, $p < 0.01$).

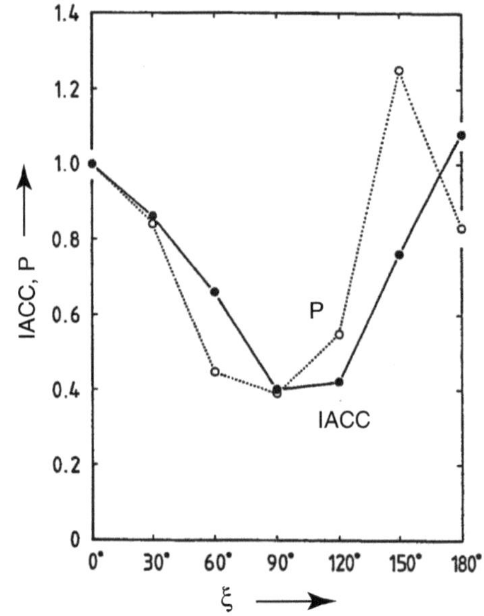

Fig. 2.16 Values of the IACC and values of *P* obtained by Eq. (2.1). A linear relationship between the IACC and the *P* value is obtained. Note that the available data at ξ = 150° was a single listener (Ando et al. 1991)

The amplitudes of the ABRs clearly differ according to the horizontal angle of the incidence of sound relative to the listener (Fig. 2.12). In particular, it is found that the amplitudes of waves I_{Vl} and I_{Vr} are nearly proportional to the SPLs at the right and left ear entrances, respectively, when the amplitude is normalized to that in front or back (ξ = 180°).

Since neurophysiological responses to IACC has examined in the auditory brain stem (Ando et al. 1991), there is little evidence of the cortical responses to IACC. To obtain such response, the present study used MEG to examine whether or not there is an activity in the auditory cortex corresponding to the IACC variations of band-pass noise.

Band-pass noises were employed for acoustic signals. To create band-pass noises, white noises were digitally filtered between 200 and 3000 Hz (Chebychev band-pass: order 18). The IACC of the stimuli was controlled by mixing in-phase diotic band-pass and dichotic independent band-pass noises in appropriate ratios (Blauert 1996). The frequency range of these noises was always kept the same. Figure 2.17 shows examples of the power spectrum, the temporal waveform, and the IACF of the stimulus used in this study. The interaural level difference was set to 0 dB, and there was no interaural delay between ear signals, i.e., the maximum of the IACC was always at $\tau = 0$.

Nine listeners with normal hearing (22–28 years, all right-handed) took part in the experiment. The AEFs were recorded using a 122 channel whole-head DC-SQUID magnetometer (Neuromag-122™; Neuromag Ltd., Helsinki, Finland) in a magnetically shielded room (Hämäläinen et al. 1993). Combinations of a reference stimulus (IACC = 1.0) and test stimuli (IACC = 0.85, 0.6, 0.2) were presented alternately at a constant 0.5-s interstimulus interval. During the experiment, listeners

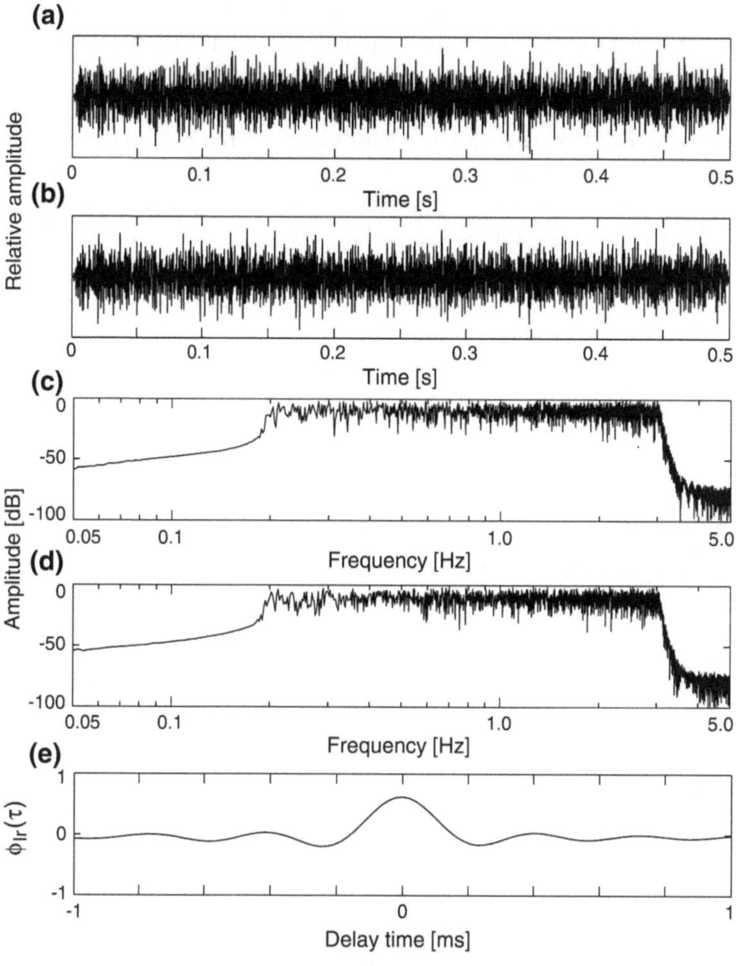

Fig. 2.17 Examples of the temporal waveform at the left (**a**) and right (**b**) ears, the power spectrum at the left (**c**) and right (**d**) ears, and the IACF (**e**) of the stimulus (IACC = 0.6) (Soeta et al. 2004b)

were asked to close their eyes to fully concentrate on the stimuli. The magnetic data were sampled at 0.4 kHz after being band-pass filtered between 0.03 and 100 Hz and averaged more than 50 times. Any responses coinciding with magnetic signals exceeding 3000 fT/cm were rejected from further analysis. The averaged responses were digitally filtered between 1.0 and 30.0 Hz. The analysis time was 0.7 s from 0.2 s prior to the stimulus onset. The average of the 0.2-s prestimulus period served as the baseline. To evaluate the amplitude and the latency of the response, the RMS of ∂Bz/∂x and ∂Bz/∂y were determined as the amplitude of the responses at each recording position. The N1m peak amplitude and latency were defined as the RMS peak and the latency in the latency range from 70 to 130 ms over the right and left hemisphere. In each listener, we employed the N1m peak latency and amplitude with a channel that showed the maximum amplitude placed at each hemisphere.

To estimate the location and the strength of the underlying neural activity of the N1m wave, a single ECD was assumed as the source of the magnetic field of the N1m wave. ECDs were found by a least square search in each hemisphere at the N1m peak latencies. Calculations were based on the spherical conductor model, which takes into account the volume current within a sphere (Kaukoranta et al. 1986). All ECDs with goodness-of-fit values exceeding 80 % were used in further analyses.

Clear N1m responses were observed in both right and left temporal regions in all listeners. The N1m latencies were not systematically affected by the IACC. Figure 2.18 depicts the mean N1m amplitude (across nine listeners) as a function of the IACC. The N1m amplitudes were normalized within each listener with respect to the maximum value. Basically, a smaller IACC provided a larger N1m

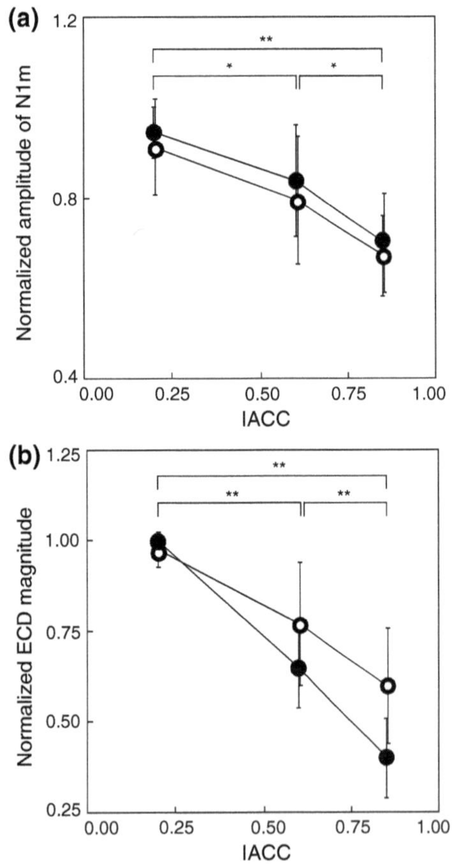

Fig. 2.18 The mean amplitude (**a**) and the mean ECD moments (**b**) of the N1m from right (*filled circle*) and left (*open circle*) hemispheres as a function of IACC. Error bars are the 95 % confidence interval. The N1m amplitudes and ECD moments are normalized within each listener with respect to the maximum value. The *asterisks* indicate statistical significance (*$P < 0.05$, **$P < 0.01$; Bonferroni test) (Soeta et al. 2004b)

amplitude. Two-way ANOVA (IACC × hemisphere) revealed a significant effect of the IACC on N1m amplitude ($P < 0.001$). Variability in the left and right hemispheres was not significantly different. The N1m ECD moment showed similar tendency with the N1m amplitude.

A previous study on auditory evoked potential (AEP) investigated the IACC using a one-third octave band-pass-filtered noise with the center frequency of 500 Hz as a sound signal (Ando et al. 1987). The results there showed that the peak-to-peak amplitude N1–P2 decreases with increasing IACC. In the present MEG study, a higher IACC provided for a significantly smaller amplitude of N1m, P2m, and the peak-to-peak amplitude N1m–P2m (Soeta et al. 2004b). A previous study on N1m response to dichotic tones of different frequencies found that N1m increased with the interaural frequency disparity (Yvert et al. 1998). Significantly, smaller AEPs or AEFs amplitudes were obtained with binaural stimulation, compared to those obtained with monaural contralateral stimulation, indicating some kind of interference between ipsilateral and contralateral pathways ("binaural interaction") (Pantev et al. 1986; McPherson and Starr 1993). It is considered that the more sounds differed arriving at both ears, the larger the N1m amplitude might be.

Blauert and Lindemann (1986) investigated the broadening and splitting of auditory events in dichotic listening condition with various degrees of the IACC by using a psychoacoustical mapping method. The response task of the listeners was to map the simple or multiple auditory events that they perceived during the presentation of each signal. The number of partial auditory events decreased with increasing IACC. This could result in larger magnetic field responses with a less interaurally coherent sound.

A starting point for modeling efforts in binaural hearing is the model of Jeffress (1948), which produces estimates of cross-correlation functions of its two inputs. Several theories regarding binaural systems rely on a cross-correlator to act as a comparator element for signals arriving at both the left and right ears (Colburn 1977; Lindemann 1986; Osman 1971; Osman et al. 1975; Blauert 1996). It has been shown that the central binaural neurons perform an operation very similar to the cross-correlation of the inputs (Ando et al. 1991; Palmer et al. 1999; Saberi et al. 1998; Yin et al. 1987). These inputs are transformed from the actual acoustic signal by the peripheral auditory system; and these transformations are reflected in the properties of the cross-correlations. Among various approaches to incorporating the evaluation of two interaural cues, namely interaural level difference (ILD) and interaural time difference (ITD or τ_{IACC}), into one consolidated model; a binaural cross-correlation model by contralateral inhibition was proposed (Lindemann 1986). A model using an IACF analysis of the ear input signals should be able to filter out the components of the ear input signals that are interaurally coherent (Lindemann 1986; Blauert 1996). Therefore, sounds that are interaurally more coherent could lead to more binaural inhibition, which causes the decrease of the strength of the N1m response.

To evaluate the effects of ITDs in relation to IACCs in human auditory cortex, the AEFs elicited by noises with different ITDs and IACCs were recorded and analyzed. The most important cues for sound localization in human are the differences in ITD and ILD of the sound waves received at the two ears. ITDs can be measured by the IACF between two sound signals received at both the left and right ears, that is, τ_{IACC}. The psychological responses to ITDs in relation to IACCs have been obtained in humans (Jeffress et al. 1962; Zimmer and Macaluso 2005), and the neurophysiological responses have been limited to animal studies (e.g., Yin et al. 1987; Yin and Chan 1990; Albeck and Konishi 1995; Keller and Takahashi 1996; Saberi et al. 1998; D'Angelo et al. 2003; Shackleton et al. 2005).

IACC was controlled by the same method with the previous research (Soeta et al. 2004b). The stimulus duration used in the experiment was 0.5 s, including rise and fall ramps of 10 ms, which were cut out of a 10-s long band-pass filtered noise with varying IACC and ITD. For stimulus lateralization, two cues were available to listeners: envelope ITD and ongoing ITD. In this experiment, the envelope ITD was zero for all stimuli, and the ongoing ITD was varied, as shown in Fig. 2.19. Here, "envelope" refers to the shape of a gating function with 10-ms linear ramps at the onset and offset. Stimuli were presented binaurally to the left and right ears through plastic tubes and earpieces inserted into the ear canals. All signals were presented at 60 dB SPL, and the ILD was set to 0 dB.

Figure 2.20 shows the N1m ECD moment as a function of ITD. When the IACC of the stimulus was 0.95, the effect of ITD on the N1m amplitude, namely ECD moments, was significant. The N1m amplitude increased with increasing ITD in the right hemisphere in the case of a left-leading stimulus and in both the left and right hemispheres in the case of a right-leading stimulus. This result is consistent with previous findings (Sams et al. 1993; McEvoy et al. 1993; Palomäki et al. 2005). The N1m amplitude increased slightly with increasing ITDs in the hemisphere contralateral to the ITDs when the IACC of the stimulus was 0.5; however, the main effect of ITDs on the N1m amplitude was not significant. Lateralization performance worsens with decreasing IACCs (Jeffress et al. 1962; McEvoy et al. 1991; Zimmer

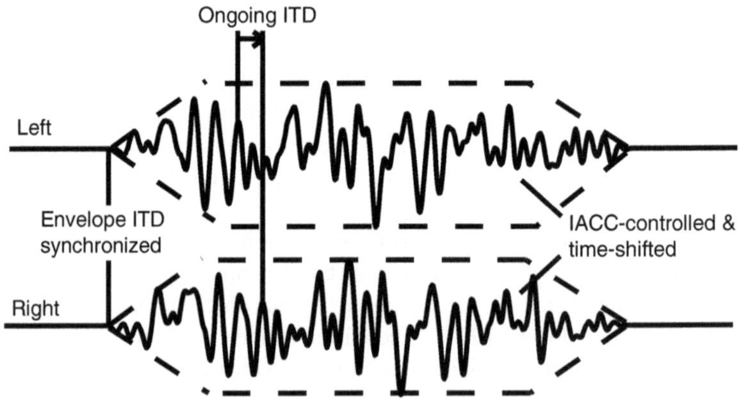

Fig. 2.19 Illustration of the stimuli used in the experiments. The fine structure (IACC controlled) of the stimulus was interaurally delayed, while the envelopes were synchronized between the ears

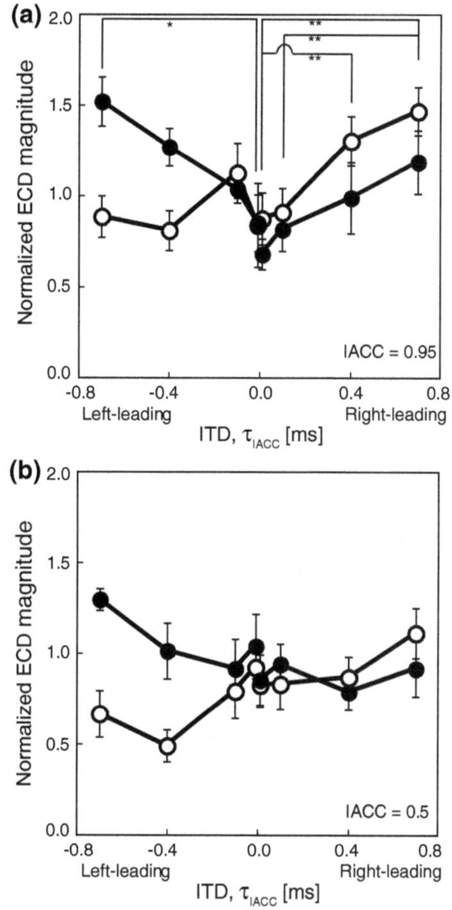

Fig. 2.20 Mean ECD moment of the N1m (±SEMs) as a function of the τ_{IACC} from the right (*filled circle*) and left (*open circle*) hemispheres when the IACC was 0.95 (**a**) and 0.5 (**b**). *Asterisks* indicate statistical significance (*$P < 0.05$, **$P < 0.01$; Post hoc Newman-Keuls test) (Soeta and Nakagawa 2009)

and Macaluso 2005); therefore, the present results may indicate that lateralization performance is reflected in N1m amplitudes. Put another way, there is a close relationship between the N1m amplitudes and the IACCs and ITDs of the stimuli.

Previous research has indicated that the N1m amplitudes significantly decreased with increasing IACCs when the ITD of the stimulus was 0.0 ms (Soeta et al. 2004b). This result contradicts the present findings. The cortical response to the change in IACCs has recently been analyzed (Chait et al. 2005), and the results indicated that the strength of the response increases with the IACC difference and that the brain responses are more sensitive to the transition from an IACC of 1.0 than to the transition from an IACC of 0.0. That is, there are two factors that influence the cortical response regarding IACC: the transition from correlated or uncorrelated noise and the size of the IACC transition. Thus, the reason the N1m amplitudes did not decrease with increasing IACCs when the ITD of the stimulus

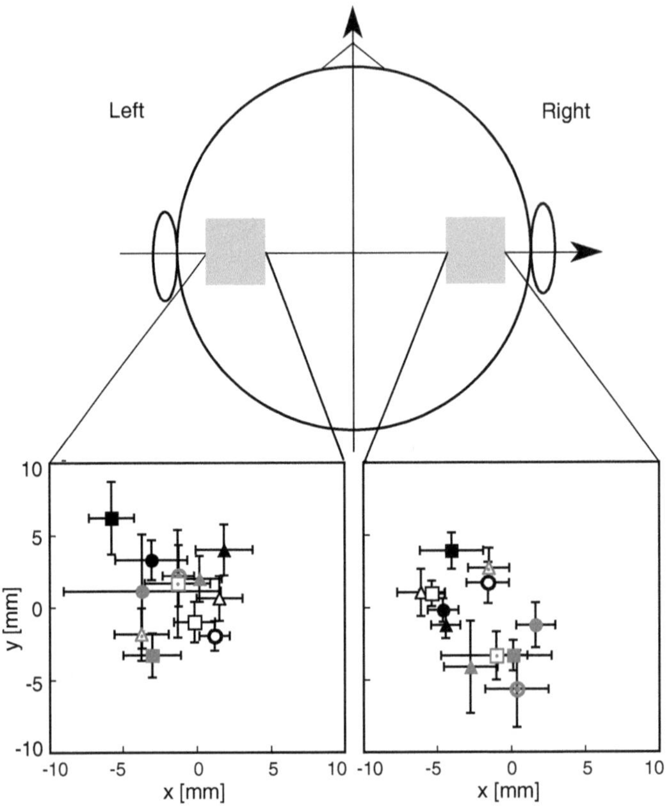

Fig. 2.21 Mean ECD location (±SEM) of all listeners in the *left* and *right* temporal planes when the IACC was 0.95 and 0.5. The ECD locations were normalized within each listener with respect to the position of ITD = 0.0 ms (Soeta and Nakagawa 2009)

was 0.0 ms in the present study could be due to an asymmetry between the adapting effect of the uncorrelated reference stimulus (IACC = 0.0) used in the present study, compared to the correlated reference stimulus (IACC = 1.0) used in the previous one (Soeta et al. 2004b).

Figure 2.21 shows the averaged ECD locations in the left and right hemispheres. The location of the ECDs underlying the N1m responses did not vary as a function of ITD or IACC, a finding in agreement with previous MEG results (McEvoy et al. 1993; Sams et al. 1993; Soeta et al. 2004b). As for Functional magnetic resonance imaging (fMRI), similarly, little evidence exists for segregated representations of specific ITDs or IACCs in auditory cortex (Woldorff et al. 1999; Maeder et al. 2001; Budd et al. 2003; Krumbholz et al. 2005; Zimmer and Macaluso 2005). Stimuli with different ITDs or IACCs may excite somewhat different neuronal populations, although the cortical source location did not differ systematically as a function of ITD or IACC. Therefore, we conclude that the present data do not show an orderly representation of ITD or IACC in the human auditory cortex that can be resolved by MEG.

Recently, it has been suggested that ITDs may be coded by the activity level in two broadly tuned hemispheric channels (McAlpine et al. 2001; Brand et al. 2002; McAlpine and Grothe 2003; Stecker et al. 2005). The present study showed that the N1m amplitude varies with the ITD; however, the location of the ECDs underlying the N1m responses did not vary with the ITD. This could suggest that different ITDs are coded nontopographically but by response level. Thus, the current data seem to be more consistent with a two-channel model (McAlpine et al. 2001; Brand et al. 2002; McAlpine and Grothe 2003; Stecker et al. 2005) rather than a topographic representation model (e.g., Jeffress 1948).

2.4 Signal Processing Model of Human Auditory System

Based on the above-mentioned physiological responses, a central auditory signal processing model has been proposed (Ando 1998). The model consists of the autocorrelation mechanisms and the interaural cross-correlation mechanisms between the two auditory pathways for temporal and spatial factors of the sound field as shown in Fig. 2.22. In this figure, a sound source $p(t)$ is located at r_0 in a three-dimensional space, and a listener is sitting at r, which is defined by the location of the center of the head, $h_{l,r}(r|r_{0,t})$, being the impulse responses between r_0 and the left and right ear canal entrances. The impulse responses of the external ear canal and the bone chain are $e_{l,r}(t)$ and $c_{l,r}(t)$, respectively. The velocity of the basilar membrane is expressed by $V_{l,r}(x, \omega)$, with x being the position along the membrane.

The action potentials from the hair cells are conducted and transmitted to the cochlear nuclei, the SOC including the MSO, the LSO, and the trapezoid body, and to the higher level of the two cerebral hemispheres. The input power density spectrum of the cochlea $I(x')$ can be roughly mapped at a certain nerve position x' (Katsuki et al. 1958; Kiang et al. 1965) as a temporal activity. Amplitudes of waves (I–IV) of the ABR reflect the SPLs at both ears as a function of the horizontal angle of incidence to a listener. Such neural activities include sufficient information to attain the ACF at the LLN as indicated by $\Phi_{ll}(\sigma)$ and $\Phi_{rr}(\sigma)$. In fact, the time domain analysis of firing rate from the auditory nerve of cat revealed a pattern of ACF, but the frequency domain analysis did not (Secker-Walker and Searle 1990). Pooled interspike interval distributions resemble the short time or the running ACF for the low-frequency component as shown in Fig. 2.23 (Cariani 2001). It traces a change of the pitch as a function of the time. The pooled interval distributions for sound stimuli consisting of the high-frequency component resemble the envelope to running ACF (Cariani and Delgutte 1996a, b). From a viewpoint of the missing

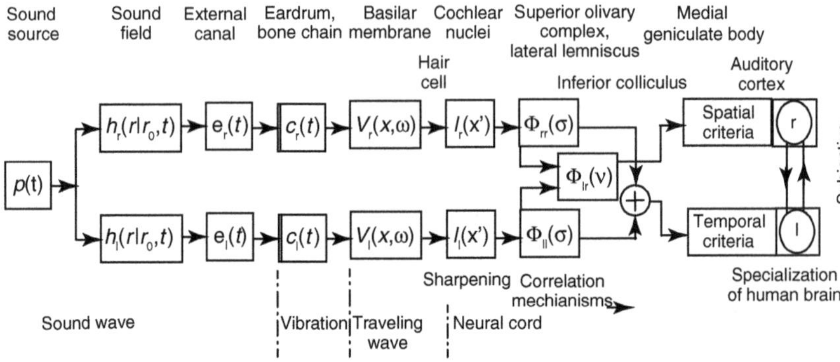

Fig. 2.22 Central auditory signal processing model for subjective responses. $p(t)$, source sound signal in the time domain; $h_{l,r}(r/r_0, t)$, head-related impulse responses from a source position of r_0 to the left and right ear entrances of a listener at r; $e_{l,r}(t)$, impulse responses of left and right external canals from the left and right ear entrances to the left and right eardrums; $c_{l,r}(t)$, impulse responses for vibration of left and right bone chains from the eardrums to *oval* windows, including transformation factors into vibrational motion at the eardrums; $V_{l,r}(x, \omega)$, traveling wave forms on the basilar membranes, where x is the position along the left and right basilar membrane measured from the *oval* window; and $I_{l,r}(x')$, sharpening in the cochlear nuclei corresponding with roughly the power spectra of input sound (i.e., responses of a single pure tone ω tend to approach a limited region of nuclei). These neural activities may be enough to convert into activities similar to the ACF. $\Phi_{ll}(\sigma)$ and $\Phi_{rr}(\sigma)$: ACF mechanisms in the left and right auditory pathways, respectively. Symbol $\oplus$ signifies that signals are combined. $\Phi_{lr}(\nu)$, IACF mechanism (Ando 1985); r and l, specialization for temporal and spatial factors of the left and right human cerebral hemispheres, respectively. Temporal sensations and spatial sensations may be processed in the left and right hemisphere according to the temporal factors extracted from the ACF and the spatial factors extracted from the IACF, respectively

fundamental or pitch of the complex tone judged by humans, the running ACF must be processed in the frequency components below about 5 kHz. The missing fundamental or pitch may be perceived less than about 1.2 kHz (Inoue et al. 2001), which may cover most musical signals. A tentative model of the running ACF processor is illustrated in Fig. 2.24. The output of the ACF processor may be dominantly connected with the left cerebral hemisphere.

As is also discussed, the neural activity (wave V together with waves IV_l and IV_r) may correspond to the IACC as shown in Fig. 2.16. Thus, the interaural cross-correlation mechanism may exist at the IC. It is concluded that the output signal of the interaural cross-correlation mechanism including the IACC may be dominantly connected to the right hemisphere. Also, the SPL expressed by a geometric average of the ACFs for the two ears at the origin of time ($\sigma = 0$) and in fact appears in the latency at the IC, and may be processed in the right hemisphere. The neural process has been developed realizing a minimum of effort and a maximum of efficiency, so that only information of criteria extracted from the ACF and IACF are transmitted into the left and right hemispheres, respectively.

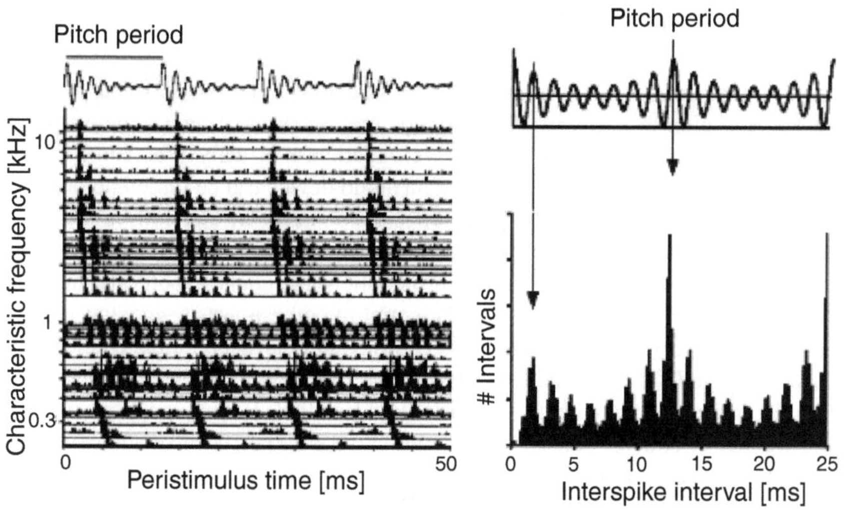

Fig. 2.23 *Top* stimulus waveform, single formant vowel, F0 = 80 Hz, F1 = 640 Hz, 60 dB SPL, 100 presentations/fiber. *Left* prestimulus time histograms of the responses of 53 auditory nerve fibers of Dial-anesthetized cats, arranged by fiber characteristics frequency. *Top right* stimulus ACF. *Bottom right* global ensemble-wide distribution of all-order interspike intervals. The most frequent interval in the distribution is 12.5 ms, which corresponds to the stimulus fundamental period (1/F0 =1/80 Hz) and the period of the low pitch that is heard (Cariani 2001)

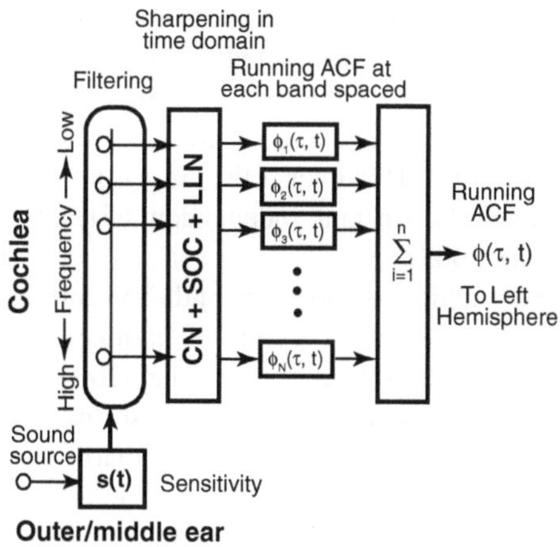

Fig. 2.24 A tentative two-dimensional model of the ACF processor in the auditory pathways. In the filtering process, however, there is no scientific evidence. The cutoff frequencies fixed, for example, might be arbitrary. Thus, it is considered that a kind of running window-filtering process in the frequency domain like the ACF analysis in the time domain might exist. In order to identify this process, a three-dimensional model might be considered, because the neuron distributes in the three-dimensional space

2.5 Brain Response in Relation to Loudness

The subjective aspect of sound intensity is loudness. Loudness is the attribute of auditory sensation in terms of which sounds may be ordered on a scale extending from quiet to loud. Loudness is what we experience in daily life. Someone can speak loudly or softly, and the volume of an audio device can be turned up or down. The mechanisms underlying the perception of loudness are not fully understood. The idea that loudness is simply proportional to the total number of action potentials fired by all auditory nerve neurons (the spike count hypothesis) has been investigated in animal studies. The spike count hypothesis was tested and justified that the rate of growth of both loudness and the auditory nerve spike count agreed over a wide range of tone intensity (Zwislocki 1965; Goldstein 1974; Lachs et al. 1984). However, disagreement also exists (Pickles 1983; Relkin and Doucet 1997).

The effect of sound intensity in the auditory cortex has been previously investigated by MEG, and AEF in response to stimulus intensity has been examined (e.g., Reite et al. 1982; Bak et al. 1985; Pantev et al. 1989; Vasama et al. 1995; Soeta and Nakagawa 2009). These results indicate that the N1m amplitude of the AEFs increases up to a stimulus intensity of 50–60 dB SPL, but then remains more or less constant or even decreases for higher intensities. The intensity dependence of the ECD location in human auditory cortex has also been examined. Pantev et al. (1989) reported that the higher the stimulus intensity is, the more superficial is the locus of cortical excitation. Vasama et al. (1995), however, failed to find any systematic variation of the N1m source locations as a function of intensity.

In previous studies on MEG response as a function of frequency, the ECD locations, that is, tonotopic organization, have been studied in considerable detail. Relatively, little research has been conducted on amplitude response as a function of frequency. The N1m response amplitude as a function of frequency has been examined using pure tones, and the results showed that the N1m amplitude peaks at 1000 Hz (Pantev et al. 1995). However, other researches have indicated that the N1m amplitude remains fairly constant, independent of test frequency (Pantev et al. 1988; Lütkenhöner et al. 2003).

Regarding the representation of noise in auditory cortex, auditory single-unit responses in the superior temporal gyrus of monkeys have been examined (Rauschecker et al. 1995; Rauschecker and Tian 2000; Lakatos et al. 2005). Responses of lateral neurons to band-pass noise are stronger than responses to pure tones. fMRI has showed that pure tones activate primarily the core, whereas band-pass noises activate preferably the belt areas in human auditory cortex (Wessinger et al. 2001). Relatively, little is known about the response amplitude in human auditory cortex as a function of frequency and bandwidth.

Changes in sound level are highly correlated with loudness changes; however, the relationship is not perfect. That is, changes in the frequency and bandwidth also affect the perceived loudness (Yost 2000). To evaluate the relationship between loudness and the activities of the auditory cortex, AEFs in response to the frequency and bandwidth were investigated (Soeta et al. 2006; Soeta and Nakagawa 2008b).

Octave band, 1/3 octave band, and 130 Hz bandwidth noises with center frequencies of 0.25, 0.5, 1, 2, 4, and 8 kHz were used as stimuli in Experiment 1 (Soeta et al. 2006). Pure tones, 1/6 octave band noise, and 1/3 octave band noise with center frequencies of 0.063, 0.125, 0.25, 0.5, 1, 2, 4, 8, 12, and 16 kHz were used in Experiment 2 (Soeta and Nakagawa 2008b). The octave band noises have fixed bandwidth in a logarithmic frequency scale. The 130 Hz bandwidth noises have fixed bandwidth irrespective of center frequency, such as pure tone. To make band-pass noises, the white noises were filtered using fourth-order Butterworth filters. Center frequencies of the noise bands are given as the geometric means of low- and high-frequency cutoffs. The stimulus duration was 500 ms, including rise and fall ramps of 10 ms. Stimuli were presented monaurally in Experiment 1 and binaurally in Experiment 2 through plastic tubes and earpieces inserted into the ear canals.

The AEFs were recorded using a 122-channel whole-head DC-SQUID (Neuromag-122^{TM}) in a magnetically shielded room (Hämäläinen et al. 1993). To maintain a constant level of attention, listeners were instructed to concentrate on a self-selected silent movie projected on a screen in front of them and to ignore the auditory stimuli. The magnetic data were sampled at 400 Hz after being band-pass filtered between 0.03 and 100 Hz and then averaged. Responses were rejected if the magnetic field exceeded 3000 fT/cm in any channel. The averaged responses were digitally filtered between 1.0 and 30.0 Hz. The analysis time was 0.7 s from 0.2 s prior to the stimulus onset. The average of the 0.2-s prestimulus period served as the baseline. To evaluate the latency of the N1m peak, the RMSs of $\partial Bz/\partial x$ and $\partial Bz/\partial y$ were determined as the amplitude of the responses at each recording position. The amplitude and latency with maximum peak amplitude in the latency range from 70 to 160 ms over each left and right hemisphere were defined as the N1m amplitude and latency in each listener.

A psychophysical loudness experiment was carried out individually in an anechoic and soundproof room. A two-interval, two-alternative, forced-choice, adaptive procedure was employed to measure the point of subjective equality (PSE) in the loudness balance task (Levitt 1971; Jesteadt 1980; Schlauch and Wier 1987). In these two sequences, the upper sequence used a 2-down, 1-up decision rule to track the 71 % response level, while the lower sequence used a 1-down, 2-up decision rule to track the 29 % response level in Experiment 1. A simple 1-up, 1-down rule was adopted in Experiment 2. In each trial, a standard sound with a fixed level and a comparison sound with a variable level that was selected randomly from one of two sequences were presented randomly in either the first interval or the second interval. The comparison stimulus was the 1000-Hz tone, and the two intervals were separated by 500 ms. The listener's task was to indicate which sound was louder by pressing a key. Each sequence was terminated by having reached 10 or 12 reversals. The level at which a reversal occurred was recorded, and the last four reversals were averaged at the end of the run to estimate the 71 and 29 % response levels. The PSE was estimated as the average of the 71 % and the 29 % levels in Experiment 1. The arithmetic mean of the last four of the levels was used to estimate the PSE in Experiment 2.

Figure 2.25 shows mean N1m amplitude plotted as a function of center frequency in Experiment 1. The main effect of the center frequencies on the N1m amplitude was significant for all bandwidth. Regarding 1/3 octave band noise, middle frequency range stimuli (500–2000 Hz) appear to be associated with larger amplitude. Regarding 130 Hz bandwidth noise, the maximum amplitude was found in the center frequency of 1000 Hz.

Figure 2.26 shows mean N1m latency plotted as a function of center frequency in Experiment 1. Middle frequency range stimuli (1000–4000 Hz) appear to be associated with shorter N1m latencies for all bandwidth. Lower and higher frequency stimuli give rise to relatively delayed N1m responses.

Figure 2.27 shows mean N1m amplitude plotted as a function of center frequency in Experiment 2. The N1m amplitude showed peaks at around 0.5 and 1 kHz, and decreased with decreasing center frequency at lower frequencies and decreased with increasing center frequency at higher frequencies. This tendency is consistent with previous studies using pure tones (Pantev et al. 1995; Fujioka et al. 2003) and band-pass noise (Soeta et al. 2006). The main effect of the center frequency on the N1m amplitude was significant ($p < 0.05$).

Figure 2.28 shows mean N1m latency plotted as a function of center frequency in Experiment 2. The latency of the N1m had minima between around 1 and 4 kHz. Middle frequency range sounds (1–4 kHz) appear to be associated with shorter N1m latencies. Lower and higher frequency stimuli had relatively delayed N1m responses. This is consistent with previous studies using pure tones (Roberts and Poeppel 1996; Stufflebeam et al. 1998; Lütkenhöner et al. 2001) and band-pass noise (Soeta et al. 2006).

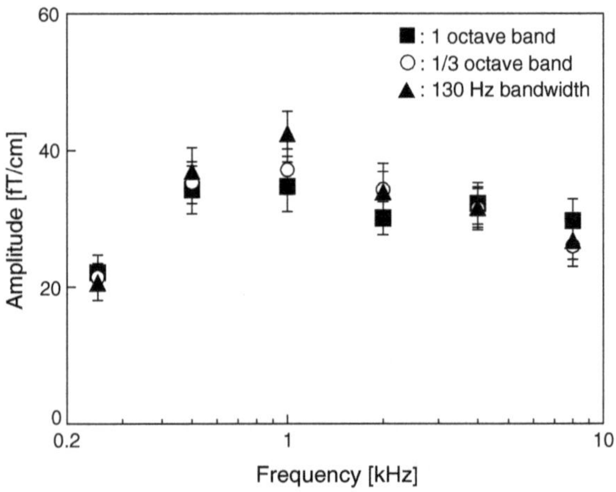

Fig. 2.25 Mean amplitude of the N1m (±SEMs) from both left and right hemispheres in response to (*filled square*) 1 octave band, (*open circle*) 1/3 octave band, and (*filled triangle*) 130 Hz bandwidth noises as a function of the center frequencies

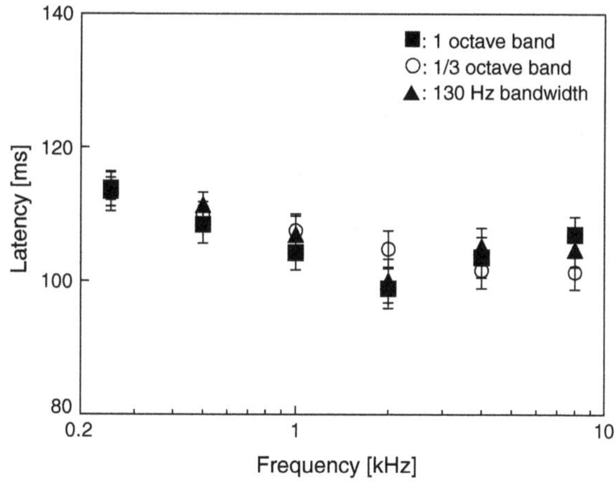

Fig. 2.26 Mean latency of the N1m (±SEMs) from both left and right hemispheres in response to (*filled square*) 1 octave band, (*open circle*) 1/3 octave band, and (*filled triangle*) 130 Hz bandwidth noises as a function of the center frequencies

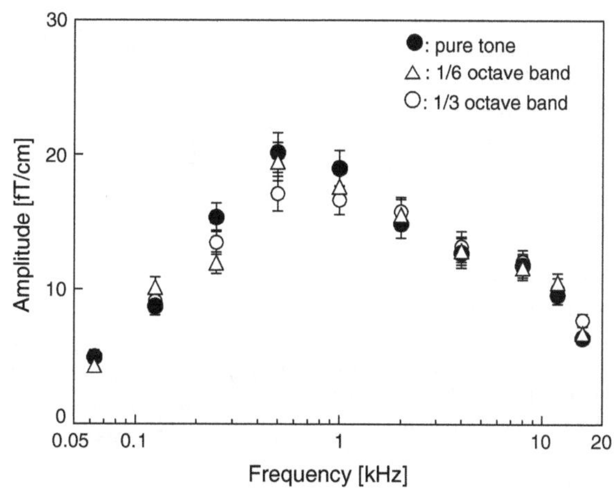

Fig. 2.27 Mean amplitude of the N1m (±SEMs) from both left and right hemispheres as a function of the center frequency. The bandwidth indicated by (*filled circle*) pure tone, (*open triangle*) 1/6 octave band, and (*open circle*) 1/3 octave band

The averaged data for the loudness balance experiment are shown in Fig. 2.29. For comparison, loudness, which is drawn based on an equal loudness contour from Suzuki and Takeshima (2004) when SPL is 60 dB, is reproduced as the solid line. Correlation coefficients between loudness values and mean N1m amplitudes are listed in Table 2.1. The mean N1m amplitude correlated well to the loudness values in the center frequency below 2 kHz in both Experiments 1 and 2. The phase-locking might be important for the perception of loudness and the precision of phase-locking decreases with increasing frequency above 1–2 kHz (Carlyon and Moore 1984). The low correlation between loudness values and N1m amplitude in

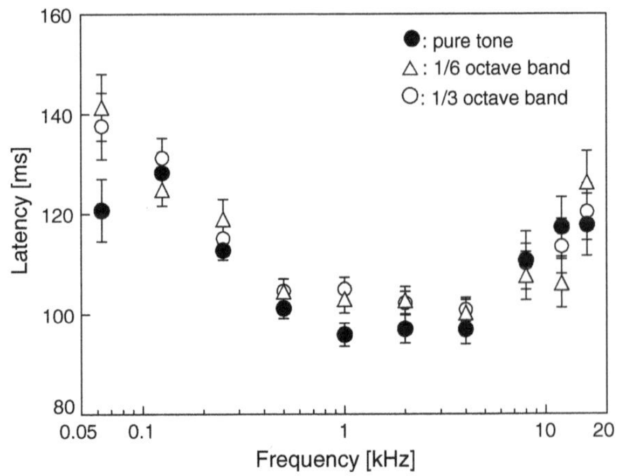

Fig. 2.28 Mean latency of the N1m (±SEMs) from both left and right hemispheres as a function of the center frequency. The bandwidth indicated by (*filled circle*) pure tone, (*open triangle*) 1/6 octave band, and (*open circle*) 1/3 octave band

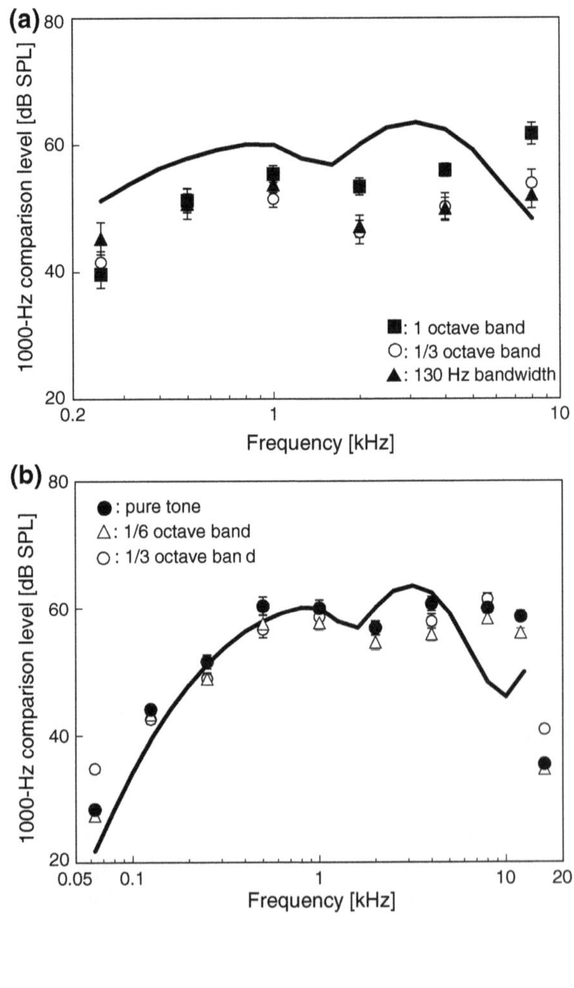

Fig. 2.29 Averaged loudness balance data between a 1000-Hz comparison tone (*y*-axis) and a standard sound as a function of the center frequency (*x*-axis) for (*filled square*) 1 octave band, (*open circle*) 1/3 octave band, and (*filled triangle*) 130 Hz bandwidth stimuli in (**a**) Soeta et al. (2006) and for (*filled circle*) pure tone, (*open triangle*) 1/6 octave band, and (*open circle*) 1/3 octave band stimuli in (**b**) Soeta and Nakagawa (2008b). Loudness when sound pressure level is 60 dB, which is drawn based on an equal loudness contour (Suzuki and Takeshima 2004), is represented by the *solid line*

Table 2.1 Correlation coefficient between loudness values and N1m amplitudes

	All frequency range	Lower frequency range (−2 kHz)	Middle frequency range (2–8 kHz)	High frequency range (8 kHz–)
Experiment 1	0.47	0.78	−0.48	
Experiment 2	0.78	0.97	−0.71	0.96

the center frequency between 2 and 8 kHz might be due to the loss of phase-locking. Scatter-plot of loudness as a function of the N1m amplitude is shown in Fig. 2.30. In the high-frequency range (>8 kHz), the amplitude of the N1m also correlated well with loudness. However, no AEF responses correlated with loudness in the frequency around 2 and 4 kHz. The activation of neurons in human auditory cortices shows a more rapid growth with SPL for a low-frequency tone

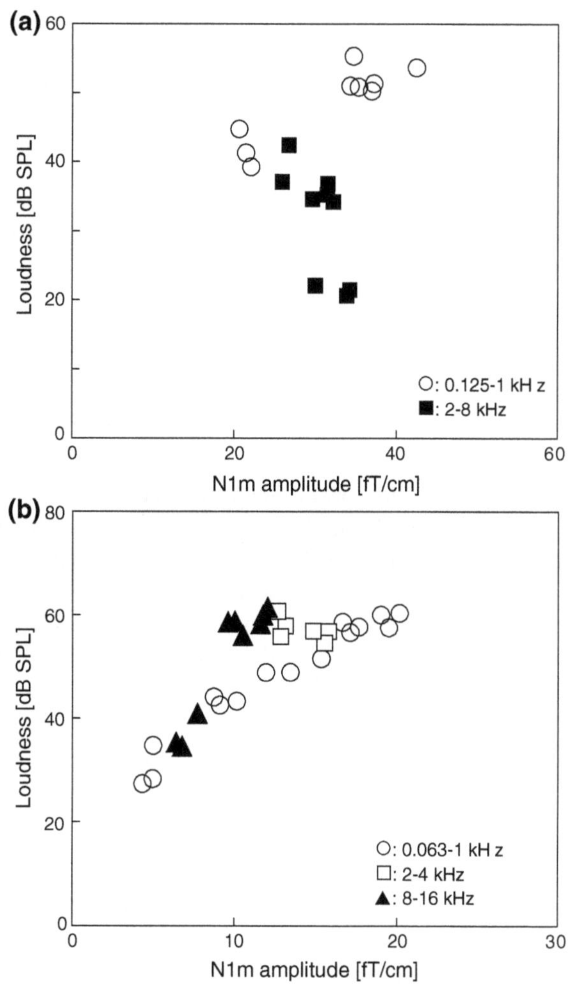

Fig. 2.30 Scatter plot of loudness as a function of the N1m amplitude in the center frequency range of **a** 250–2000 Hz (Soeta et al. 2006) and **b** 63–16,000 Hz (Soeta and Nakagawa 2008b)

than for a high-frequency tone (Hart et al. 2003). The present results indicated that the amplitudes of the N1m show a more rapid growth with loudness for lower frequency stimuli (<2 kHz) than for higher frequency stimuli (>8 kHz). The transition of the growth in activation of the human auditory cortex from low to high frequency might be between 2 and 8 kHz.

The ear does not transmit all frequencies equally. That is, it does not have a flat frequency response. The outer ear modifies the sound waves in transferring the acoustic vibrations to the eardrum. The resonances of the external ear increase the sound pressure at the eardrum, particularly in frequency range of 2–7 kHz (Wiener and Ross 1946; Shaw and Teranishi 1968; Shaw 1974; Mehrgardt and Mellert 1977). In other words, the transfer function from a sound source to the eardrum, which can be obtained by multiplying the HRTF between a sound source and the ear entrance, $H(\omega)$, and the function from the entrance to the eardrum, $E(\omega)$, together, has a broad peak at approximately 2.5 kHz. The middle ear apparatus then transfers the sound vibrations from eardrum to the cochlea. The transfer function of the human middle ear, $C(\omega)$, can also be calculated and measured (Onchi 1961; Zwislocki 1962; Rubinstein et al. 1966; Nedzelnitsky 1980; Wada et al. 1992). It has band-pass characteristics, with greatest transmission being seen around 1 kHz. The transfer function between a sound source and the cochlea may be represented by $S(\omega) = H(\omega)E(\omega)C(\omega)$. It has a broad maximum near 3 kHz, and the pattern of the transfer function agrees with the ear sensitivity for people with normal hearing ability (Ando 1998). The N1m amplitudes derived from the 1/3 octave band and 130 Hz bandwidth noises (Fig. 2.25) appear similar to the transfer function of the human middle ear, $C(\omega)$, (Fig. 2.3).

2.6 Brain Response Corresponding to Annoyance

Basically, psychoacoustic annoyance depends on the loudness, timbre, and temporal structure of sounds. Loudness and annoyance have been distinguished and defined (Berglund et al. 1975; Hellman 1982). These authors relate annoyance to an individual's reaction to noise within the context of a given situation and loudness directly to sound intensity. Environmental noise has been related to annoyance in several studies (e.g., Rylander et al. 1972, 1980, 1986; Björkman and Björkman 1997; Rylander and Björkman 1997). It is generally recognized that loudness is the most influential determinator for annoyance and can be predicted by the SPL. Previous studies have concluded that perceived loudness remains constant with increasing noise bandwidth until the bandwidth reaches the critical band. Loudness then increases with increasing bandwidth at the same SPL (Zwicker et al. 1957; Greenwood 1961a, b; Scharf 1962). However, the loudness of a sharply filtered noise increases as the effective duration of the ACF, τ_e, increases, even when the bandwidth of the signal is within the critical band (Merthayasa et al. 1994; Sato et al. 2002; Soeta et al. 2004c). The τ_e represents repetitive features within the signal itself and increases as the filter bandwidth decreases. In addition, a sound is perceived to be annoying although the SPL was only about 35 dBA in a given situation (Kitamura et al. 2002). This demonstrates that annoyance cannot be predicted by sound intensity alone.

To investigate the relationship between the human brain and the environment, studies were made using EEG and MEG. To investigate the relationship between the EEG responses and subjective preferences for a sound field, a method was developed using the ACF of EEG (Ando and Chen 1996; Chen and Ando 1996; Sato et al. 2003). The effective duration of the normalized ACF, τ_e, was analyzed with variation in the time delay of the single echo, Δt_1, reverberation time (RT), and IACC of sound fields. Their results showed that the τ_e is significantly longer in preferred conditions for the factors, Δt_1, RT, and IACC. It has also found that the τ_e and the maximum amplitude of the cross-correlation function (CCF), $|\phi(\tau)|_{max}$, of MEG between 8 and 13 Hz is correlated with subjective preference for Δt_1 and IACC of speech signal (Soeta et al. 2002, 2003).

We investigated the responses of the human brain that correspond to noise annoyance (Soeta et al. 2004c). The scale values of annoyance for each listener were obtained by paired-comparison tests. MEG measurements and analyses by the ACF and CCF were made. The relationship between the scale value of annoyance to band-pass noise and the factors extracted from the ACF and CCF of MEG in the brain's magnetic responses were investigated.

Pure tone and band-pass noises with a center frequency of 1000 Hz were used as auditory signals. The bandwidth of the signal was set to 0, 40, 80, 160, or 320 Hz with a 2000 dB/octave sharp filter, obtained by a digital FFT filter, to control the ACF of the source signal (Sato et al. 2002). The filter bandwidth of 0 Hz was the only slope component. The auditory stimuli were binaurally delivered through plastic tubes and earpieces inserted into the ear canals. The sound pressure was measured with an ear simulator, including a microphone and a preamplifier, and an adaptor connected to the earpiece. All stimuli were fixed at the same SPL (74 dBA). The signals were characterized by ACF factors, τ_e, τ_1, and ϕ_1. The measured τ_1 of all signals were 1.0 ms, which correspond to the center frequency of band-pass noise. The measured ϕ_1 and τ_e increased as the filter bandwidth decreased with a certain degree of coherence between ϕ_1 and τ_e.

Seven listeners took part in the experiment, 22–28 year old with normal hearing. They were seated in a dark soundproof room, with a comfortable thermal environment, and were presented the sound stimuli. A paired-comparison tests were performed for all combinations of the pairs of pure tone and band-pass noise, i.e., 15 pairs ($N(N-1)/2$, $N = 6$) of stimuli with interchange of the order of each pair per session and random presentation of the pairs. Ten sessions was conducted for each listener. The duration of the stimuli was 2.0 s, the rise and fall times were 10 ms, the silent interval between the stimuli was 1.0 s, and the interval between pairs was 4.0 s, which was the allowed time for the listeners to respond by pushing one of two buttons. They were asked to judge which of the two sound stimuli was more annoying. The scale values of the annoyance were calculated according to Case V of Thurstone's theory (Thurstone 1927; Gullikson 1956), and the model of Case V for all data was confirmed by the goodness-of-fit test (Mosteller 1951).

The same listeners joined in the annoyance tests took part in the recording of MEG responses. The magnetic responses were measured in a magnetically shielded room and recorded (passband 0.03–100 Hz, sampling rate 400 Hz) with a 122 channel whole-head DC-SQUID magnetometer. For the measurements, the listeners were seated in a dark soundproof room with a comfortable thermal environment and were asked to close their eyes and fully concentrate on the sound stimulus. The paired-auditory stimuli were presented in the same way as in the subjective annoyance test. Combinations of a reference stimulus (pure tone) and test stimuli (band-pass noise) were presented alternately 30 times at a constant 2.0-s inter-stimulus interval and MEGs recorded. Eighteen channels that were located around the temporal area in each hemisphere were selected for ACF and CCF analysis (Fig. 2.31). This resulted of 36 channels selected to be analyzed. Each response, corresponding to one stimulus, was analyzed by ACF and CCF for each listener. The relationship between the degree of annoyance and the averaged τ_e values at 18 sites, measured at two tangential derivatives, was investigated.

The ACF provides the same information as the power spectral density of a signal. Figure 2.32a shows an example of a measured ACF. A normalized ACF can be expressed by:

$$\phi(\tau) = \frac{\Phi(\tau)}{\Phi(0)}, \tag{2.3}$$

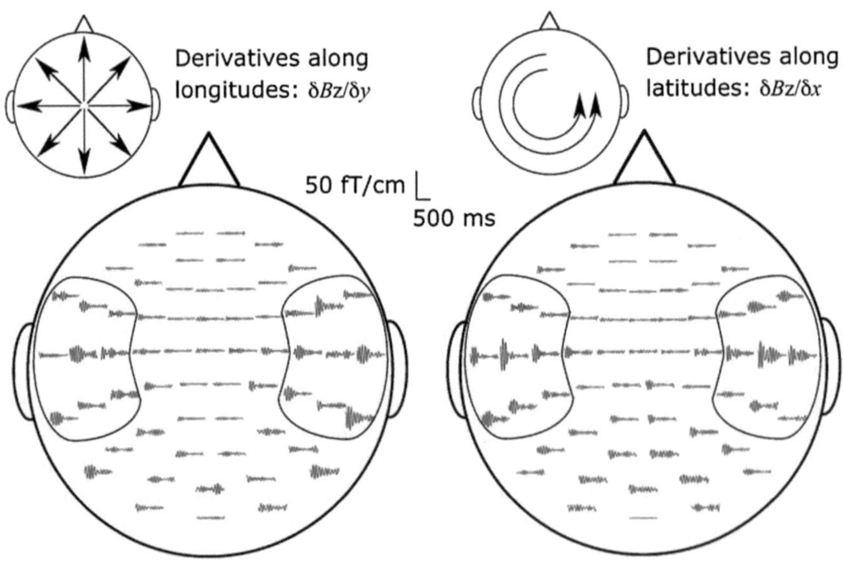

Fig. 2.31 Examples of recorded MEG responses to band-pass noise. The passband is 8–13 Hz. 36 channels that were located around the left and temporal area were selected for the ACF and CCF analysis (Soeta et al. 2004c)

where

$$\Phi(\tau) = \frac{1}{2T}\int_{0}^{2T} \alpha(t)\alpha(t+\tau)\mathrm{d}t, \tag{2.4}$$

where $2T$ is the integral interval, τ is the time delay, and $\alpha(t)$ is the MEG between 8 and 13 Hz. Figure 2.32b shows the absolute value of the ACF in a logarithmic form as a function of the time delay, τ. To calculate the degree of the ACF envelope decay, the effective duration, τ_e, is determined. As shown in Fig. 2.32b, a straight-line regression of the ACF can only be made by using the initial declining portion, 0 dB > 10 log|$\phi(\tau)$| > −5 dB (Ando and Chen 1996). In most cases, the envelope decay of the initial part of the ACF may fit a straight line. The values of τ_e were analyzed at $2T = 2.0$ s. Given the two signals are $\alpha_1(t)$ and $\alpha_2(t)$, then the CCF is defined by:

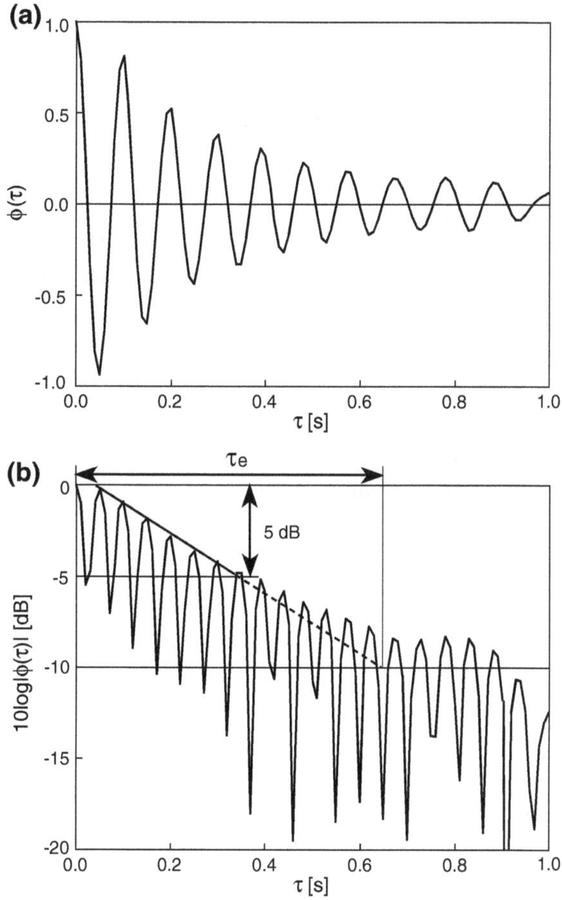

Fig. 2.32 **a** Examples of normalized ACF of MEGs between 8 and 13 Hz. **b** Examples of determining the effective duration of ACF, τ_e (Soeta et al. 2004c)

$$\Phi_{12}(\tau) = \frac{1}{2T}\int_{-T}^{+T} \alpha_1(t)\alpha_2(t+\tau)\mathrm{d}t. \tag{2.5}$$

The normalized CCF is given by:

$$\phi_{12}(\tau) = \frac{\Phi_{12}(\tau)}{\sqrt{\Phi_{11}(0)\Phi_{22}(0)}}, \tag{2.6}$$

where $\Phi_{11}(0)$ and $\Phi_{22}(0)$ are the ACFs of $\alpha_1(t)$ and $\alpha_2(t)$ at $\tau = 0$, respectively. The normalized CCF between the MEG responses was recorded at the reference channels, with 18 channels for each hemisphere, and those recorded at the 35 test channels (with the exception of the reference channel) were calculated. Examples of a normalized CCF and the definition of the maximum value of the CCFs, $|\phi(\tau)|_{max}$, are shown in Fig. 2.33. The values of $|\phi(\tau)|_{max}$ were analyzed at $2T = 2.0$ s.

The values of τ_e for the most annoying stimuli were significantly shorter than those for the least annoying stimuli in six listeners ($p < 0.05$, one-way ANOVA). Figure 2.34a shows the relationship between the ratio of averaged values of τ_e of band-pass noise to those of a pure tone, and the difference between the scale values of band-pass noise and those of a pure tone. The ratio of τ_e increases as the difference of scale values of annoyance decreased (except for one listener). This indicates that the value of τ_e became shorter during the presentation of an annoying stimulus. The correlation coefficient between the ratio of τ_e values and the difference in scale values of annoyance was −0.83 ($p < 0.01$).

The results from the reference channel with the highest correlation between the scale values of annoyance and averaged $|\phi(\tau)|_{max}$ values of all test channels showed a significant effect of the stimulus on $|\phi(\tau)|_{max}$ values ($p < 0.001$). The values of $|\phi(\tau)|_{max}$ for the most annoying stimuli were significantly smaller than those for the

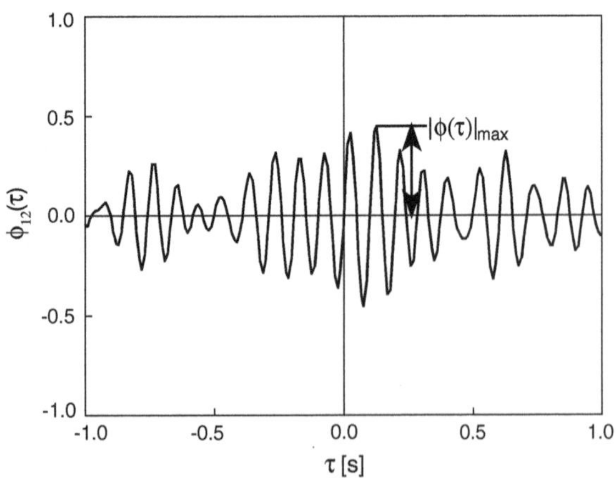

Fig. 2.33 Examples of normalized CCF of MEGs between 8 and 13 Hz and the definitions of the maximum value of CCF, $|\phi(\tau)|_{max}$ (Soeta et al. 2004c)

least annoying stimuli for six listeners ($p < 0.001$, one-way ANOVA). The results indicate that the ratio of $|\phi(\tau)|_{max}$ increases as the difference of scale values of annoyance decrease (except for one listener), as shown in Fig. 2.34b. This indicates that the value of $|\phi(\tau)|_{max}$ becomes smaller during the presentation of an annoying stimulus. The correlation coefficient between the ratio of $|\phi(\tau)|_{max}$ value and the difference in scale values of annoyance was −0.72 ($p < 0.01$).

The value of τ_e becomes shorter and the values of $|\phi(\tau)|_{max}$ becomes smaller during presentation of an annoying stimulus. The τ_e is the degree of similar repetitive features included in MEG between 8 and 13 Hz, and the $|\phi(\tau)|_{max}$ signifies the degree of similar repetitive features that appear in MEG between 8 and 13 Hz recorded at two different channels. Thus, the brain is unstable over a wider range, in both space and time during annoying conditions. Previous studies on EEG and MEG between 8 and 13 Hz show that the τ_e becomes significantly longer and $|\phi(\tau)|_{max}$ significantly larger in preferred sound fields (Ando and Chen 1996; Chen and Ando 1996; Sato et al. 2003; Soeta et al. 2002, 2003). This indicates that the

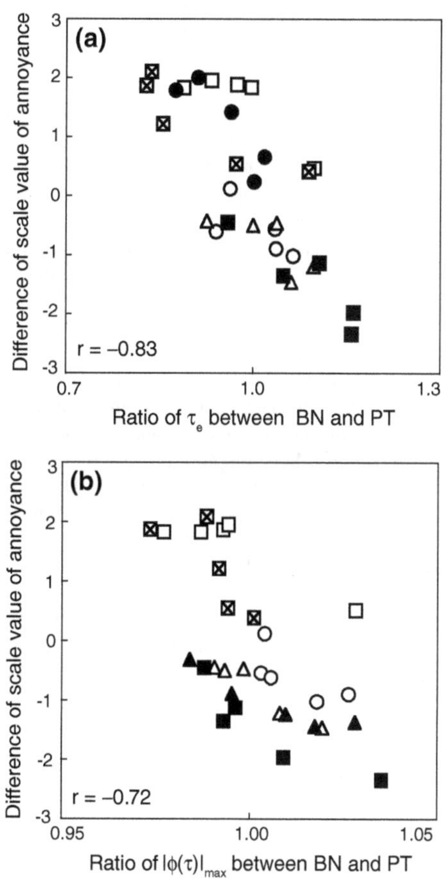

Fig. 2.34 Relationship between the difference of scale values (*SV*) [SV (band-pass noise) − SV (pure tone)] and the ratio of **a** τ_e and **b** $|\phi(\tau)|_{max}$ values of band-pass noise (*BN*) to those of a pure tone (*PT*). Each *symbol* represents one listener

brain repeats a similar rhythm over a wider range, in both space and time in preferred conditions. These are considered to be consistent with the present results.

Alpha activity is commonly defined as fluctuations between 8 and 13 Hz that can be detected on the occipital scalp (Chapman et al. 1984). Similar oscillatory activity, seen over the auditory cortex, is called τ rhythm (Tiihonen et al. 1991; Dinse et al. 1997). It is this τ rhythm that is analyzed by the ACF and CCF in this study. The high correlation between annoyance and the values of τ_e and $|\phi(\tau)|_{max}$ was not found in other frequency bands, such as delta (1–4 Hz), theta (4–8 Hz), and beta (13–30 Hz).

References

Aibara R, Welsh JT, Puria S, Goode RL (2001) Human middle-ear sound transfer function and cochlear input impedance. Hear Res 152:100–109

Albeck Y, Konishi M (1995) Responses of neurons in the auditory pathway of the barn owl to partially correlated binaural signals. J Neurophysiol 74:1689–1700

Ando Y (1985) Concert hall acoustics. Springer, Heidelberg

Ando Y (1998) Architectural acoustics: blending sound sources, sound fields, and listeners. AIP Press/Springer, New York

Ando Y (2001) A theory of primary sensations and spatial sensations measuring environmental noise. J Sound Vib 241:3–18

Ando Y (2002) Correlation factors describing primary and spatial sensations of sound fields. J Sound Vib 258:405–417

Ando Y, Chen C (1996) On the analysis of autocorrelation function of a-waves on the left and right cerebral hemispheres and in relation to the time delay of single sound reflection. J Architec Plan Env Eng 488:67–73

Ando Y, Hosaka I (1983) Hemispheric difference in evoked potentials to spatial sound field stimuli. J Acoust Soc Am 74(S1):S64–S65

Ando Y, Kang SH, Nagamatsu H (1987) On the auditory-evoked potentials in relation to the IACC of sound field. J Acoust Soc Jpn (E) 8:183–190

Ando Y, Yamamoto K, Nagamastu H, Kang SH (1991) Auditory brainstem response (ABR) in relation to the horizontal angle of sound incidence. Acoust Lett 15:57–64

Ando Y, Sato S, Sakai H (1999) Fundamental subjective attributes of sound fields based on the model of auditory brain system. In: Sendra JJ (ed) Computational acoustics in architecture. WIT Press, Southampton, pp 63–99

Bak CK, Lebech J, Saermark K (1985) Dependence of the auditory evoked magnetic field (100 msec signal) of the human brain on the intensity of the stimulus. Electroenceph Clin Neurophysiol 61:141–149

Berglund B, Berglund U, Lindvall T (1975) Scaling loudness, noisiness, and annoyance of aircraft noise. J Acoust Soc Am 57:930–934

Bilsen FA (1966) Repetition pitch: monaural interaction of a sound with the repetition of the same, but phase shifted sound. Acustica 17:295–300

Bilsen FA, Ritsma RJ (1969) Repetition pitch and its implication for hearing theory. Acustica 22:63–73

Bilsen FA, ten Kate JH, Buunen TJF, Raatgever J (1975) Responses of single units in the cochlear nucleus of the cat to cosine noise. J Acoust Soc Am 58:858–866

Björkman M, Rylander R (1997) Maximum noise levels in city traffic. J Sound Vib 205:513–516

Blauert J (ed) (1996) Spatial hearing. The MIT Press, Cambridge

Blauert J, Lindemann W (1986) Spatial mapping of intracranical auditory events for various degrees of interaural coherence. J Acoust Soc Am 79:806–813

Brand A, Behrend O, Marquardt T, McAlpine D, Grothe B (2002) Precise inhibition is essential for microsecond interaural time difference coding. Nature 417:543–547

Budd TW, Hall DA, Gonçalves MS, Akeroyd MA, Foster JR, Palmer AR, Head K, Summerfield AQ (2003) Binaural specialisation in human auditory cortex: an fMRI investigation of interaural correlation sensitivity. Neuroimage 20:1783–1794

Butler RA, Belundiuk K (1977) Spectral cues utilized in the location of sound in the median sagittal plane. J Acoust Soc Am 61:1264–1269

Cansino S, Ducorps A, Ragot R (2003) Tonotopic cortical representation of periodic complex sounds. Hum Brain Mapp 20:71–81

Cariani PA (2001) Neural timing nets. Neural Netw 14:737–753

Cariani PA, Delgutte B (1996a) Neural correlates of the pitch of complex tones. I. Pitch and pitch salience. J Neurophysiol 76:1698–1716

Cariani PA, Delgutte B (1996b) Neural correlates of the pitch of complex tones. II. Pitch shift, pitch ambiguity, phase invariance, pitch circularity, rate pitch, and the dominance region for pitch. J Neurophysiol 76:1717–1734

Carlyon RP, Moore BCJ (1984) Intensity discrimination: a severe departure from Weber's law. J Acoust Soc Am 76:1369–1376

Chait M, Poeppel D, Cheveigne A, Simon JZ (2005) Human auditory cortical processing of changes in interaural correlation. J Neurosci 25:8518–8527

Chandrasekaran B, Kraus N (2010) The scalp-recorded brainstem response to speech: neural origins and plasticity. Psychophysiology 47:36–246

Chapman RM, Ilmoniemi RJ, Barbanera S, Romani GL (1984) Selective localization of alpha brain activity with neuromagnetic measurements. Electroencephalogr Clin Neurophysiol 58:569–572

Chen C, Ando Y (1996) On the relationship between the autocorrelation function of a-waves on the left and right cerebral hemispheres and subjective preference for the reverberation time of music sound field. J Architec Plan Env Eng 489:73–80

Colburn HS (1977) Theory of binaural interaction based on auditory-nerve data. II. Detection of tones in noise. J Acoust Soc Am 61:525–533

D'Angelo WR, Sterbing SJ, Ostapoff EM, Kuwada S (2003) Effects of ampli-tude modulation on the coding of interaural time differences of low-frequency sounds in the inferior colliculus. II. Neural mechanisms. J Neurophysiol 90:2827–2836

Denham S (2005) Pitch detection of dynamic iterated rippled noise by humans and a modified auditory model. Biosystems 79:199–206

Dinse HR, Krüger K, Akhavan AC, Spengler F, Schüoner G, Schreiner CE (1997) Low-frequency oscillations of visual, auditory and somatosensory cortical neurons evoked by sensory stimulation. Int J Psychophysiol 26:205–227

Elberling C, Bak C, Kofoed B, Lebech J, Sarmark G (1982) Auditory magnetic fields from the human cerebral cortex: location and strength of an equivalent current dipole. Acta Neurol Scand 65:553–569

Fastl H, Stoll G (1979) Scaling of pitch strength. Hear Res 1:293–301

Fay RR, Yost WA, Coombs S (1983) Psychophysics and neurophysiology of repetition noise processing in a vertebrate auditory system. Hear Res 12:31–55

Fujioka T, Ross B, Okamoto H, Takeshima Y, Kakigi R, Pantev C (2003) Tonotopic representation of missing fundamental complex sounds in the human auditory cortex. Eur J Neurosci 18:432–440

Gardner MB, Gardner RS (1973) Problem of localization in the median plane: Effect of pinna cavity occlusion. J Acoust Soc Am 53:400–408

Goldstein JL (1974) Is the power law simply related to the driven spike response rate from the whole auditory nerve? In: Moskowitz HR, Scharf B, Stevens JC (eds) Sensation and measurement. Reidel, Dordrecht, pp 223–229

Greenwood DD (1961a) Auditory masking and the critical band. J Acoust Soc Am 33:484–502

Greenwood DD (1961b) Critical bandwidth and the frequency of the basilar membrane. J Acoust Soc Am 33:1344–1356
Gullikson H (1956) A least squares solution for paired comparisons with incomplete data. Psychometrika 21:125–134
Hämäläinen MS, Hari R, Ilmoniemi RJ, Knuutila J, Lounasmaa OV (1993) Magnetoencephalography? Theory, instrumentation, and applications to noninvasive studies of the working human brain. Rev Mod Phys 65:413–497
Hart HC, Hall DA, Palmer AR (2003) The sound-level-dependent growth in the extent of fMRI activation in Heschl's gyrus is different for low- and high-frequency tones. Hear Res 179:104–112
Hellman RP (1982) Loudness, annoyance and noisiness produced by single-tone-noise complexes. J Acoust Soc Am 72:62–73
Inoue M, Ando Y, Taguti T (2001) The frequency range applicable to pitch identification based upon the autocorrelation function model. J Sound Vib 241:105–116
ISO 226:2003 Acoustics—normal equal-loudness-level contours
Jeffres LA (1948) A place theory of sound localization. J Comp Physiol Psych 61:468–486
Jeffress LA, Blodgett HC, Deatherage BH (1962) Effects of interaural correlation on the precision of centering a noise. J Acoust Soc Am 34:1122–1123
Jesteadt W (1980) An adaptive procedure for subjective judgments. Percept Psychophys 28:85–88
Katsuki Y, Sumi T, Uchiyama H, Watanabe T (1958) Electric responses of auditory neurons in cat to sound stimulation. J Neurophysiol 21:569–588
Kaukoranta E, Hämäläinen M, Sarvas J, Hari R (1986) Mixed and sensory nerve stimulations activate different cytoarchitectonic areas in the human primary somatosensory cortex SI: Neuromagnetic recordings and statistical considerations. Exp Brain Res 63:60–66
Keller CH, Takahashi TT (1996) Binaural cross-correlation predicts the responses of neurons in the owl's auditory space map under conditions simulating summing localization. J Neurosci 16:4300–4309
Kiang NYS, Watanabe T, Thomas EC, Clark LF (1965) Discharge patterns of single fibers in the cat's auditory nerve. MIT Press, Cambrdge
Kitamura T, Sato S, Shimokura R, Ando Y (2002) Measurement of temporal and spatial factors of a flushing toilet noise in a downstairs bedroom. J Temporal Des Arch Environ 2:13–19
Knuutila J, Ahonen A, Hämäläinen M, Kajola M, Laine P, Lounasmaa O, Parkkonen L, Simola J, Tesche C (1993) A 122-channel whole cortex SQUID system for measuring the brain's magnetic fields. IEEE Trans Magn 29:3315–3320
Krishnan A (2007) Human frequency following response. In Burkard RF, Don M, Eggermont JJ (eds) Auditory evoked potentials: basic principles and clinical application, Lippincott Williams & Wilkins, Baltimore, pp 313–335
Krishnan A, Bidelman GM, Gandour JT (2010) Neural representation of pitch salience in the human brainstem revealed by psychophysical and electrophysiological indices. Hear Res 268:60–66
Krishnan A, Xu Y, Gandour J, Cariani P (2005) Encoding of pitch in the human brainstem is sensitive to language experience. Brain Res Cogn Brain Res 25:161–168
Krumbholz K, Patterson RD, Seither-Preisler A, Lammertmann C, Lütkenhöner B (2003) Neuromagnetic evidence for a pitch processing center in Heschl's gyrus. Cereb Cortex 13:765–772
Krumbholz K, Schönwiesner M, von Cramon DY, Rübsamen R, Shah NJ, Zilles K, Fink GR (2005) Representation of interaural temporal information from left and right auditory space in the human planum temporale and inferior parietal lobe. Cereb Cortex 15:317–324
Lachs G, Al-Shaikh R, Bi Q, Saia RA, Teich M (1984) A neural counting model based on physiological characteristics of the peripheral auditory system. V. Applications to loudness estimation and intensity discrimination. IEEE Trans Syst Man Cybern SMC 14:819–836
Lakatos P, Pincze Z, Fu KG, Javitt DC, Karmos G, Schroeder CE (2005) Timing of pure tone and noise-evoked responses in macaque auditory cortex. NeuroReport 16:933–937

Langner G, Sams M, Heli P, Schulze H (1997) Frequency and periodicity are represented in orthogonal maps in the human auditory cortex: evidence from magnetoencephalography. J Comp Physiol A 181:665–676
Levitt H (1971) Transformed up-down procedures in psychophysics. J Acoust Soc Am 49:467–477
Licklider JCR (1951) A duplex theory of pitch perception. Experimenta 7:128–134
Lindemann W (1986) Extension of a binaural cross-correlation model by means of contralateral inhibition, I: Simulation of lateralization of stationary signals. J Acoust Soc Am 80:1608–1622
Lütkenhöner B, Lammertmann C, Knecht S (2001) Latency of auditory evoked field deflection N100m ruled by pitch or spectrum? Audiol NeuroOtol 6:263–278
Lütkenhöner B, Krumbholz K, Seither-Preisler A (2003) Studies of tonotopy based on wave N100 of the auditory evoked field are problematic. Neuroimage 19:935–949
Maeder PP, Meuli RA, Adriani M, Bellmann A, Fornari E, Thiran JP, Pittet A, Clarke S (2001) Distinct pathways involved in sound recognition and localization: a human fMRI study. Neuroimage 14:802–816
McAlpine D, Grothe B (2003) Sound localization and delay lines—do mammals fit the model? Trends Neurosci 13:347–350
McAlpine D, Jiang D, Palmer AR (2001) A neural code for low-frequency sound localization in mammals. Nature Neurosci 4:396–401
McEvoy LK, Picton TW, Champagne SC (1991) The timing of the processes underlying lateralization: psychophysical and evoked potential measures. Ear Hear 12:389–398
McEvoy L, Hari R, Imada T, Sams M (1993) Human auditory cortical mechanisms of sound lateralization: II. Interaural time differences at sound onset. Hear Res 67:98–109
McPherson DL, Starr A (1993) Binaural interaction in auditory evoked potentials: Brainstem, middle- and long-latency components. Hear Res 66:91–98
Meddis R, Hewitt M (1991) Virtual pitch and phase sensitivity of a computer model of the auditory periphery. I: Pitch identification. J Acoust Soc Am 89:2866–2882
Mehrgardt S, Mellert V (1977) Transformation characteristics of the external human ear. J Acoust Soc Am 61:1567–1576
Merthayasa IN, Hemmi H, Ando Y (1994) Loudness of a 1 kHz pure tone and sharply (1080 dB/Oct.) filtered noises centered on its frequency. Mem Grad School Sci Tech Kobe Univ 12A:147–156
Mosteller F (1951) Remarks on the method of paired comparisons III. Psychometrika 16:207–218
Nedzelnitsky V (1980) Sound pressure in the basal turn of the cat cochlea. J Acoust Soc Am 68:1676–1689
Onchi Y (1961) Mechanism of the middle ear. J Acoust Soc Am 21:794–805
Osman E (1971) A correlation model of binaural masking level differences. J Acoust Soc Am 50:1494–1511
Osman E, Tzuo HY, Tzuo PL (1975) Theoretical analysis of detection of monaural signals as a function of interaural noise correlation and signal frequency. J Acoust Soc Am 57:939–942
Palmer AR, Jiang D, McAlpine D (1999) Desynchronizing responses to correlated noise: a mechanism for binaural masking level differences at the inferior colliculus. J Physiol 417:722–734
Palomäki K, Tiitinen H, Mäkinen V, May PJC, Alku P (2005) Spatial pro-cessing in human auditory cortex: the effects of 3D, ITD, and ILD stimulation techniques. Cogn Brain Res 24:364–379
Pantev C, Lütkenhörner B, Hoke M, Lehnertz K (1986) Comparison between simultaneously recorded auditory-evoked magnetic fields and potentials elicited by ipsilateral, contralateral, and binaural tone burst stimulation. Audiology 25:54–61
Pantev C, Hoke M, Lehnertz K, Lütkenhöner B, Anogianakis G, Wittkowski W (1988) Tonotopic organization of the human auditory cortex revealed by transient auditory evoked magnetic fields. Electroenceph Clin Neurophysiol 69:160–170
Pantev C, Hoke M, Lehnertz K, Lütkenhöner B (1989) Neuromagnetic evidence of an amplitopic organization of the human auditory cortex. Electroenceph Clin Neurophysiol 72:225–231

Pantev C, Bertrand O, Eulitz C, Verkindt C, Hampson S, Schuierer G, Elbert T (1995) Specific tonotopic organizations of different areas of the human auditory cortex revealed by simultaneous magnetic and electric recordings. Electroenceph Clin Neurophysiol 94:26–40

Patterson RD, Allerhand M, Giguere C (1995) Time-domain modelling of peripheral auditory processing: A modular architecture and a software platform. J Acoust Soc Am 98:1890–1894

Pickles JO (1983) Auditory-nerve correlates of loudness summation with stimulus bandwidth, in normal and pathological cochleae. Hear Res 12:239–250

Pickles JO (ed) (2008) An introduction to the physiology of hearing, 3rd edn. Academic Press, London

Puria S, William T, Peake WT, Rosowski JJ (1997) Sound-pressure measurements in the cochlear vestibule of human-cadaver ears. J Acoust Soc Am 101:2754–2770

Rauschecker JP, Tian B (2000) Mechanisms and streams for processing of "what" and "where" in auditory cortex. Proc Natl Acad Sci USA 97:11800–11806

Rauschecker JP, Tian B, Hauser M (1995) Processing of complex sounds in the macaque nonprimary auditory cortex. Science 268:111–114

Reite M, Zimmerman JT, Edrich J, Zimmerman JE (1982) Auditory evoked magnetic fields: response amplitude vs. stimulus intensity. Electroenceph Clin Neurophysiol 54:147–152

Relkin EM, Doucet JR (1997) Is loudness simply proportional to the auditory nerve spike count? J Acoust Soc Am 101:2735–2740

Roberts TPL, Poeppel D (1996) Latency of auditory evoked M100 as a function of tone frequency. NeuroReport 7:1138–1140

Romani GL, Williamson SJ, Kaufman L (1982) Tonotopic organization of the human auditory cortex. Science 216:1339–1340

Rubinstein M, Feldman B, Fischler F, Frei EH, Spira D (1966) Measurement of stapedial-footplate displacements during transmission of sound through the middle ear. J Acoust Soc Am 44:1420–1426

Rylander R, Björkman M (1997) Annoyance by aircraft noise around small airports. J Sound Vib 205:533–538

Rylander R, Sörensen S, Kajland A (1972) Annoyance reactions from aircraft noise exposure. J Sound Vib 24:419–444

Rylander R, Björkman M, Åhrlin U, Sörensen S, Kajland A (1980) Aircraft noise contours: importance of overflight frequency and noise level. J Sound Vib 69:583–595

Rylander R, Björkman M, Åhrlin U, Arntzen U, Solberg S (1986) Dose–response relationships for traffic noise and annoyance. Arch Environ Health 41:7–10

Saberi K, Takahashi Y, Konishi M, Albeck Y, Arthur BJ, Farahbod H (1998) Effects of interaural decorrelation on neural and behavioral detection of spatial cues. Neuron 21:789–798

Sams M, Hämäläinen M, Hari R, McEvoy L (1993) Human auditory cortical mechanisms of sound lateralization: I. Interaural time differences within sound. Hear Res 67:89–97

Sato S, Kitamura T, Ando Y (2002) Loudness of sharply (2068 dB/Octave) filtered noises in relation to the factors extracted from the autocorrelation function. J Sound Vib 250:47–52

Sato S, Nishio K, Ando Y (2003) Propagation of alpha waves corresponding to subjective preference from the right hemisphere to the left with changes in the IACC of a sound field. J Temporal Des Arch Environ 3:60–69

Scharf B (1962) Loudness summation and spectrum shape. J Acoust Soc Am 34:228–233

Schlauch RS, Wier CC (1987) A method for relating loudness matching and intensity discrimination data. J Speech Hear Res 30:13–20

Secker-Walker HE, Searle C (1990) Time domain analysis of auditory-nerve-fiber firing rates. J Acoust Soc Am 88:1427–1436

Seither-Preisler A, Krumbholz K, Lutkenhoner B (2003) Sensitivity of the neuromagnetic N100m deflection to spectral bandwidth: A function of the auditory periphery? Audiol Neurootol 8:322–337

Seither-Preisler A, Krumbhol K, Patterson RD, Seither S, Lutkenhoner B (2004) Interaction between the neuromagnetic responses to sound energy onset and pitch onset suggests common generators. Eur J Neurosci 19:3073–3080

Shackleton TM, Arnott RH, Palmer AR (2005) Sensitivity to interaural correlation of single neurons in the inferior colliculusof guinea pigs. J Assoc Res Otolaryngol 6:244–259
Shaw EAG (1974) Transformation of sound pressure level from the free field to the eardrum in the horizontal plane. J Acoust Soc Am 56:1848–1861
Shaw EAG, Teranishi R (1968) Sound pressure generated in an external-ear replica and real human ears by a nearby point source. J Acoust Soc Am 44:240–249
Shofner WP (1991) Temporal representation of rippled noise in the anteroventral cochlear nucleus of the chinchilla. J Acoust Soc Am 90:2450–2466
Shofner WP (1999) Responses of cochlear nucleus units in the chinchilla to iterated rippled noises: analysis of neural autocorrelograms. J Neurophysiol 81:2662–2674
Soeta Y, Nakagawa S (2008a) The effect of pitch and pitch strength on an auditory-evoked N1m. NeuroReport 19:783–787
Soeta Y, Nakagawa S (2008b) Relationship between loudness and auditory evoked N1m. Interdisci Res Explor, Biomagnetism, pp 95–97
Soeta Y, Nakagawa S (2009) Level-dependent growth on auditory evoked N1m for low- and high-frequency tones. NeuroReport 20:548–552
Soeta Y, Nakagawa S, Tonoike M, Ando Y (2002) Magnetoencephalographic responses corresponding to individual subjective preference of sound fields. J Sound Vib 258:419–428
Soeta Y, Nakagawa S, Tonoike M, Ando Y (2003) Spatial analyses of magnetoencephalographic activities in relation to subjective preference of a sound field. J Temporal Des Arch Environ 3:28–35
Soeta Y, Maruo T, Ando Y (2004a) Annoyance of bandpass filtered noises in relation to the factor extracted from autocorrelation function. J Acoust Soc Am 116:3275–3278
Soeta Y, Hotehama T, Nakagawa S, Tonoike M, Ando Y (2004b) Auditory evoked magnetic fields in relation to the inter-aural cross-correlation of bandpass noise. Hear Res 96:109–114
Soeta Y, Nakagawa S, Tonoike M, Ando Y (2004c) Magnetoencephalographic responses corresponds to individual annoyance of bandpass noise. J Sound Vib 277:479–489
Soeta Y, Nakagawa S, Tonoike M (2005a) Auditory evoked magnetic fields in relation to bandwidth variations of bandpass noise. Hear Res 202:47–54
Soeta Y, Nakagawa S, Tonoike M (2005b) Auditory evoked magnetic fields in relation to the iterated rippled noise. Hear Res 205:256–261
Soeta Y, Nakagawa S, Matsuoka K (2006) The effect of center frequency and bandwidth on the auditory evoked magnetic field. Hear Res 218:64–71
Stecker GC, Harrington IA, Middlebrooks JC (2005) Location coding by opponent neural populations in the auditory cortex. PLoS Biol 3:520–528
Stufflebeam SM, Poeppel D, Rowley HA, Roberts TPL (1998) Peri-threshold encoding of stimulus frequency and intensity in the M100 latency. NeuroReport 9:91–94
Suzuki Y, Takeshima H (2004) Equal-loudness counters for pure tones. J Acoust Soc Am 116:918–933
ten Kate JH, van Bekkum MF (1988) Synchrony-dependent autocorrelation in eighth-nerve-fiber response to rippled noise. J Acoust Soc Am 84:2092–2102
Thurstone LL (1927) A law of comparative judgement. Psychol Rev 34:273–289
Tiihonen J, Hari RM, Kajola M, Karhu J, Ahlfors S, Tissari S (1991) Magnetoencephalographic 10-Hz rhythm from the human auditory cortex. Neurosci Lett 129:303–305
Vasama JP, Mäkelä JP, Tissari SO, Hämäläinen MS (1995) Effects of intensity variation on human auditory evoked magnetic fields. Acta Otolaryngol (Stockh) 115:616–621
Wada H, Metoki T, Kobayashi T (1992) Analysis of dynamic behavior of human middle-ear using a finite-element method. J Acoust Soc Am 92:3157–3168
Wessinger CM, VanMeter J, Tian B, Van Lare J, Pekar J, Rauschecker JP (2001) Hierarchical organization of the human auditory cortex revealed by functional magnetic resonance imaging. J Cogn Neurosci 13:1–7
Wiener FM, Ross DA (1946) The pressure distribution in the auditory canal in a progressive sound field. J Acoust Soc Am 18:401–408
Wightman FL (1973) The pattern-transformation model of pitch. J Acoust Soc Am 54:407–416

Winter IM, Wiegrebe L, Patterson RD (2001) The temporal representation of the delay of iterated rippled noise in the ventral cochlear nucleus of the guinea-pig. J Physiol 537:553–566

Woldorff MG, Tempelmann C, Fell J, Tegeler C, Gaschler-Markefski B, Hinrichs H, Heinze H, Scheich H (1999) Lateralized auditory spatial perception and the contralaterality of cortical processing as studied with functional magnetic resonance imaging and magnetoencephalography. Hum Brain Mapp 7:49–66

Yin TC, Chan JC (1990) Interaural time sensitivity in medial superior olive of cat. J Neurophysiol 64:465–488

Yin TC, Chan JCK, Carney LH (1987) Effects of interaural time delays of noise stimuli on low-frequency cells in the cat's inferior colliculus. III. Evidence for cross-correlation. J Neurophysiol 58:562–583

Yost WA (1996) Pitch strength of iterated rippled noise. J Acoust Soc Am 100:3329–3335

Yost WA (ed) (2000) Fundamentals of hearing: an introduction. Academic Press, San Diego

Yost WA, Hill R (1979) Models of the pitch and pitch strength of ripple noise. J Acoust Soc Am 66:400–410

Yost WA, Patterson R, Sheft S (1996) A time domain description for the pitch strength of iterated ripple noise. J Acoust Soc Am 99:1066–1078

Yvert B, Bertrand O, Pernier J, Ilmoniemi RJ (1998) Human cortical responses evoked by dichotically presented tones of different frequencies. NeuroReport 9:1115–1119

Zimmer U, Macaluso E (2005) High binaural coherence determines successful sound localization and increased activity in posterior auditory areas. Neuron 47:893–905

Zwicker E, Flottorp G, Stevens SS (1957) Critical bandwidth in loudness summation. J Acoust Soc Am 29:548–557

Zwislocki JJ (1962) Analysis of middle ear function. Part I: input impedance. J Acoust Soc Am 35:1514–1523

Zwislocki JJ (1965) Analysis of some auditory characteristics. Handb Math Psycol 3:1–97

Chapter 3
Noise Measurement Method Based on the Model

Abstract To evaluate environmental noise, we need to use methods based on functioning of our auditory system. In this chapter, basic concept of noise measurement is described. First, the correlation between two variables, correlation function as a function of time, such as autocorrelation function (ACF) and cross-correlation function (CCF), and factors extracted from the ACF and CCF are described. ACF and CCF are related monaural and binaural criteria, respectively. Second, system of noise measurement based on the ACF and IACF factors is described.

Keywords Correlation · Autocorrelation function · Interaural cross-correlation function · System of noise measurement

3.1 Correlation Function

A correlation is one of the most common and most useful statistics. The correlation coefficient, r, between two variables x and y is defined as

$$r = \frac{\sum_{i=1}^{n}(x_i - \bar{x})(y_i - \bar{y})}{\sqrt{\sum_{i=1}^{n}(x_i - \bar{x})^2}\sqrt{\sum_{i=1}^{n}(y_i - \bar{y})^2}}. \tag{3.1}$$

It measures the strength and direction of a linear relationship between two variables and lies between −1 and 1. Figure 3.1 shows some examples of relationship between two variables, white noises, pure tones with the same frequency, pure tones with the different frequency, human voices and its time delayed one, environmental noises and its time delayed one, and environmental noises obtained from left and right ears.

When a signal is a time series, it is characterized by periodicity or randomness as a function of time. Figure 3.2 shows some examples of the relationship between a signal and the delayed version. The signals are white noise, pure tone, human voice,

Y. Soeta and Y. Ando, *Neurally Based Measurement and Evaluation of Environmental Noise*, Mathematics for Industry 20,
DOI 10.1007/978-4-431-55432-5_3

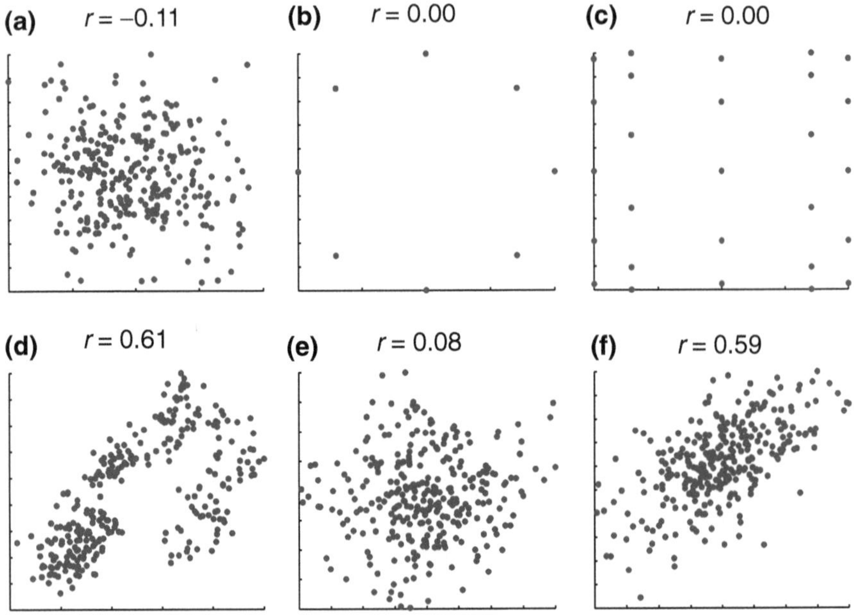

Fig. 3.1 Relationship between two variables. **a** White noises, **b** pure tones with the same frequency, **c** pure tones with the different frequency, **d** human voice and its time delayed one, **e** environmental noises and its time delayed one, and **f** environmental noises obtained from left and right ears

and train noise. The correlation coefficients can change as a function of time and it can be observed by autocorrelation function (ACF). The ACF is a set of correlation coefficients between the series and lags of itself over time. The ACF is a time domain function that is a measure of how much a signal shape, or waveform, resembles a delayed version of itself. For example, white noise is random, so the ACF has quite low value. Pure tone is completely periodic, so the ACF is also periodic. Human voice and environmental noise have periodic components, and the ACF has high values as shown in Fig. 3.3.

In a similar way, the relationship between two time series data is characterized by the correlation coefficient as a function of time, that is, cross-correlation function (CCF). The CCF is a measure of similarity of two waveforms as a function of a time lag applied to one of them. Figure 3.4 shows examples of the CCF between pure tones with different frequencies, human voice, and environmental noises obtained from left and right ears. The CCF between signals obtained from left and right ears is called interaural cross-correlation function (IACF).

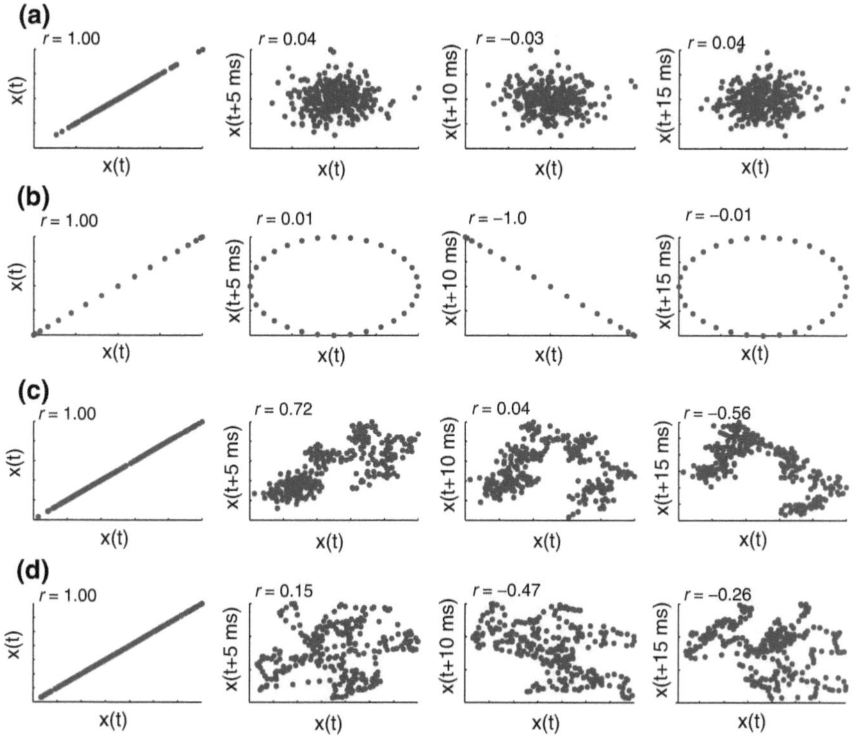

Fig. 3.2 Relationship between signal and its delayed one. **a** White noise, **b** pure tone, **c** human voice, and **d** train noise

3.2 Temporal Factors Extracted from the ACF

The ACF/IACF factors of the noise are proposed for describing the sound quality (Ando 2001). The normalized ACF for the signals received at each ear from the microphones, $p_{l(r)}(t)$, as a function of the running step, s, is defined by

$$\phi_{ll(rr)}(\tau; s, T) = \frac{\Phi_{ll(rr)}(\tau; s, T)}{\sqrt{\Phi_{ll(rr)}(0; s, T)\Phi_{ll(rr)}(0; s+\tau, T)}} \quad (3.2)$$

where

$$\Phi_{ll(rr)}(\tau; s, T) = \frac{1}{2T}\int_{s-T}^{s+T} p'_{l(r)}(t)p'_{l(r)}(t+\tau)\mathrm{d}t. \quad (3.3)$$

The ACF is normalized by the geometric mean of the energy at s and at $s + \tau$ and should not be normalized by the energy only at s. This ensures that the normalized

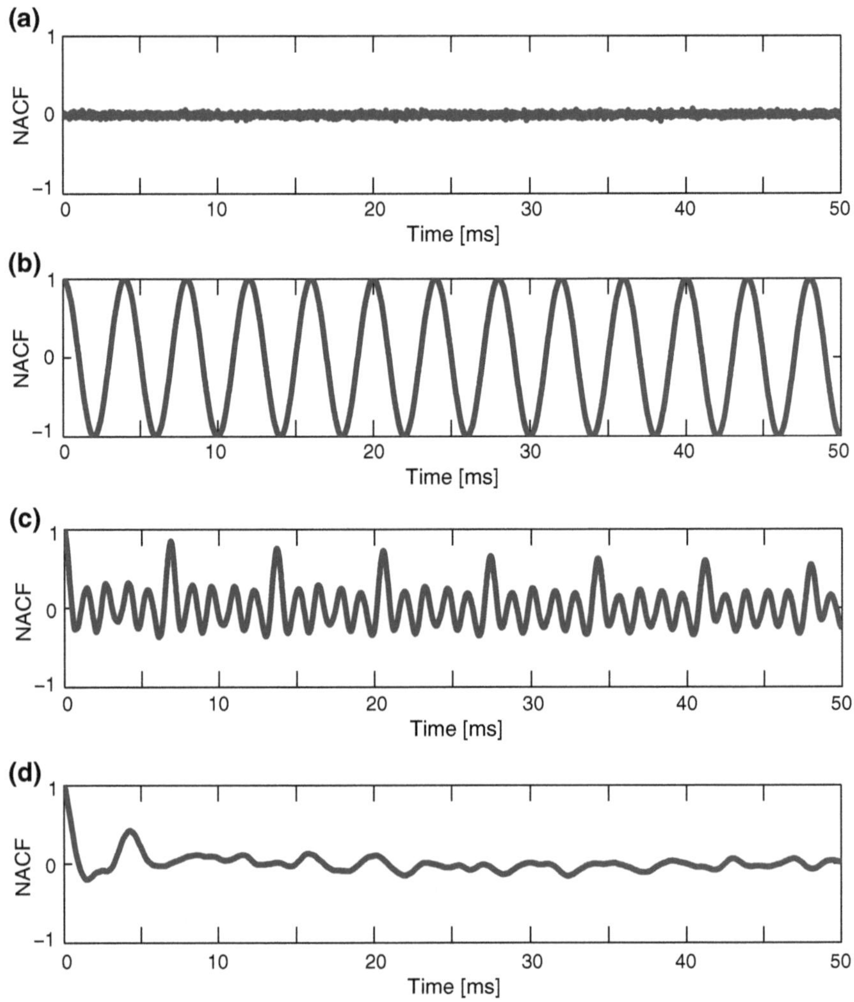

Fig. 3.3 Examples of the normalized ACF (NACF) of **a** white noise, **b** pure tone, **c** human voice, and **d** train noise

ACF satisfies $\phi_{ll(rr)}(0) = 1$ and $\phi_{ll(rr)}(\tau) \leq 1$ at $\tau > 0$. Here, $\Phi_{ll}(0)$ and $\Phi_{rr}(0)$ are the ACFs at $\tau = 0$ for the left and right ears, $2T$ is the integration interval, and $p'_{l(r)}(t) = p_{l(r)}(t) * s_e(t)$, $s_e(t)$ is the ear sensitivity. For convenience, we chose $s_e(t)$ as the impulse response of an A-weighted network, which is formed by the transfer function of the physical system from the free field to the oval window of the cochlea as shown in Fig. 2.4 (Ando 2001; Ando and Cariani 2009). Rigorously, individual sensitivity represented by $s_e(t)$ is different, for the purpose of practical use; however, $s_e(t)$ may be chosen as the impulse response of an A-weighted network, which is commonly used. The auditory temporal window $2T$ is described in the following section.

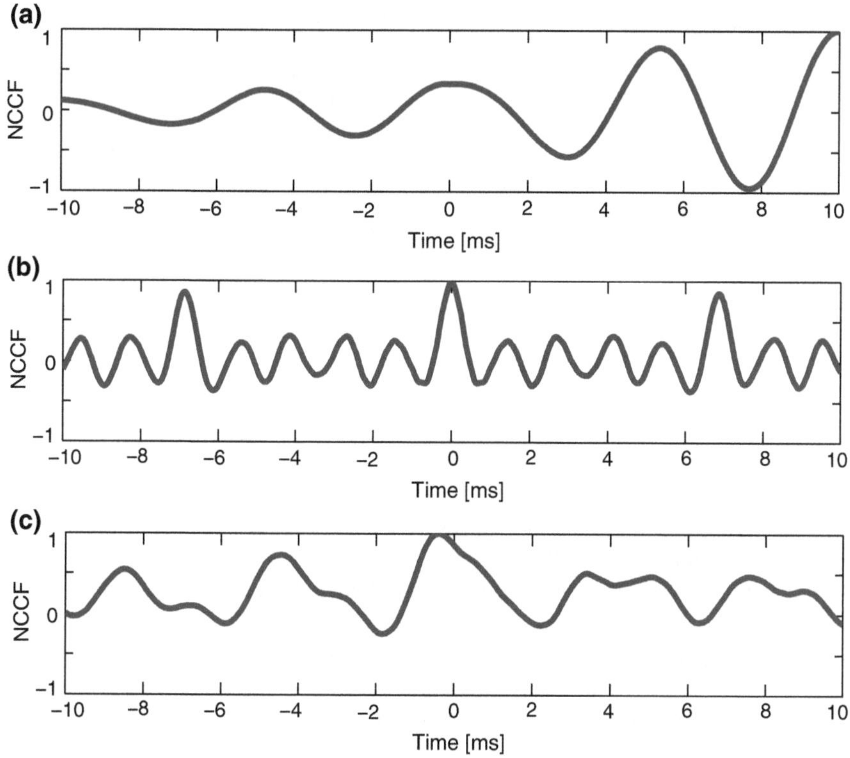

Fig. 3.4 Examples of the normalized CCF (NCCF) of **a** pure tone, **b** human voice, and **c** train noise

The ACF and the power density spectrum mathematically contain the same information. However, after a rough analysis of the power spectrum at the basilar membrane, the neural system with the synaptic delay and the summation have activities similar to the ACF as discussed in the previous section. It is considered that the correlation mechanisms in the neural system may work in such a manner, "the minimum effort conveying the maximum information to the brain."

From the ACF analysis, five temporal factors are extracted, which may play important role for any temporal sensations:

1. Energies at two ear entrances represented are given by $\Phi_{(ll,\ rr)}(0)$. The geometrical mean of the sound energies arriving at the two ears yields the binaural listening level, which corresponds to the A-weighted equivalent continuous sound pressure level (SPL), L_{Aeq}, that is given by as a function of the running step, s,

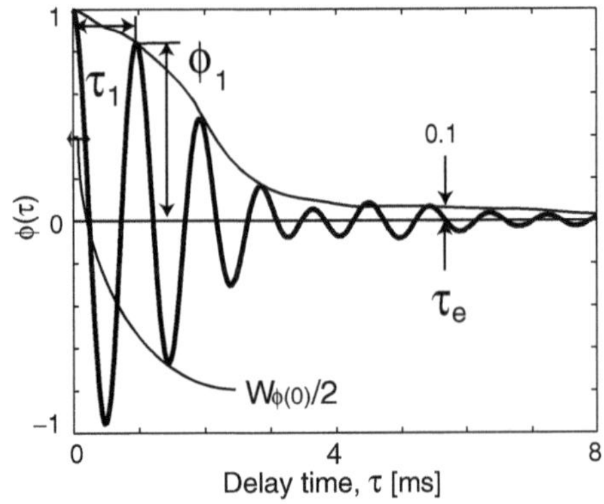

Fig. 3.5 The definition of the ACF factors, τ_1, ϕ_1, τ_e, and $W_{\phi(0)}$

$$L_{\mathrm{Aeq}}(s,T) = 10\log\sqrt{\Phi_{ll}(0;s,T)\Phi_{rr}(0;s,T)}\Big/\Phi_{\mathrm{ref}}(0). \tag{3.4}$$

where $\Phi_{\mathrm{ref}}(0)$ is the reference corresponding to the pressure 20 μPa that is the reference value of the SPL.

In the fine structures including peaks and delays, for instance, τ_1 and ϕ_1 are the delay time and the amplitude of the first maximum of the ACF. Then, there are τ_n and ϕ_n being the delay time and the amplitude of the nth local peak ($n > 1$). Usually, there are certain correlations between τ_n and τ_{n+1} and between ϕ_n and ϕ_{n+1}; thus, the most significant factor can be obtained at the first maximum, so that:

2. Factor τ_1.
3. Factor ϕ_1.
4. The effective duration of the normalized ACF, τ_e, is defined by the ten-percentile delay and represents a repetitive feature containing the sound source itself (Ando 1985, 1998). For convenience, τ_e is obtained by fitting a straight line for extrapolation of delay time by using only the initial declining portion, 0 dB > 10 log $|\phi_{ll(rr)}(\tau)| > -5$ dB, if the initial envelope of the ACF decays exponentially, as shown in Fig. 2.32b (Ando 1985, 1998). In most cases, the envelope decay of the initial part of the ACF can fit a straight line.
5. As shown in Fig. 3.5, the width of the amplitude $\phi_{ll(rr)}(\tau)$ around the origin of the delay time defined at a value of 0.5 is $W_{\phi(0)}$, according to the fact that $\phi_{ll(rr)}(\tau)$ is an even function.

These factors are usually analyzed by as a function of the running step, s, because characteristics of source signals change as a function of time. The concept of the running analysis is shown in Fig. 3.6.

Some examples of ACF factor analysis are shown here. Analyzed sound sources are pure tone, 1 and 1/3 octave band noises, white noise, and pink noise. To obtain sharp filtered 1 and 1/3 octave band noises, the white noises, which each had a 10 s duration, were repeatedly digitally filtered 50 times by the 8th-order Butterworth

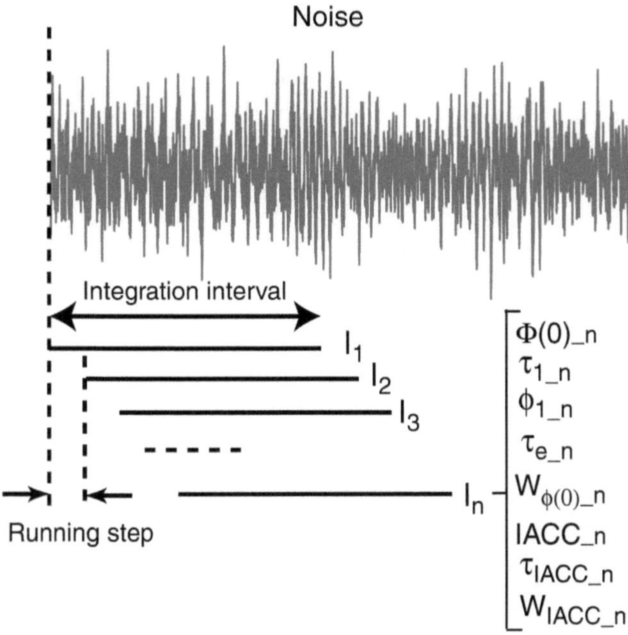

Fig. 3.6 Concept of running analysis

filter. Figure 3.7 shows the ACF factors for pure tone, 1 and 1/3 octave band noises, white noise, and pink noise. τ_1 and $W_{\phi(0)}$ decrease with increasing the frequency of pure tones and the center frequency of 1 and 1/3 octave band noises. ϕ_1 increases with decreasing the bandwidth. Figure 3.8 shows the relationship between $1/\tau_1$ and the frequency of pure tones or the center frequency of 1 and 1/3 octave band noises. τ_1 corresponds to the frequency. The spectral centroid, C_t, is defined as the center of gravity of the magnitude spectrum of the short-time Fourier transform

$$C_t = \frac{\sum_{n=1}^{N} n \cdot |M_t[n]|^2}{\sum_{n=1}^{N} |M_t[n]|^2}, \tag{3.5}$$

where $M_t[n]$ is the magnitude of the Fourier transform at frame t and frequency bin n. The centroid is a measure of spectral shape, and higher centroid values correspond to "brighter" textures with more high frequencies. Figure 3.9 shows the relationship between the C_t and $1/\{3 \times W_{\phi(0)}\}$. The correlation coefficient between $1/\{3 \times W_{\phi(0)}\}$ and the spectral centroid, C_t, is 0.98.

In Chap. 4, four temporal sensations (i.e., loudness, pitch, timbre, and duration) are described by means of these factors extracted from the ACF.

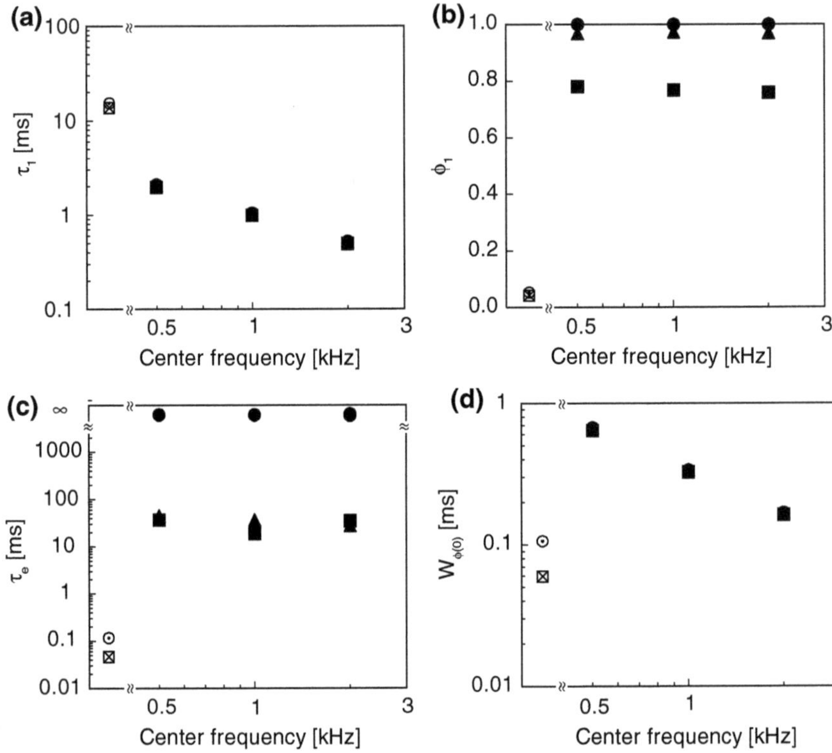

Fig. 3.7 ACF factors **a** τ_1, **b** ϕ_1, **c** τ_e, and **d** $W_{\phi(0)}$ for (*filled circle*) pure tone, (*filled triangle*) 1 and (*filled square*) 1/3 octave band noises, (*squared times*) white noise, and (*circled dot*) pink noise

Fig. 3.8 Relationship between $1/\tau_1$ and the frequency of pure tones or the center frequency of 1 and 1/3 octave band noises

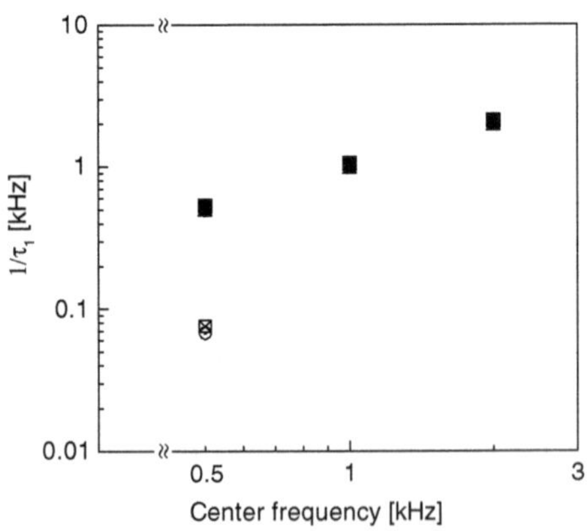

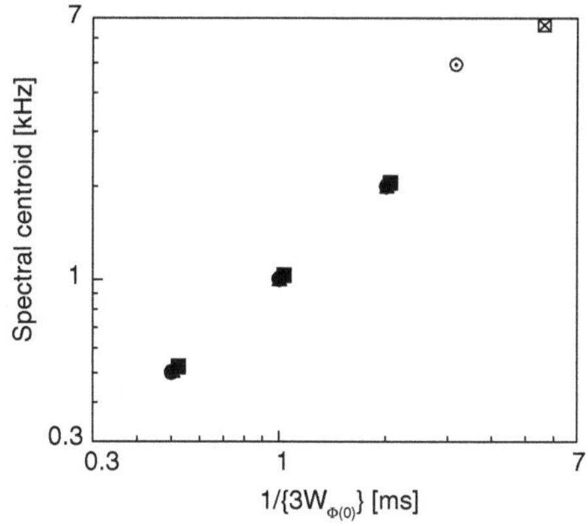

Fig. 3.9 Relationship between the spectral centroid, C_t and $1/\{3 \times W_{\phi(0)}\}$

3.3 Temporal Window of the ACF Processing

In analysis of the running ACF, the so-called auditory temporal window $2T$ in Eq. (3.3) must be carefully determined. The initial part of the ACF within the effective duration τ_e of the ACF contains important information of the signal. To determine the auditory temporal window, successive loudness judgments in pursuit of the running SPL have been conducted (Mouri et al. 2001).

Telephone ringing, the fan noise of the air conditioner, the sound of keypunch, the fan noise of the personal computer, human voice, music B (Beethoven, Symphonie No. 6, F - dur, op. 68, Pastorale Andante molto mosso), music M (Mozart, String Quintet, No. 4, first movement), Motif A, Motif B (Ando 1977), aircraft noise A, and aircraft noise B (Fujii et al. 2001) were analyzed by the ACF as a function of time. The $2T$ was set at 0.1, 0.2, 0.4, 1.0, and 2.0 s with the running intervals of 0.1 s. Figure 3.10 shows the results of $\Phi(0)$ for music B as a function of time.

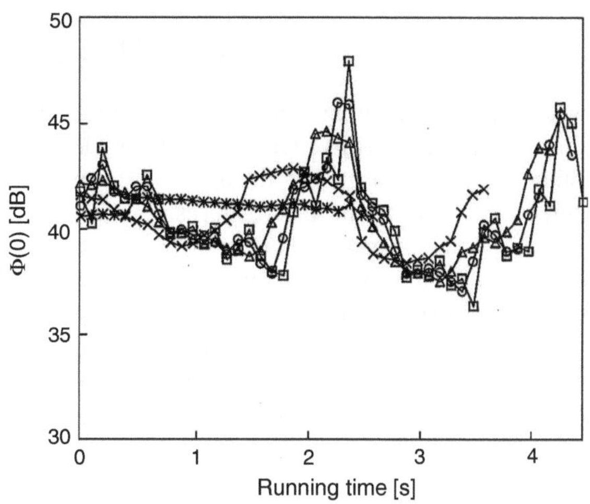

Fig. 3.10 Running $\Phi(0)$ of music B for each to $2T$: (*open square*) $2T = 0.1$ s; (*open circle*) $2T = 0.2$ s; (*open triangle*) $2T = 0.4$ s; (*times*) $2T = 1.0$ s; and (*division times*) $2T = 2.0$ s (Mouri et al. 2001)

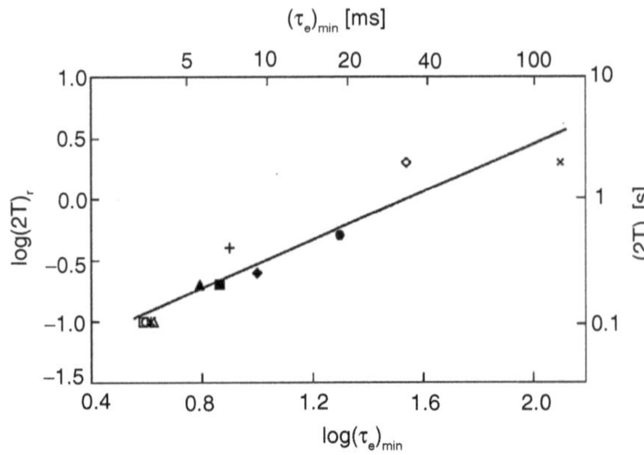

Fig. 3.11 Relationship between $(2T)_r$ and the $(\tau_e)_{min}$ for each sound source: (*filled triangle*) telephone ringing; (*open circle*) the fan noise of the air conditioner; (*division times*) the sound of the keypunch; (*open triangle*) the fan noise of the personal computer; (*open square*) human voice; (*plus*) music H; (*filled square*) music I; (*times*) Motif A; (*open diamond*) Motif B; (*filled circle*) aircraft noise A; and (*filled diamond*) aircraft noise B (Mouri et al. 2001)

The waveform of $\Phi(0)$, τ_1, ϕ_1, and τ_e for each $2T$ was compared to its hearing impression, respectively. The relationship between the waveform and hearing impression was observed in $\Phi(0)$ and τ_1; however, it was not found in ϕ_1 and τ_e because these factors may be masked by $\Phi(0)$ and τ_1. For the fan noise of the air conditioner, the sound of keypunch, the fan noise of the personal computer, and human voice, the hearing impression matched the waveform of $\Phi(0)$ for $2T$ of 0.1 s. This $2T$ is defined as recommended signal duration, $(2T)_r$. For the telephone ringing, music B, and music I, the $(2T)_r$ was 0.2 s.

The relationship between $(2T)_r$ and τ_e is shown in Fig. 3.11. A linear relation is observed between log $(2T)_r$ and log $(\tau_e)_{min}$, and its regression line is approximately expressed by the equation

$$(2T)_r \approx 30(\tau_e)_{min}. \tag{3.6}$$

where $(\tau_e)_{min}$ is the minimum value of τ_e obtained by analyzing the running ACF. This signifies an adaptive temporal window depending on the temporal characteristics of the sound signal in the auditory system. Therefore, the temporal window differs according to the music pieces [$(2T)_r$ = 0.5–5 s], the vowels [$(2T)_r$ = 50–100 ms], and consonants [$(2T)_r$ = 5–10 ms] in the continuous speech signals. For example, the time constant represented by "fast" or "slow" in the current system of sound level meter should be replaced by this temporal window.

The running step (R_s), which signifies a degree of overlap of the signal to be analyzed, is not so critical. It may be selected as $K_2(2T)_r$, K_2 being chosen in the range of 1/4–1/2.

3.4 Spatial Factors Extracted from the IACF

The normalized IACF for the signals received at each ear from the microphones, $p_l(t)$ and $p_r(t)$, as a function of the running step, s, in the possible range of maximum interaural delay times is defined by

$$\phi_{lr}(\tau) = \phi_{lr}(\tau; s, T) = \frac{\Phi_{lr}(\tau; s, T)}{\sqrt{\Phi_{ll}(0; s, T)\Phi_{rr}(0; s+\tau, T)}}, \quad -1\,\text{ms} < \tau < +1\,\text{ms} \tag{3.7}$$

where

$$\Phi_{lr}(\tau; s, T) = \frac{1}{2T}\int_{s-T}^{s+T} p_l'(t)p_r'(t+\tau)\mathrm{d}t. \tag{3.8}$$

Here $\Phi_{ll}(0; s, T)$ and $\Phi_{rr}(0; s, T)$ are the ACFs at $\tau = 0$ for the left and right ears, $2T$ is the integration interval, and $p'_{l,\,r}(t) = p_{l,\,r}(t) * s_e(t)$, $s_e(t)$ is the ear sensitivity. Practically, $s_e(t)$ may be approximated by the inverse Fourier transform of the A-weighting network.

From the IACF analysis, four spatial factors are determined as shown in Fig. 3.12, which may play important role for any spatial sensations:

1. The magnitude of the interaural cross-correlation is defined by

$$\text{IACC} = |\phi_{lr}(\tau)|_{\max} \tag{3.9}$$

for the possible maximum interaural time delay, -1 ms $< \tau <$ +1 ms.

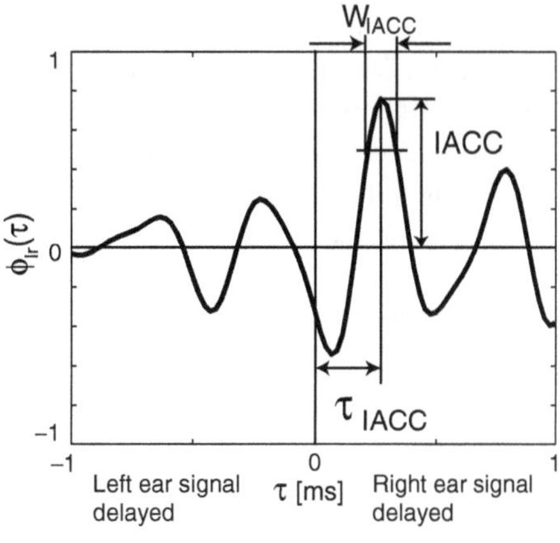

Fig. 3.12 The definition of the IACF factors, IACC, τ_{IACC}, and W_{IACC}

2. The interaural delay time at which IACC is defined is τ_{IACC}.
3. The width of the IACF defined by the interval of delay time at a value of δ below the IACC that may correspond to the just noticeable difference of the IACC is given by W_{IACC}. This value mainly depends on the frequency component of the source signals. A well-defined horizontal localization associated with the interaural time delay τ_{IACC} is perceived when listening to sounds with a sharp peak in the IACF and with a short value of $W_{IACC.}$ On the other hand, when listening to a sound field with a low value of the IACC < 0.15, subjectively diffuse sound is perceived (Damaske and Ando 1972).
4. The denominator of Eq. (3.7) defined by the SPL is the geometric mean of the sound energies arriving at the two ears.

In Chap. 5, the spatial sensations of the sound field are described by means of these factors extracted from the IACF. These spatial sensations may be judged immediately when we come into a sound field, because our binaural system may process the IACF in the short temporal window as discussed later in this chapter. This is quite different from the adaptive temporal window for the sound signals, which varies due to the effective duration of the ACF, τ_e, of the sound source signal.

3.5 Auditory Temporal Window for the IACF Processing

When a sound signal is moving in the horizontal direction, we must identify a suitable "temporal window" $2T$ in analyzing the running IACF, which corresponds to the moving image of the sound localization. The range of τ_{IACC} extracted from the IACF can describe the range of such a moving image. It is obvious that the range of τ_{IACC} disappears when the integration interval ($2T$) of the IACF is longer than the period of movement, and the value greatly fluctuates when $2T$ is shorter than the possible maximum value of $\tau_{IACC} < 1$ ms. For a sound source moving sinusoidal in the horizontal plane with less than 0.2 Hz, $2T$ may be selected in a range from 30 to 1000 ms, and when a sound source is moving below 4.0 Hz, $2T = 30$–100 ms is acceptable (Mouri 2003). In order to obtain reliable results, it is recommended for such a temporal window that the IACF covering a wide range of movement velocity in the horizontal localization be fixed at about 30 ms.

For the sound source fixed on the stage in a concert hall, for example, the value of ($2T$) may be selected longer than 1.0 s for the measurement of spatial factors at each audience seat in a room.

3.6 System of Noise Measurement

Based on the auditory-brain model, the noise measurement system has been developed (Ando 2001; Sakai et al. 2002; Fujii et al. 2004). Figure 3.13 is a block diagram of the measurement system. The system consists of binaural microphones, a laptop computer, and software calculating the ACF and IACF factors from a

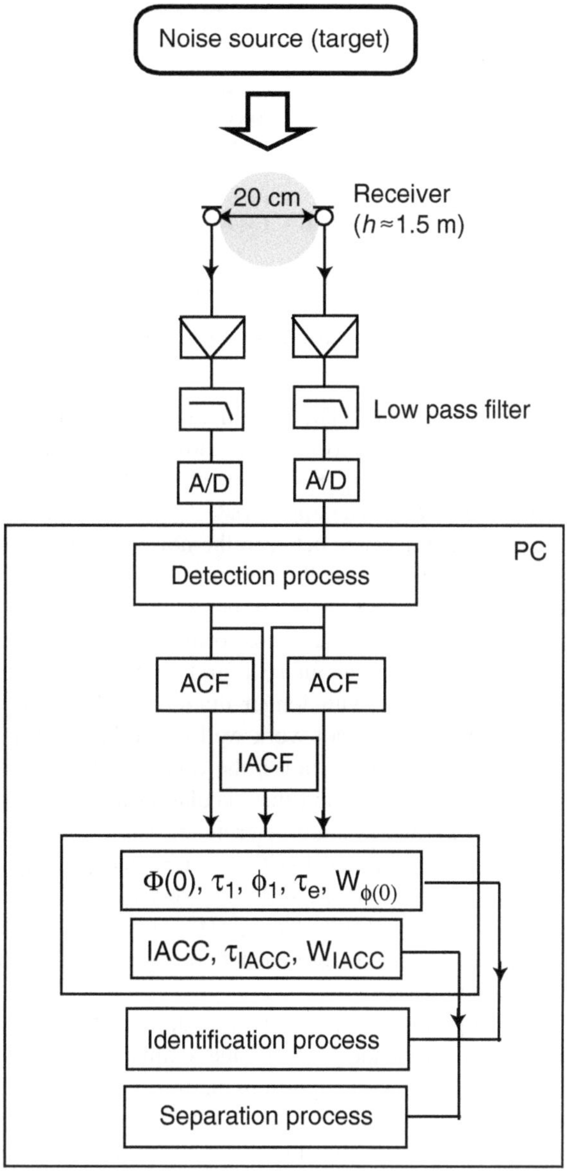

Fig. 3.13 A flowchart of the measurement system. The ACF and the IACF parameters are extracted after the process of automatic detection of noise event. Noise source is identified by using the ACF parameters

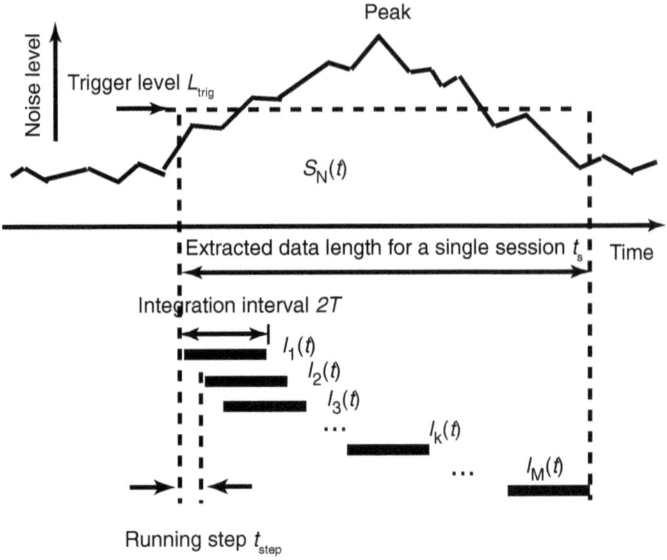

Fig. 3.14 Process of the extraction of noise event and the running ACF analysis

real-time noise data. The system can measure noise event automatically by using the peak detection algorithm. As shown in Fig. 3.14, the noise level is continuously monitored, and a target noise event is extracted when the noise level exceeds the trigger level, L_{trig}. The appropriate L_{trig} value varies according to the kind of target noise and the distance between the noise source and the receiver. It must therefore be determined by a preliminary measurement. The noise data with duration of t_s centered on its peak level are recorded on the hard disk as a single session. The duration t_s should be set so as to include the peak level after exceeding L_{trig}. For example, it is set to about 10 s for aircraft noises. This value is different between a steady-state noise with longer duration and an intermittent noise. Note that the current system does not work well when there are interfering noises.

For each session of noise, the running ACF and IACF are calculated with the integration time ($2T$) and the calculation interval (running step: t_{step}). Appropriate values of $2T$ and t_{step} are determined before the measurement. As you can see in Fig. 3.14, the ACF and IACF are calculated in every step (n = 1, 2, …, M) for the data length of $2T$ which shifts in every t_{step}, as $\{(0, 2T), (t_{step}, t_{step} + 2T), (2t_{step}, 2t_{step} + 2T), \ldots, ((M - 1)t_{step}, (M - 1)t_{step} + 2T)\}$. Acoustical factors described before are extracted from each step of the ACF and IACF and used for the identification process.

In the proposed system, noise source is identified based on the similarity of the acoustical parameters by using the template-matching algorithm. The basic concept of the identification algorithm in this system is illustrated in Fig. 3.15. Three dimensions in the figure represent the acoustical factors considered. A fundamental assumption here is that the noises from the same source have similar acoustical

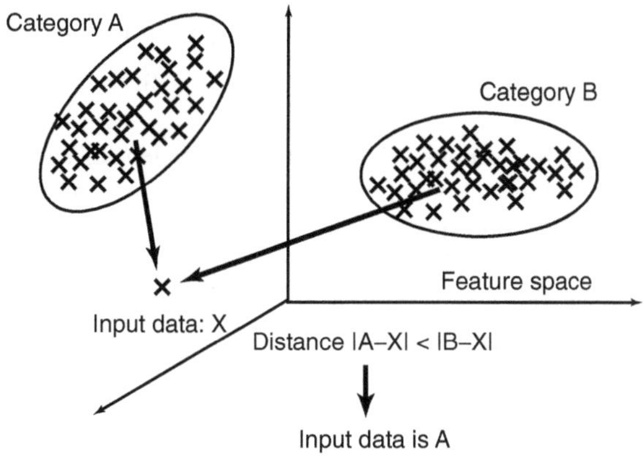

Fig. 3.15 Basic concept of the identification algorithm

qualities and therefore, they are mapped into the same cluster in the feature space. Now, suppose that there are two noise sources, and they are represented as the factor set *A* and *B*. When new data *X* are measured, the distance of factors |*A*–*X*| and |*B*–*X*| is calculated. Distance represents the similarity of factors. Thus, if the distance |*A*–*X*| is smaller than |*B*–*X*|, the input data *X* are categorized in *A*.

The acoustical factors used in the identification algorithm are derived from the running ACF. As described above, timbre information of noise is extracted from the ACF. Distance $D(x)$ (x: $\Phi(0)$, τ_1, ϕ_1, τ_e, and $W_{\phi(0)}$) between the unknown target data (indicated by symbol *a* in Eqs. (3.10)–(3.14)) and the template (already calculated data of noise sources, indicated by symbol *b*) is calculated by the following equations:

$$D(\Phi(0)) = \left|\log(\Phi(0))^a - \log(\Phi(0))^b\right|, \tag{3.10}$$

$$D(\tau_1) = \left|\log(\tau_1)^a - \log(\tau_1)^b\right|, \tag{3.11}$$

$$D(\phi_1) = \left|\log(\phi_1)^a - \log(\phi_1)^b\right|, \tag{3.12}$$

$$D(\tau_e) = \left|\log\{(\tau_e)_{\min}\}^a - \log\{(\tau_e)_{\min}\}^b\right|, \tag{3.13}$$

$$D(W_{\phi(0)}) = \left|\log(W_{\phi(0)})^a - \log(W_{\phi(0)})^b\right|. \tag{3.14}$$

$D(x)$ is calculated at the maximum SPL of measured noise, because the noise is considered as steady state at this point in many cases. The total distance D is calculated as the sum of the right-hand terms of Eqs. (3.10)–(3.14), so

$$D(x) = W^{\Phi(0)}D(\Phi(0)) + W^{\tau_1}D(\tau_1) + W^{\phi_1}D(\phi_1) + W^{\tau_e}D(\tau_e) + W^{W_{\phi(0)}}D(W_{\phi(0)}) \tag{3.15}$$

where $W^{(x)}$ shows the weighting coefficient of each parameter. The template with the smallest D is defined as the noise source of the target data. Weighting coefficients $W^{(x)}$ in Eq. (3.15) are obtained by using statistical values $S_1^{(x)}$ and $S_2^{(x)}$, as $(S_2/S_1)^{1/2}$. Here, $S_1^{(x)}$ is the arithmetic mean of the standard deviations (SD) for all categories (i.e., SD within a certain category), and $S_2^{(x)}$ is the SD of the arithmetic means in each category (i.e., SD between categories), and category means a set of data for the same noise source. It means that a parameter (x) becomes important when the SD between categories becomes large and the SD among a category becomes small.

The performance of the identification algorithm was tested for several kinds of sounds, such as car engine sound, aircraft noise, motorcycle noise, and human voice. Good identification performance was obtained in the measurement of the idling car engine sound. It is considered that the identification worked well because the acoustical quality of idling sound is almost constant during measurement and the ACF factors differed for different sounds. The proposed algorithm is thus applicable for the steady-state noise. Noise with nonstationary sound qualities was sometimes misidentified. It is because the algorithm calculates the distance by using sound factors only at the maximum SPL point. This problem would be solved by using the whole time pattern of the ACF factors for identification. A more important problem in the present algorithm is that it cannot deal with the mixed noise. Separation of sound source is indispensable in order to apply the measurement system in the real environment. It has been found that the binaural localization cues are very effective in separating a target sound from an interfering sound (Roman et al. 2003).

Further research is planned to deal with the fluctuating noise and mixed noise for applying the system in the real environment by ambisonics and independent component analysis. In addition, to increase the accuracy of the identification and evaluations of the environments, introduction of narrowband ACF (Valero and Alías 2013) and summary ACF (Sato et al. 2014) is under consideration.

References

Ando Y (1977) Subjective preference in relation to objective parameters of music sound fields with a single echo. J Acoust Soc Am 62:1439–1441

Ando Y (eds) (1985) Concert hall acoustics. Springer-Verlag, Heidelberg

Ando Y (eds) (1998) Architectural acoustics: blending sound sources, sound fields, and listeners. AIP Press, Springer-Verlag, New York

Ando Y (2001) A theory of primary sensations and spatial sensations measuring environmental noise. J Sound Vib 241:3–18

Ando Y, Cariani P (2009) Auditory and visual sensations. Springer, New York

Damaske P, Ando Y (1972) Interaural crosscorrelation for multichannel loudspeaker reproduction. Acustica 27:232–238

Fujii K, Soeta Y, Ando Y (2001) Acoustical properties of aircraft noise measured by temporal and spatial factors. J Sound Vib 241:69–78

Fujii K, Sakurai M, Ando Y (2004) Computer software for identification of noise source and automatic noise measurement. J Sound Vib 277:573–582

Mouri K (2003) On the temporal window of auditory-brain system in connection with subjective responses. Ph.D dissertation, Kobe University

Mouri K, Akiyama K, Ando Y (2001) Preliminary study on recommended time duration of source signals to be analyzed, in relation to its effective duration of the auto-correlation function. J Sound Vib 241:87–95

Roman N, Wang D, Brown GJ (2003) Speech segregation based on sound localization. J Acoust Soc Am 114:2236–2252

Sakai H, Hotehama T, Prodi N, Pompoli R, Ando Y (2002) Diagnostic system based on the human auditory-brain model for measuring environmental noise—an application to the railway noise. J Sound Vib 250:9–21

Sato S, Bidondo A, Soeta Y (2014) MATLAB program for calculating the parameters of the autocorrelation and interaural cross-correlation functions based on Ando's auditory-brain model. In: 137th AES convention paper, Los Angeles Convention Center, Los Angeles, 9–12 October 2014

Valero X, Alías F (2013) Narrow-band autocorrelation function features for the automatic recognition of acoustic environments. J Acoust Soc Am 134:880–890

Chapter 4
Temporal Primary Sensations of Noise

Abstract The basic, perceived attributes of sound can be divided into those qualities that distinguish different sounds independent of location (temporal sensations) and those related to a sound's perceived location in space (spatial sensations). Temporal sensations include pitch, loudness, timbre, and duration. They can be described in terms of temporal factors extracted from the autocorrelation function (ACF). The ACF has the same information as the power density spectrum of the signal under analysis. From the ACF, however, significant factors may be extracted, which are related to temporal sensations.

Keywords Temporal sensation · Pitch · Loudness · Timbre

4.1 Formulation of Temporal and Spatial Sensations

The model outlined in Chap. 2 leads to a division of primary sensations into two main categories: temporal sensations and spatial sensations. To begin with, we will discuss the relationship of subjective sensations to physical factors.

In psychophysical models, perceptual attributes associated with a given sensation can depend upon multiple physical factors, such that a single sensation j may not be well described only in terms of a single factor. Let c_i (i = 1, 2, …, I) be physical factors representing cues influencing any primary sensation (temporal and spatial sensations as mentioned above) S_j (j = 1, 2, …, $J < I$), where J is a number of significant physical factors and I is the total number of physical factors, a sensation S_j may be expressed by

$$S_j = f(x_1, x_2, \ldots, x_1), \quad j = 1, 2, \ldots, J \tag{4.1}$$

If physical factors are orthogonal to each other and contribute independently to a given sensation, then S_j may be expressed by a linear combination, such that

Y. Soeta and Y. Ando, *Neurally Based Measurement and Evaluation of Environmental Noise*, Mathematics for Industry 20,
DOI 10.1007/978-4-431-55432-5_4

$$S_j = f(x_1) + f(x_2) + \cdots + f(x_I), \quad j = 1, 2, \ldots, J \tag{4.2}$$

For example, let us consider the scale value of loudness, which might be described by not only the sound pressure level (SPL) but also by factors extracted from autocorrelation function (ACF), such as τ_1 corresponding to pitch and τ_e corresponding to repetitive feature of the signal.

Here, a question arises as to whether or not a single sensation is independent from other sensations. It is demonstrated by means of the simplest case with two physical factors x_1 and x_2 in Eq. (4.2) so that

$$\begin{aligned} S_1 &= f_1(x_1) + f_1(x_2), \quad j = 1 \\ S_2 &= f_2(x_1) + f_2(x_2), \quad j = 2 \end{aligned} \tag{4.3}$$

then, the correlation coefficient between S_1 and S_2 is given by

$$r_{12} = S_1 S_2 = f_1(x_1)f_2(x_1) + f_1(x_2)f_2(x_2) + f_1(x_1)f_2(x_2) + f_1(x_2)f_2(x_1) \tag{4.4}$$

and is not zero, because in general, the first and second terms of the right-hand side are not always zero. Previously, it was believed that these perceptual attributes are largely independent of each other. For example, timbre is independent and usually not much affected by loudness, pitch, or duration, within limits. And we can identify musical instruments by their timbres irrespective of how loudly and how long any arbitrary notes are being played. However, more rigorously, we shall discuss sensations in relation to the possible dimension of physical factors.

In our auditory model, we consider differences in response patterns between the cerebral hemispheres of human listeners. Here, temporal factors are more prominent in the left hemisphere, whereas spatial factors are more prominent in the right hemisphere. Models in which an internal variable associated with each hemisphere is modeled as the linear combination of hemisphere-specific factors can explain these differences. Temporal sensations S_L and spatial sensations S_R can thus be modeled in terms of the contributions of different factors that dominate in the neural response:

$$\begin{aligned} S_L &= f_L(x_{1l}) + f_L(x_{2l}) + \cdots + f_L(x_{Il}), \quad I = 1, 2, \ldots L \\ S_R &= f_R(x_{1r}) + f_R(x_{2r}) + \cdots + f_R(x_{Ir}), \quad I = 1, 2, \ldots R \end{aligned} \tag{4.5}$$

where $L + Rx \equiv J$.

Individual differences in the weighting of factors can also produce differences in sensation, preference, and annoyance. Even for such temporal and spatial sensations, there are substantial individual differences due to multiple physical factors as expressed by Eqs. (4.1) and (4.2). Individual differences can be caused both by differing individual sensitivities and/or unique responses to the various factors. These differences of sensation can be seen as characteristics of different individual listeners who have distinct auditory and visual "personalities."

Subjective responses that are related to the overall intensity of the evoked perceptual experience (e.g., preference or annoyance) can be expressed by both temporal and spatial factors, S_L and S_R, so that

$$S = S_L + S_R. \tag{4.6}$$

4.2 Loudness

4.2.1 Loudness of Sharply Filtered Noise

Previous studies on the relationship between loudness and the bandwidth of noise have concluded that for sounds having the same SPL, loudness remains constant as bandwidth increases, up until the bandwidth reaches the "critical band." For bandwidths larger than the critical band, loudness increases with bandwidth (Zwicker et al. 1957). The spectral characteristics of the filters used in those studies were not specified, except by Greenwood (1961a, b). Mathews and Pfafflin (1965) suggested that loudness of band-pass noises might differ between that using an actual filter and that using an ideal (rectangular shape) filter. An actual filter passes not only frequencies within the band defined by the 3 dB attenuation at the low and high cutoff frequencies but also at frequencies outside the band. The outside bandwidth response of the filter greatly affects the repetitive feature of the signal, represented by τ_e extracted from the ACF (Ando 1998; Ando et al. 1999). Due to the sharpening effect at least in the high-frequency range, such a sharp filter may exist in the auditory system (Katsuki et al. 1958); a roll-off of more than 1000 dB/octave is required. It is considered that loudness of a sharply (1080 dB/octave) filtered noise increases as the effective duration of the normalized ACF (τ_e) increases, even if the bandwidth of the signal is within the critical band. It is worth noting that the tendency observed that as the reverberation time (RT) of a sound field increases, the τ_e also increases (Ando 1998).

The purpose of this study was to examine the loudness of the band-pass noise in terms of factors extracted from the ACF. It is assumed that when the SPL is fixed at a constant value, the scale value of loudness S is expressed by

$$S = S_L = f_L(\tau_1, \phi_1, \tau_e, D) \tag{4.7}$$

where the factors are defined in Chap. 3; $W_{\phi(0)}$ is excluded in the above equation because the center frequency of the noise is fixed and is represented by τ_1, and D is the duration of the sound signal. It is worth noting that loudness does not depend on the interaural cross-correlation coefficients (IACCs) when the SPL at both ear entrances is fixed. This confirms the results obtained using headphone reproduction (Dubrovskii and Chernyak 1969).

White noise was produced by a random generator and it was filtered. The source signal of band-pass noises is characterized in terms of their ACFs as shown in Fig. 4.1.

Bandwidth (Δf) was changed by using a sharp filter with the cutoff slope of 2068 dB/octave, which was realized by a combination of two filters. Factors of τ_1, τ_e, and ϕ_1 analyzed are shown in Fig. 4.2. The filter bandwidth of 0 Hz included only its slope component. All source signals were the same SPL at 74 dBA. Whereas filter bandwidth has absolutely no effect on the signal's dominant periodicity (as reflected by τ_1), it has a profound effect on the slope of the ACF envelope (effective duration, as reflected by τ_e) and a lesser effect on the relative height of the peak associated with the dominant periodicity ϕ_1.

The loudness judgment was performed by the paired comparison test for which the ACF of the band-pass noise was changed. A headphone delivered the same sound signal to the two ears. Thus, the IACCs were kept constant at nearly unity. Sound signals were digitized at a sampling frequency of 48 kHz. Five listeners with normal hearing participated in the experiment. They were seated in an anechoic chamber and asked to judge which of two paired sound signals were perceived to be louder. Stimulus durations were 1.0 s, rise and fall times were 50 ms, and silent intervals between the stimuli were 0.5 s. A silent interval of 3.0 s separated each pair of stimuli, and the pairs were presented in random order.

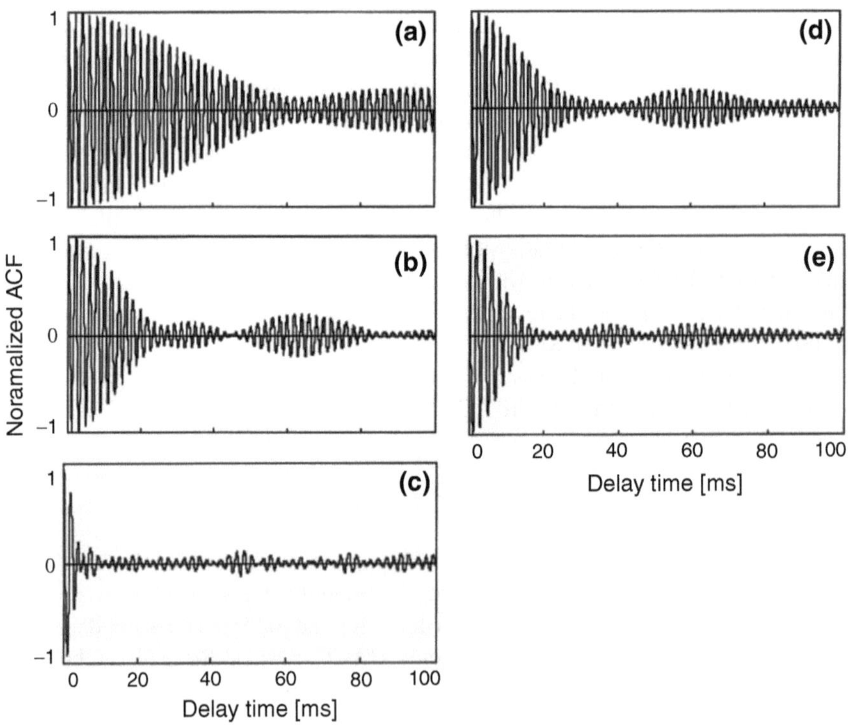

Fig. 4.1 Examples of the NACF analyzed for the center frequency of 500 Hz. The filter bandwidth Δf: **a** 0 Hz, **b** 40 Hz, **c** 80 Hz, **d** 160 Hz, **e** 320 Hz

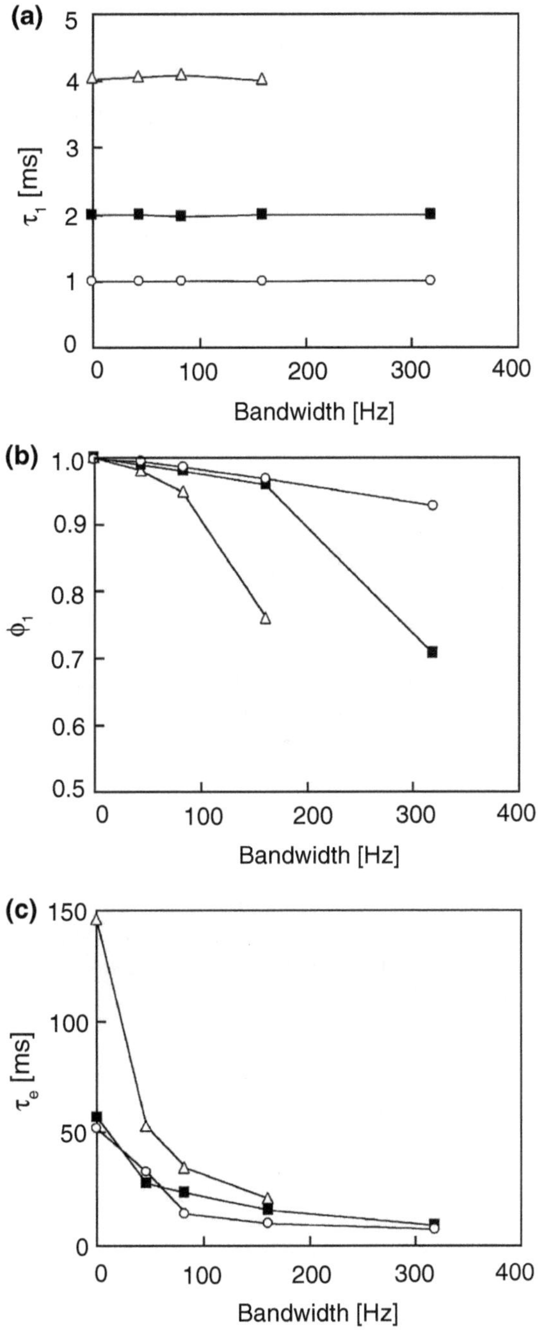

Fig. 4.2 Measured factors extracted from the ACF of the source signal as a function of the bandwidth. Different symbols indicate different frequencies: (*open triangle*) 250 Hz, (*filled square*) 500 Hz, (*open circle*) 1000 Hz. **a** Delay time of the first peak of ACF (τ_1); **b** amplitude of the first peak of ACF (ϕ_1); and **c** effective duration of ACF (τ_e)

Fifty responses (5 listeners × 10 sessions) to each stimulus were obtained. Consistency tests indicated that all listeners had a significant ($p < 0.05$) ability to discriminate loudness. The test of agreement also indicated that there was significant ($p < 0.05$) agreement among all listeners. A scale value of loudness was obtained by applying the law of comparative judgment (Thurstone's case V) and was confirmed by goodness of fit (Thurstone 1927).

The relationship between the scale value of loudness and the filter bandwidth is shown in Fig. 4.3. The scale value difference of 1.0 corresponds with about 1 dB due to the preliminary experiment. For all center frequencies (250, 500, 1000 Hz), the scale value of loudness is maximal for the pure tone with the infinite value of τ_e and large bandwidths, with minima at smaller bandwidths (40, 80, 160 Hz, respectively). From the dependence of τ_e on filter bandwidth, loudness increases with increasing τ_e almost within the critical bandwidth. Results of analysis of variance (ANOVA) for the scale values of loudness indicated that for all center frequencies tested, the scale value of loudness of pure tone was significantly larger than that of other band-pass noises within the critical band ($p < 0.01$). When the results of loudness with changes in the RT of the sound field are taken into account, the fact is that the factor τ_e, a measure of repetitive features of the sound signal, may contribute to the loudness (Merthayasa et al. 1994).

Consequently, loudness of the band-pass noise with identical SPL was not constant within the critical band. Also, loudness of the pure tone was significantly larger than that of sharply filtered noises, and loudness increased with increasing τ_e within the critical band. Therefore, Eq. (4.7) within the critical band may be reduced by

$$S = S_L = f_L(\tau_1) + f_L(\tau_e). \tag{4.8}$$

The similar loudness function has also been found in human auditory cortex. Auditory-evoked magnetic fields showed that the N1 m magnitude keeps constant or decreases with increasing bandwidth when the bandwidth is less than the critical bandwidth. Also, it increases with increasing bandwidth beyond the critical band (Soeta et al. 2005).

4.2.2 Loudness of Complex Noise

In the previous section, effects of τ_e on loudness were examined using simple band-pass noise. But the sounds are more complex in our surroundings. Thus, the loudness of spectrally complex noise was examined (Sato et al. 2001).The complex noises used consisted of multiple band-pass noises whose center frequencies were harmonics of a 1000-Hz fundamental. There were no correlations between the noise bands. The perceived pitch was centered on 1000 Hz, the "missing fundamental" of the noise bands. Perceptual judgments by listeners were compared with those for

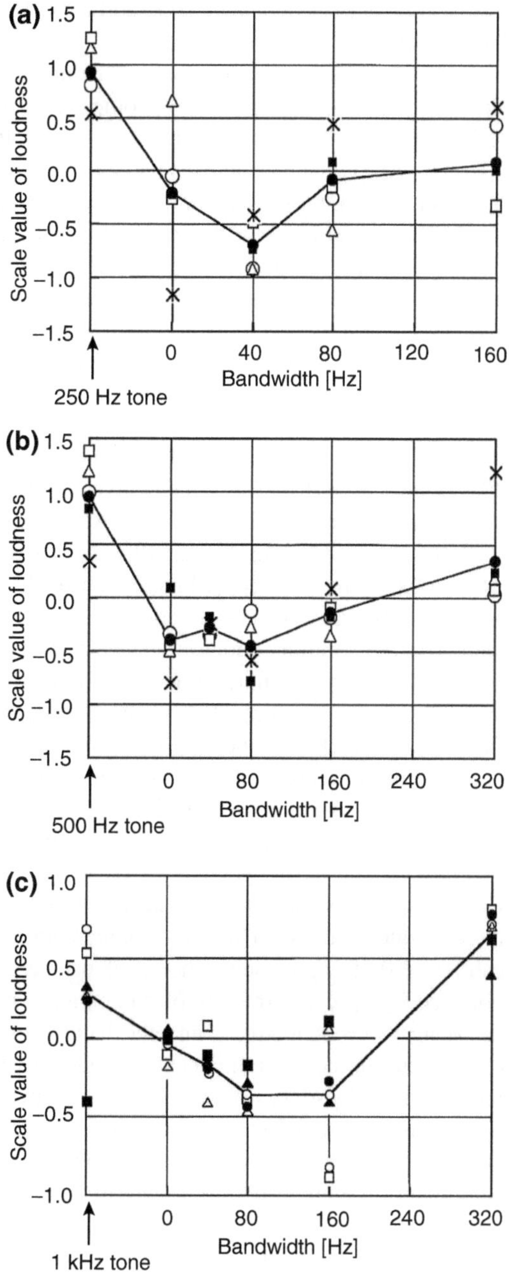

Fig. 4.3 Scale values of loudness as a function of the bandwidth of band-pass noise. f_c: **a** 250 Hz, **b** 500 Hz, and **c** 1000 Hz. Different symbols indicate the scale values obtained with different listeners

the single band-pass noise of 1000 and 2000 Hz center frequencies in terms of the factors extracted from the ACF.

Source signals in the experiments included the following: (1) a complex noise stimulus with two band-pass noise components whose center frequencies were 2000 and 3000 Hz and (2) a complex tone with pure-tone components of 2000 and 3000 Hz. All partial components had the same SPL. To control the τ_e of the ACF of the complex noise, the bandwidths of each partial noise (Δf) were changed to 0, 40, 80, 160, and 320 Hz with the cutoff slope of 2068 dB/octave. In fact, 0 Hz of the bandwidth was the only slope component of the filter used. Figure 4.4 shows the normalized ACF of the complex noise whose fundamental frequencies of 1000 Hz and the single noise component centered on 1000 Hz. As shown in Fig. 4.5a, all of the ACFs indicate the maximum peak at $\tau_1 = 1.0$ ms. Figure 4.5b, c show the measured ϕ_1 and τ_e of the source signals as a function of the bandwidth.

Loudness judgments were performed by the paired comparison test. Pairwise comparisons were made using the complex tones and five complex noises ($\Delta f = 0$, 40, 80, 160, 320 Hz). The same source signal was presented to both ears through headphones. The magnitude of the IACCs was thus kept constant at unity. All stimuli were fixed at the same SPL at 74 dBA. Input signals were digitized at 24 kHz sampling frequency. Fluctuation of the measured $\Phi(0)$ for all stimuli was within ± 0.06 dB when the duration of the signals was lengthened beyond 0.8 s; therefore, the stimulus duration was chosen at 1.0 s.

Four listeners with normal hearing ability were seated in the anechoic chamber and asked to judge which of two sound signals they perceived louder. The rise and fall times were 50 ms, and the silent interval between the stimuli was 0.5 s. Each pair of stimuli was separated by an interval of 3.0 s, and the pairs were presented in random order. A single test session consisted of 15 pairs [$N(N-1)/2$; $N = 6$] of stimuli and lasted about 1.5 min. Ten sessions were performed for each listener.

Forty responses (4 listeners × 10 sessions) to each stimulus were obtained. Consistency tests indicated that all listeners had a significant ($p < 0.01$) ability to discriminate loudness. The test of agreement also indicated that there was significant ($p < 0.05$) agreement among all listeners. The scale value of loudness was obtained by applying the law of comparative judgment. The relationship between the scale value of loudness and the bandwidth of each partial component of the complex noise with its fundamental frequency of 1000 Hz is shown in Fig. 4.6. The minimum loudness was observed for a bandwidth of 160 Hz. Loudness increased with increasing τ_e of the source signal within the bandwidth of 160 Hz for the complex noise. ANOVA for the scale values of loudness showed that there were significant differences between the pairings of a complex tone and 160 Hz, 0 and 80 Hz, 0 and 160 Hz, 40 and 80 Hz, and 40 and 160 Hz. It is remarkable that the result of this experiment is similar to the previous one that measured loudness of the single band-pass noise of the 1000 Hz center frequency (Fig. 4.3c).

As shown in Fig. 4.7, the loudness of the sharply (2068 dB/octave) filtered band-pass noise centered on 2000 Hz that was obtained by constant method is flat up to 160 Hz, although the τ_e increases with a decrease in bandwidth as shown in Fig. 4.8. Thus, loudness may be described in relation to effective duration, τ_e, for

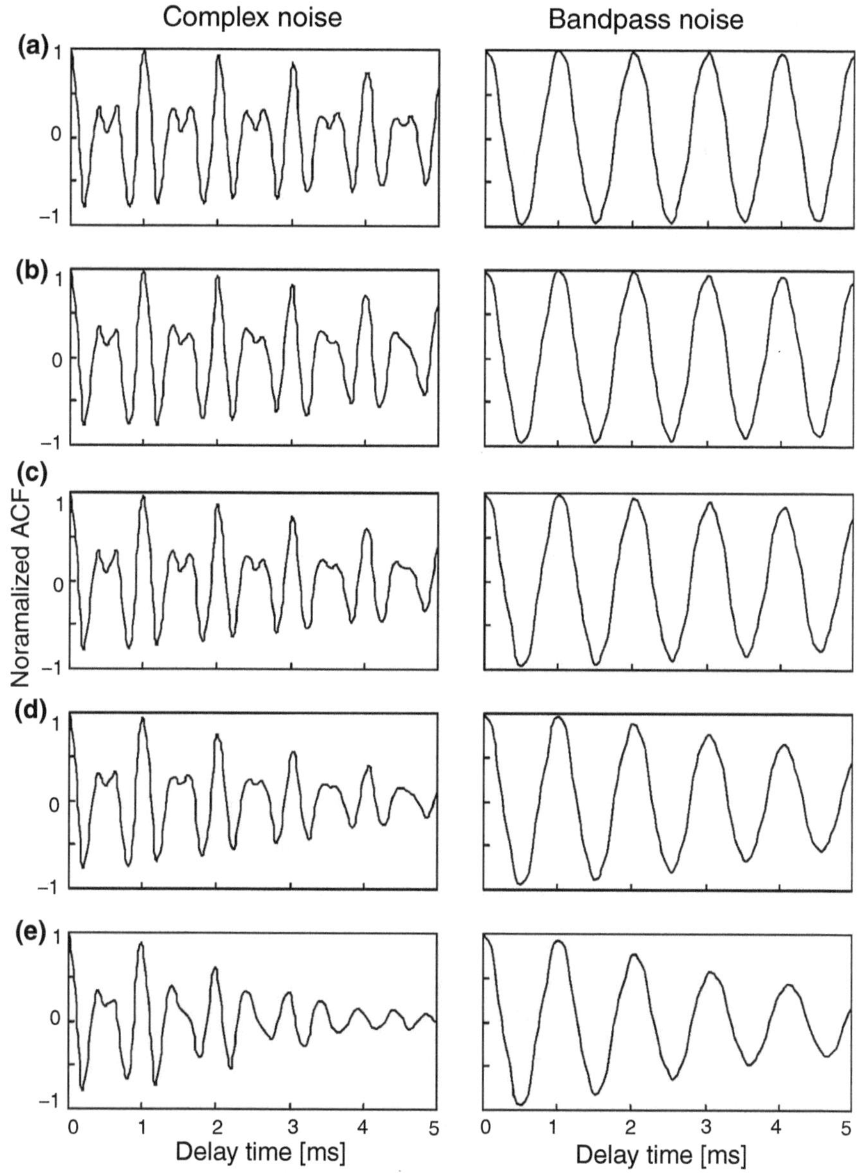

Fig. 4.4 The NACF of the complex noise with fundamental frequencies of 1000 Hz (*left-hand side*) and the band-pass noise of 1000 Hz center frequency (*right-hand side*). **a** Δf = 0 Hz, **b** Δf = 40 Hz, **c** Δf = 80 Hz, **d** Δf = 160 Hz, **e** Δf = 320 Hz

fundamental frequencies below 1200 Hz, which is the limitation of the missing fundamental phenomenon (Inoue et al. 2001). In these experiments, it was found that loudness for the complex noise with fundamental frequencies of 1000 Hz is

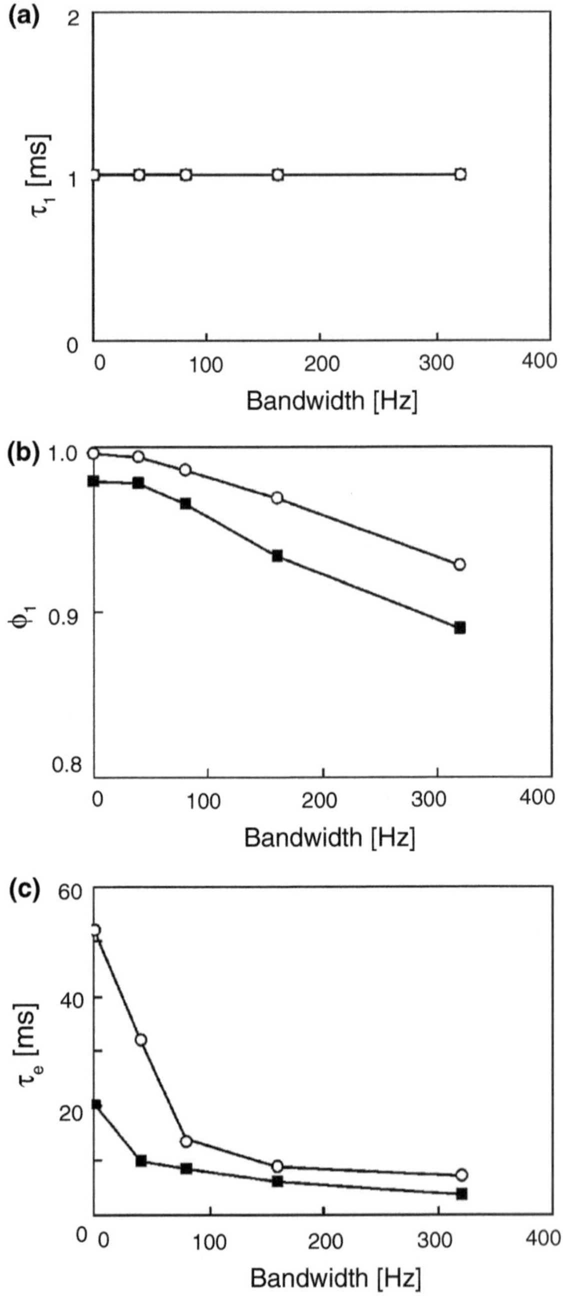

Fig. 4.5 Measured factors extracted from the ACF, **a** τ_1, **b** ϕ_1, and **c** τ_e of the signal as a function of the bandwidth: *filled square* complex noises with fundamental frequencies of 1000 Hz; *open circle* band-pass noises of 1000 Hz center frequency

similar to that of the single noise component centered on 1000 Hz. This is because both signals have the same τ_1. Also, loudness increases with the increasing value of τ_e within the critical band of 1000 Hz. However, loudness for the band-pass noise,

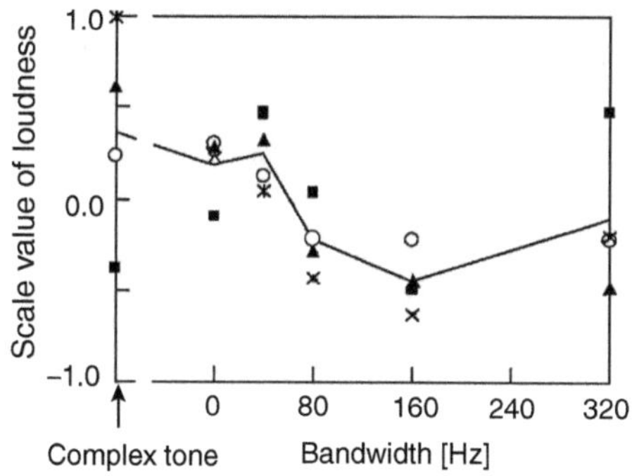

Fig. 4.6 Scale values of loudness as a function of the bandwidth for complex noises with fundamental frequencies of 1000 Hz. Different symbols indicate the scale values obtained with different listeners

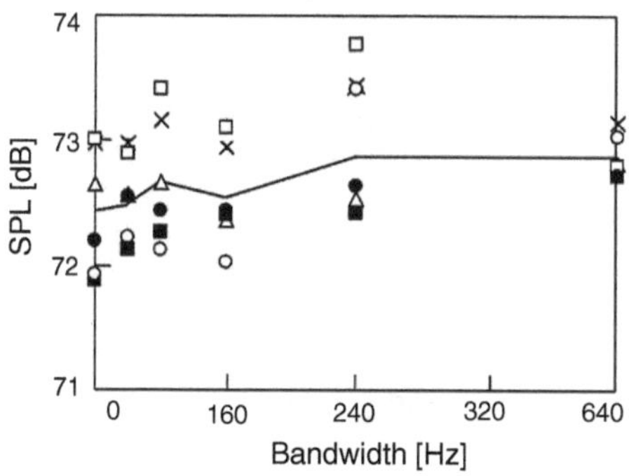

Fig. 4.7 Loudness of the band-pass noises centered at 2000 Hz obtained by the constant method comparing the 2000-Hz tone as a function of the bandwidth. Different symbols indicate the loudness obtained with different listeners

centered on 2000 Hz, is not affected by the value of τ_e because of the limitation on the ACF model.

4.2.3 Loudness of Iterated Rippled Noise

It was found that the loudness of a sharply (2068 dB/octave) filtered band-pass noise increases as the effective duration of the ACF, τ_e, increases, even when the bandwidth of the band-pass noise is within the critical band (Sato et al. 2002; Soeta et al. 2004). The τ_e represents repetitive components within the signal itself and increases as the bandwidth of a band-pass noise decreases. However, the envelope and SPL also vary as the bandwidth of a band-pass noise changes. This variation of the envelope and SPL might therefore affect the loudness of a band-pass noise

Fig. 4.8 Measured factor τ_e extracted from the ACF of the band-pass noise of 2000 Hz center frequency as a function of the bandwidth

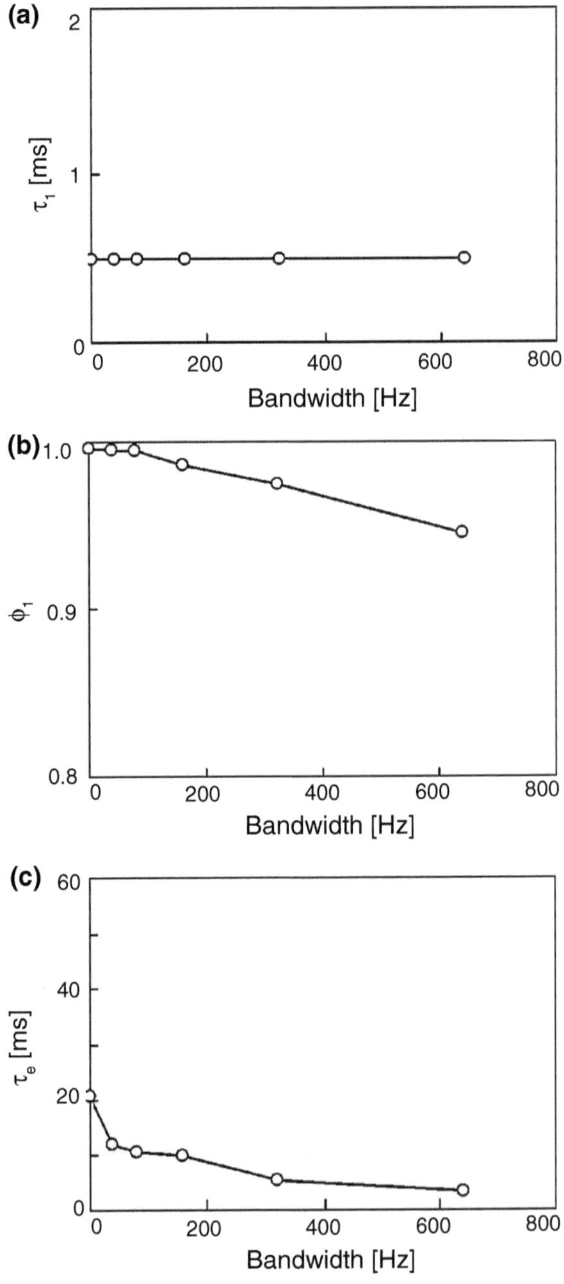

(Zhang and Zeng 1997; Moore et al. 1999). To eliminate such effects, the effects of τ_e on loudness were investigated using iterated rippled noise (IRN) (Soeta et al. 2007; Soeta and Nakagawa 2008a). IRN is produced by delaying a noise, adding it

to the original, and iterating the delay-and-add process. The reciprocal of the delay determines the pitch, and the number of iterations determines the pitch strength (Yost et al. 1996; Yost 1996). Thus, the delay time and number of iterations determine the τ_e of the ACF. In the present study, therefore, loudness matches were obtained using a two-interval, adaptive forced-choice procedure converging on the point of subjective equality (PSE) following a simple 1-up, 1-down rule.

Digitally generated white noise with a sampling rate of 48 kHz was used to produce band-pass noise. IRN was produced by a delay-and-add algorithm applied to the band-pass noise that was filtered using fourth-order Butterworth filters between 100–3500 Hz. The number of iterations of the delay-and-add process was set at 2, 4, 8, 16, and 32. The delay was set at 0.5, 1, 2, 4, 8, and 16 ms, for which the reciprocals were 2000, 1000, 500, 250, 125, and 62.5 Hz, respectively. The duration of the stimuli was 0.5 s, including linear rise and fall ramps of 10 ms. The sounds were digital-to-analogue-converted with a 16-bit sound card and sampling rate of 48 kHz. The sounds were presented diotically at a SPL of 60 dB through insert earphones with 29-cm plastic tubes and ear tips inserted into the ear canals. The passband in the transfer function of the plastic tubes approximately corresponded to the passband of the stimuli (100–3500 Hz). Figure 4.9 shows the temporal waveforms, power spectra, and ACFs of some of the stimuli. Figure 4.10 shows the τ_e and τ_1 (calculated at integration interval 0.5 s) of the stimuli used in the experiment. The integration interval was determined according to the psychological present, i.e., 0.5–5.0 s (Fraisse 1984).

Ten listeners (aged 21–37 years) with normal hearing took part in the experiment. There was no practice session because all listeners had already performed the same task to obtain loudness matches. Loudness matches were obtained using a two-interval, adaptive forced-choice procedure converging on the PSE following a simple 1-up, 1-down rule (Levitt 1971) in an anechoic and soundproof room. In each trial, the fixed (test) and variable (reference) sounds were presented in random order with equal, a priori probability, and separated by a 500-ms pause. The test sound was an IRN, and the reference sound was a 1-kHz pure tone. Each listener's task was to indicate which sound was louder by pressing a key. For each adaptive track, the overall level of the test sound was fixed at 60 dB SPL, and the starting level of the reference sound was 50 dB SPL. The level of the reference sound was controlled by an adaptive procedure: Whenever the listener judged the reference sound to be louder than the test sound, its SPL was lowered by a given amount; whenever the listener judged the test sound to be louder than the reference sound, the SPL of the reference sound was increased by that same amount. The initial step size was 5 dB; after four reversals (i.e., changes in the direction of the adaptive track), it was decreased to 2 dB. A total of 12 reversals were collected for each adaptive track; the arithmetic mean of the last four was used to estimate the PSE. One estimate was obtained for each IRN and listener. The effects of stimulus parameters (number of iterations and delay time) on the PSE for loudness were statistically analyzed by a repeated-measures of ANOVA.

Figure 4.11 shows the PSE for loudness as a function of the number of iterations of the IRN for each delay time. The main effects of the number of iterations

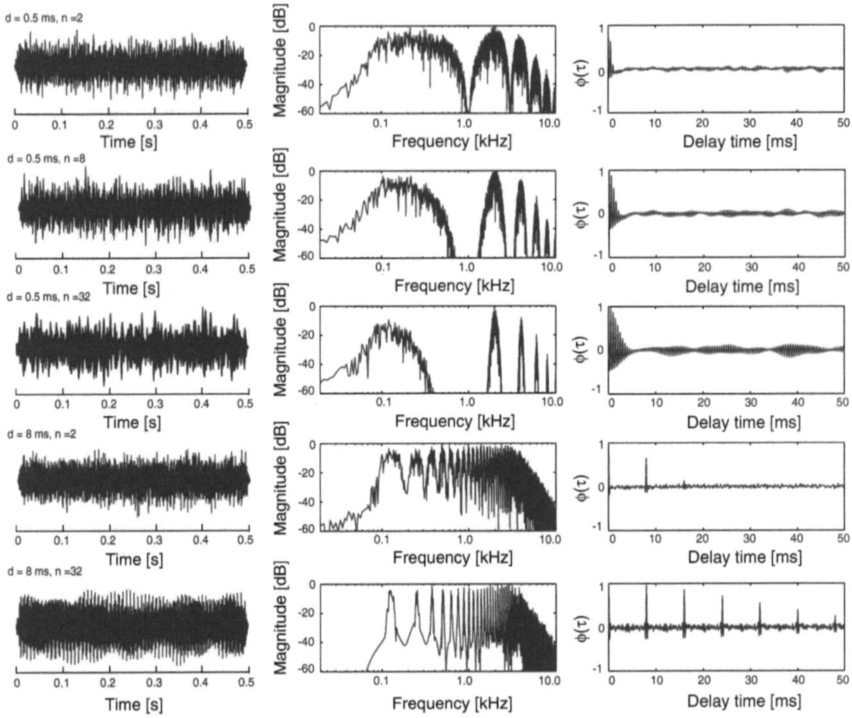

Fig. 4.9 Temporal waveforms (*left panels*), power spectra (*middle panels*), and ACF (*right panels*) of IRN with different delay times (*d*) and number of iterations (*n*). **a** $d = 0.5$ ms, $n = 2$; **b** $d = 0.5$ ms, $n = 8$; **c** d = 0.5 ms, $n = 32$; **d** $d = 8$ ms, $n = 2$; **e** $d = 8$ ms, $n = 32$

($p < 0.001$) and delay time ($p < 0.005$) were significant. The interaction was also significant ($p < 0.001$). Hence, the main effects of the number of iterations were analyzed by a one-way repeated-measures ANOVA for each delay time.

When the delay was 0.5 ms, the averaged PSE for loudness decreased as the number of iterations increased. Seven of the 10 listeners had decreasing PSE for loudness with increasing number of iterations. The main effect of the number of iterations on the PSE for loudness was significant ($p < 0.005$).

When the delay was 1 ms, the averaged PSE for loudness decreased as the number of iterations increased when the number of iterations was less than 8; however, the averaged PSE for loudness slightly increased with increasing number of iteration when the number of iterations was 16 and 32. Seven of the 10 listeners showed these tendencies. The main effect of the number of iterations on the PSE for loudness was significant ($p < 0.05$).

When the delay was 2 ms, the averaged PSE for loudness increased as the number of iterations increased, except for when the number of iterations was 2. Seven of the 10 listeners showed such tendencies. The main effect of the number of iterations on the PSE for loudness was significant ($p < 0.005$).

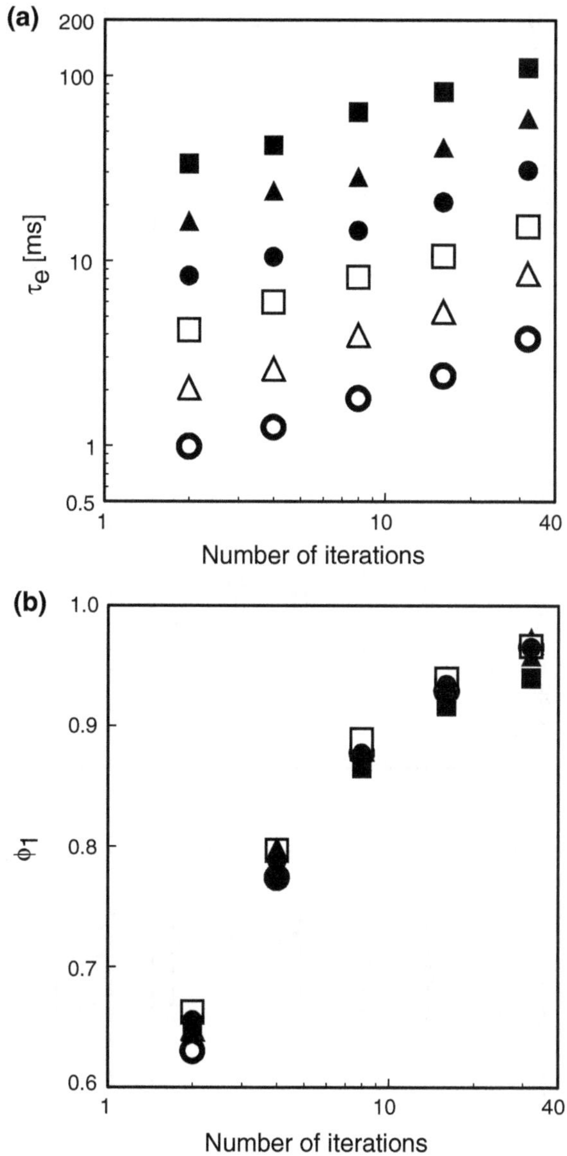

Fig. 4.10 **a** τ_e and **b** ϕ_1 of the stimuli used in the experiment as a function of the number of iterations with delays of (*open circle*) 0.5, (*open triangle*) 1, (*open square*) 2, (*filled circle*) 4, (*filled triangle*) 8, and (*filled square*) 16 ms

When the delay was 4 and 8 ms, the averaged PSE for loudness increased as the number of iterations increased, except for when the number of iterations was 32. Seven of the 10 listeners showed such tendencies. Three of the 10 listeners did not show clear PSE change as a function of number of iterations. The main effect of the number of iterations on the PSE for loudness was significant for the delay of 4 ms ($p < 0.001$) and the delay of 8 ms ($p < 0.001$).

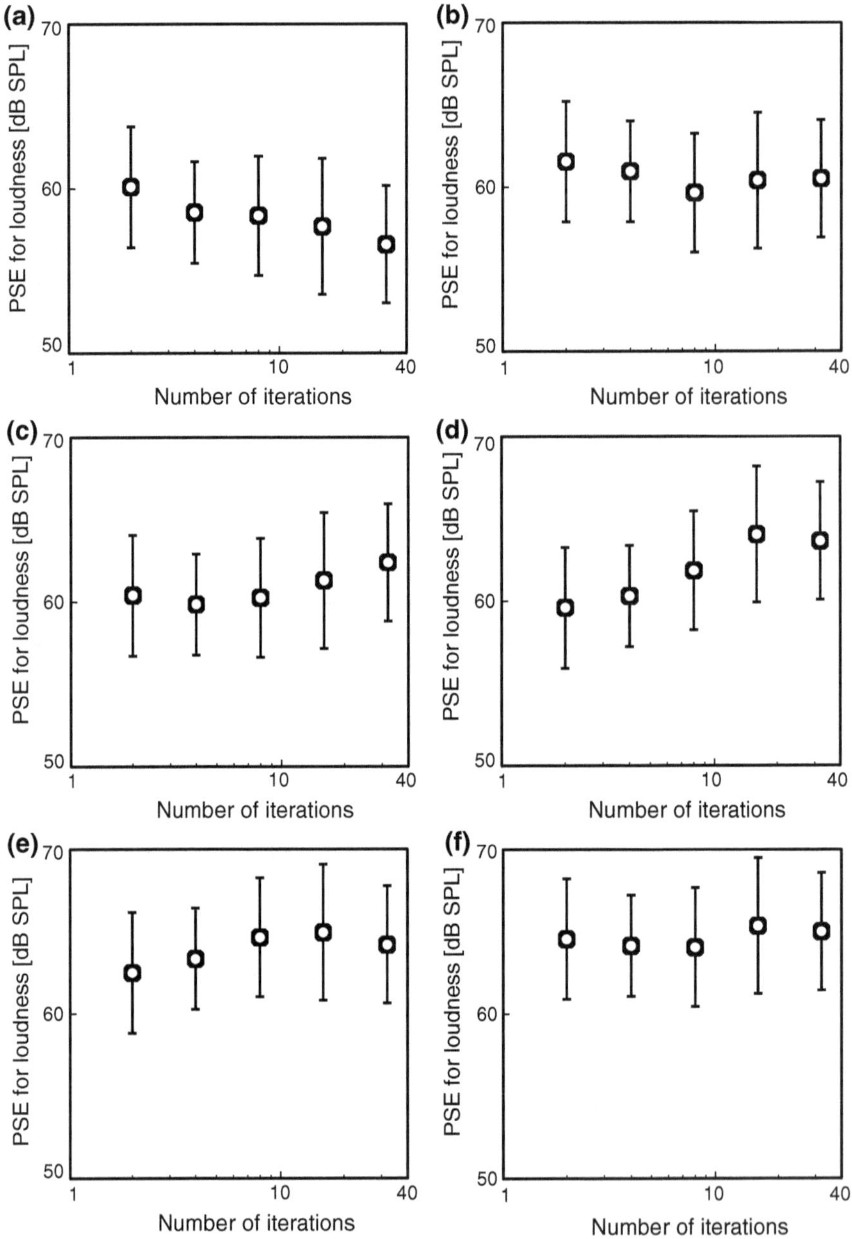

Fig. 4.11 Mean PSE for loudness as a function of **a** τ_e and **b** ϕ_1 for IRN with a delay of **a** 0.5, **b** 1, **c** 2, **d** 4, **e** 8, and **f** 16 ms. Error bars represent the standard deviation of the means

When the delay was 16 ms, the averaged PSE for loudness did not change significantly.

The relationship between the PSE for loudness, the τ_e, and the τ_1 is shown in Fig. 4.12. The τ_1 was not correlated with the perceived loudness. When the τ_e was between 10 and 100 ms, the perceived loudness increased with increasing τ_e. This is consistent with previous findings using band-pass noise (Sato et al. 2002; Soeta et al. 2004) and IRN (Soeta et al. 2007), clearly confirming that loudness is influenced by the repetitive components of sounds in the τ_e range between 10 and 100 ms. Therefore, the τ_e could be useful criteria for measuring loudness.

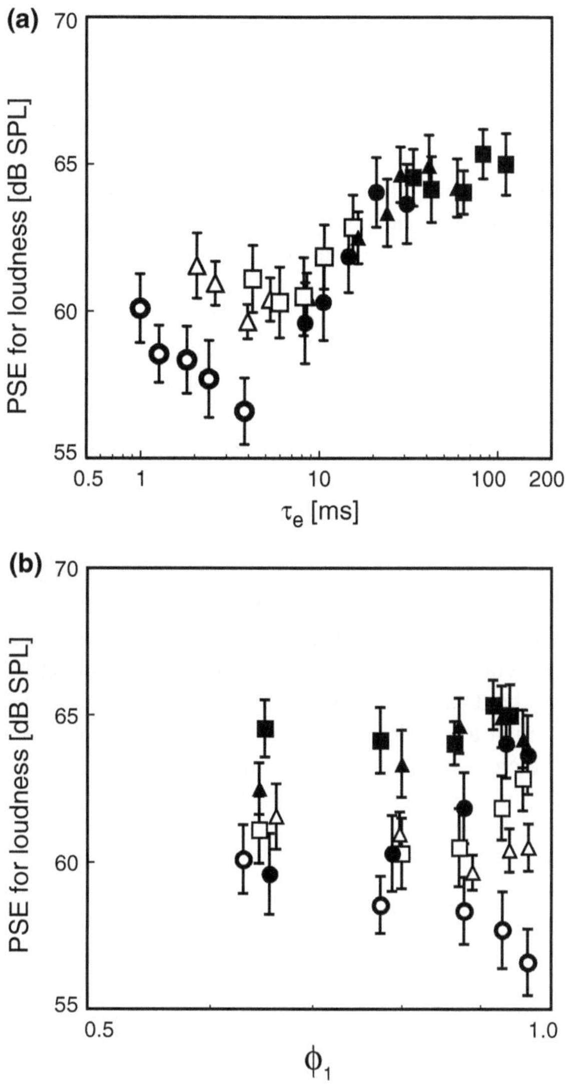

Fig. 4.12 Mean PSE for loudness as a function of the iteration number of IRN with a delay of (*open circle*) 0.5, (*open triangle*) 1, (*open square*) 2, (*filled circle*) 4, (*filled triangle*) 8, and (*filled square*) 16 ms. Error bars represent the standard deviation of the means

In addition, it was found that the increase in loudness for the τ_e values between 10 and 100 ms was approximately 5 dB.

When the τ_e was more than 100 ms, the loudness of the IRN remained constant, suggesting the contribution of perceived pitch strength. Previous studies have shown that pitch strength for the IRN is strongest for pitches of approximately 500 Hz and that stimuli had essentially no pitch strength for pitches below 50 Hz and above 2000 Hz (Yost and Hill 1979). When the τ_e was approximately 100 ms in the present study, the delay time of the IRN was 16 ms and the reciprocal of the delay was 62.5 Hz; this indicates that the perceived pitch was weak. Hence, the loudness might not have increased with increasing τ_e because of the perceived pitch weakness. In addition, the perceived pitch weakness might have also affected the loudness judgment when the delay time of the IRN was 0.5 ms and the reciprocal of the delay was 2000 Hz.

When the delay of the IRN was 0.5 ms, the loudness of the IRN decreased with increasing τ_e. When the delay of the IRN was 1 ms, the loudness of the IRN decreased with increasing τ_e up to approximately 4 ms and then increased with increasing τ_e. These tendencies might be due to the effect of the critical band, that is, loudness remains constant as the bandwidth of the noise increases up to the critical band and then increases with increasing bandwidth beyond the critical band. This can be predicted by loudness models (Moore et al. 1997; Zwicker and Fastl 1999), as shown in Fig. 4.13. The power spectrum of the IRN has a ripple with peaks at integer multiples of the reciprocal of the delay, as shown in Fig. 4.10. Figure 4.14 shows the bandwidth of the IRN stimuli as a function of the number of iterations. Bandwidth is defined as the bandwidth measured at 3 dB down from the peak. The bandwidth of the IRN stimuli decreases with an increasing number of iterations, and the bandwidth of the IRN stimuli at the delay times of 0.5 ms and 1 ms with the number of iterations of 2, 4, and 8 is larger than the critical bandwidth. This suggests that the IRN stimuli at the delay times of 0.5 ms and 1 ms with the number of iterations of 2, 4, and 8 is perceived to be louder than that with the number of iterations of 16 and 32 because of the effect of the critical band. If the bandwidth of the IRN stimuli at the delay times of 0.5 ms and 1 ms with the number of iterations of 2, 4, and 8 was smaller than the critical bandwidth at the reciprocal of the delay, then the loudness increased with an increasing number of iterations, similar to the IRN stimuli at the delay times of 2, 4, and 8 ms. Hence, the effect of the critical band on loudness might impair the effect of the repetitive component on loudness when the IRN stimuli at the delay times of 0.5 ms and 1 ms have 2, 4, and 8 iterations.

The effect of the τ_e on loudness was not seen in some of the listeners, indicating that there could have been listeners who were hardly affected by the τ_e with respect to loudness judgment. Thus, the effect of the τ_e on loudness judgment seems to partially depend upon the individual. This is consistent with previous findings using band-pass noise (Sato et al. 2002; Soeta et al. 2004) and IRN (Soeta et al. 2007).

The previously introduced loudness model (Moore et al. 1997; Zwicker and Fastl 1999) cannot predict loudness when the delay is 2 or 4 ms, which corresponds to a pitch of 500 and 250 Hz, respectively. This suggests that the τ_e is useful for

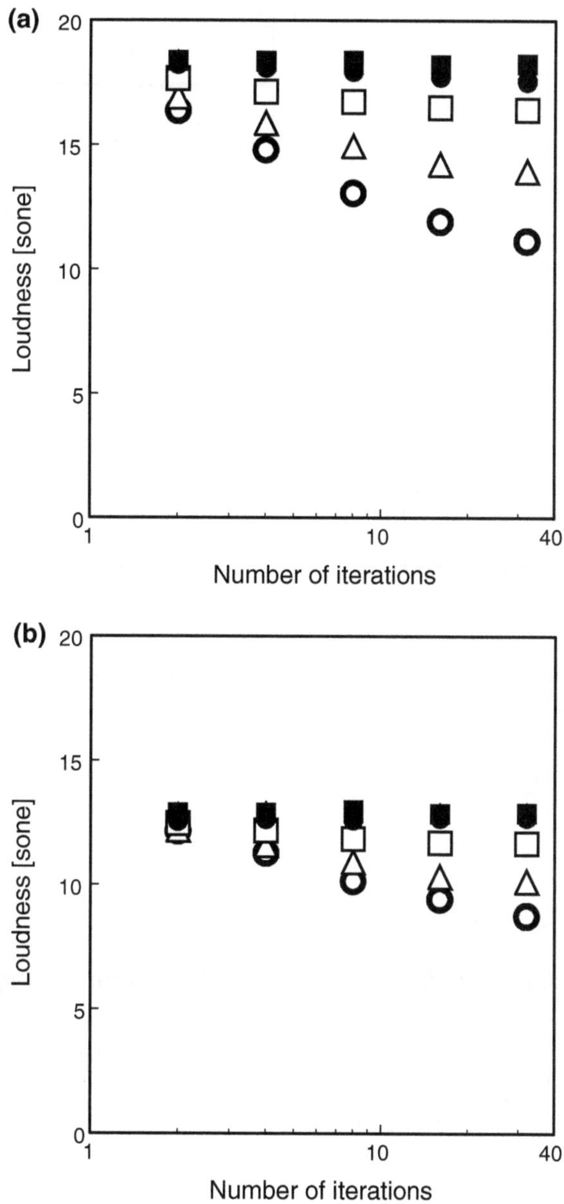

Fig. 4.13 Estimated loudness as a function of the number of iterations with delays of (*open circle*) 0.5, (*open triangle*) 1, (*open square*) 2, (*filled circle*) 4, (*filled triangle*) 8, and (*filled square*) 16 ms by the models proposed by **a** Moore and Glasberg (1997) and **b** Zwicker and Fastl (1999)

supplementing the loudness model. Loudness increases caused by a tonal component might be predictable by the τ_e in a certain range. The τ_e of various noise sources, such as airplanes (Fujii et al. 2001; Sakai et al. 2001), trains (Sakai et al. 2002), motor bikes (Fujii et al. 2002), and flushing toilets (Kitamura et al. 2002), have been analyzed, with results indicating that the τ_e varied within a range of

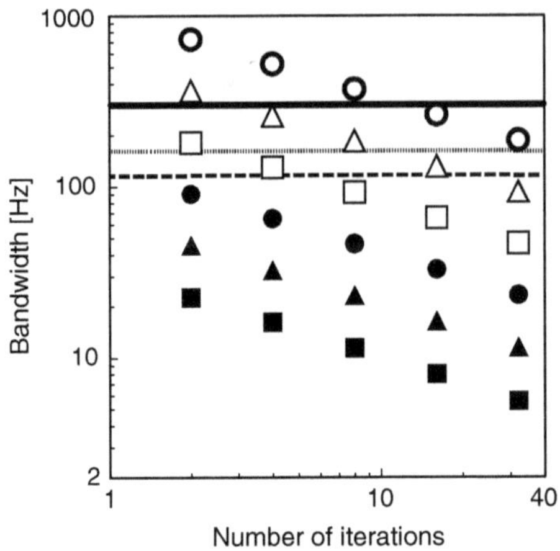

Fig. 4.14 Bandwidths of the IRN stimuli used in the experiment as a function of the number of iterations with delays of (*open circle*) 0.5, (*open triangle*) 1, (*open square*) 2, (*filled circle*) 4, (*filled triangle*) 8, and (*filled square*) 16 ms. Bandwidth is defined as the bandwidth measured at 3 dB down from the peak at the reciprocal of the delay. Solid, *short-dashed,* and *long-dashed lines* indicate critical bandwidth at a frequency of 2000, 1000, and 500 Hz, respectively (Zwicker 1961)

1–200 ms according to the type of noise source. We found that the increase in loudness for the τ_e values from 10 to 100 ms corresponds to an approximately 5 dB increase in SPL. Hence, τ_e values extracted from the ACF should be useful for supplementing the current loudness model when the τ_e value is between 10 and 100 ms.

4.3 Pitch

4.3.1 Pitches of Complex Tones and "Complex Noise"

As is widely known, a phenomenon of the missing fundamental represents a characteristic of pitch perception. The pitch of harmonic components without a fundamental frequency is perceived as being the same as the pitch of a pure tone of the fundamental frequency. The frequency perceived from such harmonic components is called a residue pitch, a periodicity pitch, a subjective pitch, or a virtual pitch. This phenomenon cannot be explained by the spectrum of source signals.

Some ACF-based models for predicting a residue pitch have been proposed. The ACF model of pitch perception was originally a "duplex" model Licklider (1951) and was improved by adding new rules (Meddis and Hewitt 1991a, b). Three famous models called "pattern recognition" models have been generally accepted since the 1970s (Goldstein 1973; Wightman 1973a; Terhardt 1974). Because the phase relation does not influence the pitch perception, a pattern-transformation model was proposed (Wightman 1973a, b). A pitch transformer based on an ACF detects the locations of peaks from the output waveform in each frequency band. Identically, the pitch strength can be estimated from the height of the maximum peak extracted from the ACF form. The effectiveness of the pattern-transformation model was examined in order to evaluate the validity in the peripheral weighting model using IRNs (Yost et al. 1978, 1996; Yost 1996). The time delay of maximum peaks of an ACF for a sound source was used as a significant parameter for predicting the pitch in their pitch-matching study. A cancelation model involving an array of delay lines and inhibitory gating neurons has also been proposed as an extension of the autocorrelation model (de Cheveigne 1998). It seems reasonable to suppose, as Ohgushi (1978) proposed, that both spatial and temporal cues relate to the perception of a residue.

From physiological point of view, temporal models for pitch that were based on first-order interspike intervals (times between successive spikes produced by a given neuron) in the auditory nerve were proposed in the beginning of the 1980s (van Noorden 1982). In these models, interspike interval information was pooled together from all regions of the auditory nerve to form a temporal population code for frequency and periodicity. By the end of the decade, temporal ACF models for pitch were tested using computer simulations of the cochlea and auditory nerve (Meddis and Hewitt 1991a, b). These ACF models are based on all-order interspike intervals (times between all spikes produced by a neuron, consecutive, and non-consecutive) rather than first-order intervals. Soon after, studies of temporal discharge patterns in the cat auditory nerve were conducted to test the temporal models (Cariani and Delgutte 1996a, b; Cariani 1999, 2001). Taken together, these studies showed that the temporal ACF models based on interspike interval distributions could predict a very wide range of pitch phenomena: pitch of the missing fundamental, pitch equivalence between pure and complex tones, level and phase invariance, pitch shift of inharmonic complex tones, pitch dominance, octave similarity, and the nonspectral pitch of amplitude-modulated noise.

Here, the fine structures of ACF, that is, the first major delay τ_1 and its amplitude ϕ_1, are adopted to predict a residue pitch of complex tones and complex noises (in the next session). The main purpose of the current experiment is to confirm whether the time delay of the first major peak of the ACF, τ_1, can predict the pitch of harmonic components, which has in-phase and random-phase conditions, without a fundamental frequency.

First, a pitch-matching test, comparing pitches of pure and complex tones, was performed to reconfirm previous results. The test signals were all complex tones consisting of harmonics 3–7 of a 200-Hz fundamental. All tone components had the same amplitudes, as shown in Fig. 4.15. As test signals, the two waveforms of complex tones, (a) in phases and (b) random phases, were applied as shown in Fig. 4.16. Starting phases of all components of the in-phase stimuli were set at zero. The phases of the components of random-phase stimuli were randomly set to avoid any periodic peaks in the real waveforms. As shown in Fig. 4.17, the normalized ACF (NACF) of these stimuli was calculated at the integration interval $2T = 0.8$ s. Though the waveforms differ greatly from each other, as shown in Fig. 4.17, their

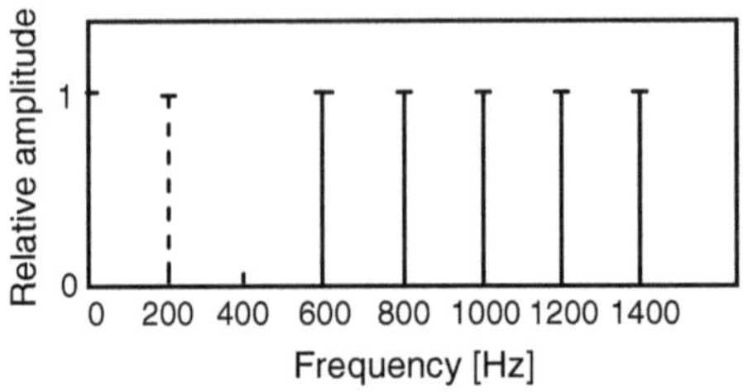

Fig. 4.15 Complex tone presented with pure-tone components of 600, 800, 1000, 1200, and 1400 Hz without the fundamental frequency of 200 Hz

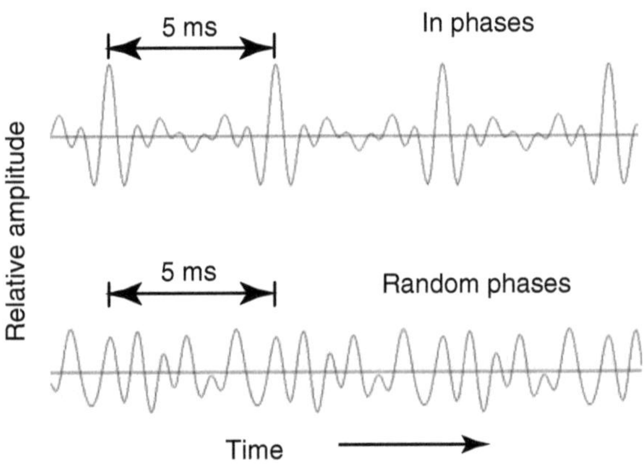

Fig. 4.16 Real waveforms of the complex-tone in-phase components (*above*) and random-phase components (*below*)

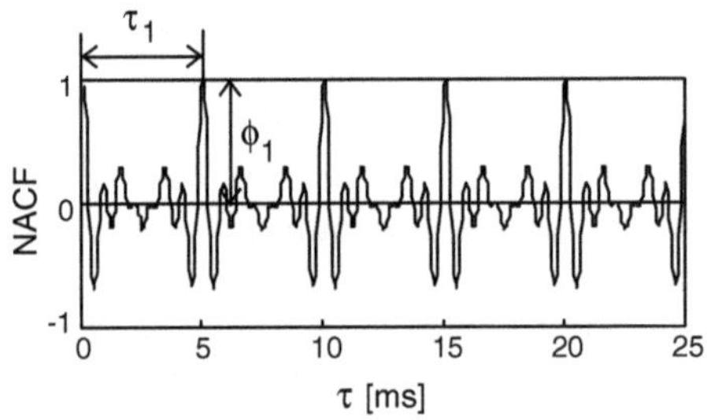

Fig. 4.17 Normalized autocorrelation function (NACF) of the two complex tones with different phase components, τ_1 = 5 ms (200 Hz)

NACFs are identical. The time delay at the first maximum peak of the NACF, τ_1, equals 5 ms (200 Hz), corresponding to the fundamental frequency. Five 20- to 26-year-old musicians participated as listeners in the experiment. Test signals were produced from the loudspeaker in front of each listener in a semi-anechoic chamber.

The SPL of each complex tone at the center position of the listener's head was fixed at 74 dB. The distance between a listener and the loudspeaker was 0.8 m ± 1 cm.

Probability of matching frequencies counted for each 1/12 octave band (chromatic scale) of the in-phase stimuli and random-phase stimuli is shown in Fig. 4.18. The dominant pitch of 200 Hz is included neither in the spectrum nor in the real waveform of random phases. However, it is obviously included in the period in the NACF. For both in-phase and random-phase conditions, about 60 % of the responses clustered within a semitone of the fundamental. There are no fundamental differences in the distributions of pitch-matching data between the two conditions. In fact, the pitch strength remains the same under both conditions as confirmed by $\phi_1 = 1$. Thus, pitch of complex tones can be predicted from the time delay at the first maximum peak of the NACF, τ_1. This result reconfirmed those obtained by Yost (1996), who demonstrated that pitch perception of IRN is greatly affected by the first major ACF peak of the stimulus signal.

From Eq. (4.6), pitch as one of temporal sensations may be expressed by

$$S = S_L = f_L(\tau_1) \approx 1/\tau_1(\text{Hz}), \tag{4.9}$$

when $\phi_1 = 1$.

Individual differences in pitch perception were also found. The results for each listener are indicated in Fig. 4.19. Listeners B and D matched only around the fundamental frequency (200 Hz). About 20 % of the responses were clustered around 400 Hz, and the NACF has a distinct dip at $\tau = 2.5$ ms (Fig. 4.17). However, an octave shift for a phase change (Lundeen and Small 1984) was not observed in the results obtained from these listeners. Listeners A and E matched at the fundamental frequency and at the frequency an octave higher. This octave change might be caused by a similarity for the octave relation. The time delay of the ACF for this pitch is 2.5 ms, so this pitch cannot be predicted because of a dip in the ACF structure. None of the listeners matched at $\tau_1 = 10$ ms (100 Hz), which is an octave lower than the fundamental frequency, though there is a peak at $\tau_1 = 10$ ms (Fig. 4.17).

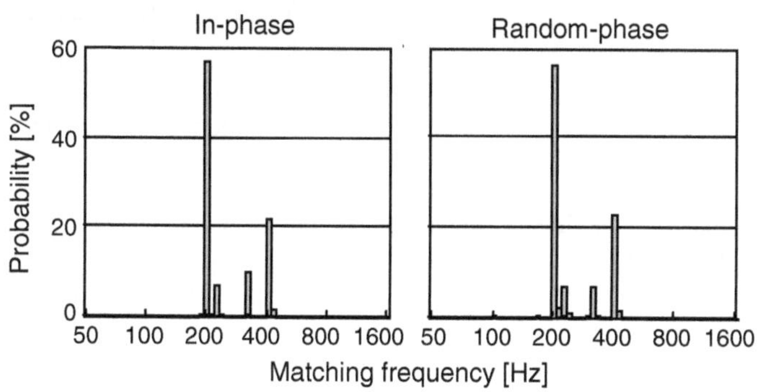

Fig. 4.18 Results of pitch-matching tests of the two complex tones, $\tau_1 = 5$ ms (five listeners)

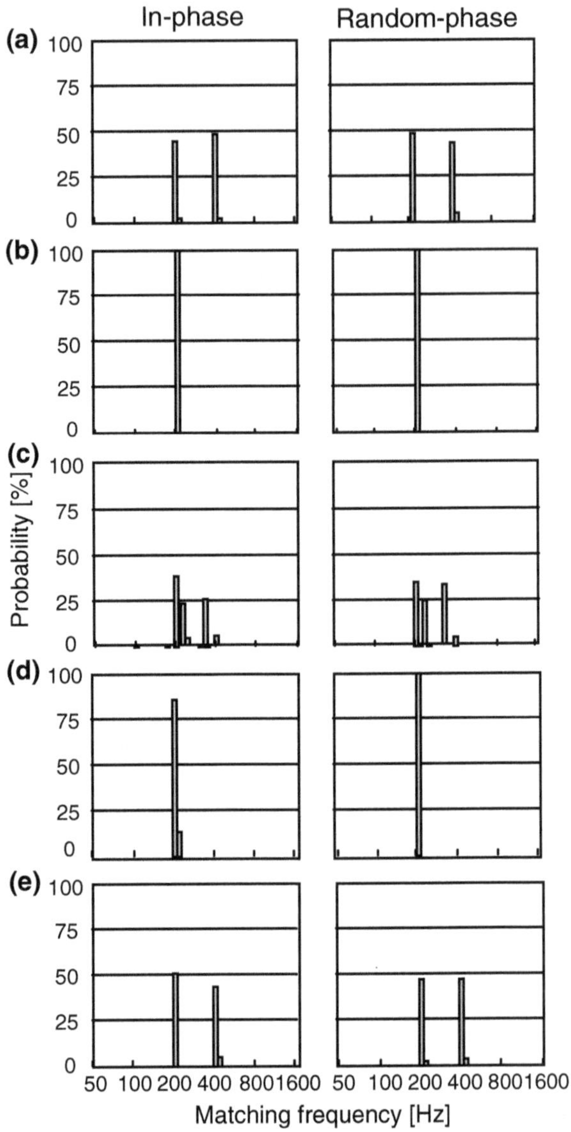

Fig. 4.19 Results of the pitch-matching tests for each of five listeners. **a–e** Illustrations represent results of different listeners

Listener C matched in three categories of center frequencies (200.0, 224.5, and 317.5 Hz). An E-flat note corresponds to the center frequency of 317.5 Hz, and the G note corresponds to 200.0 Hz. This is a musical interval (E-flat: G = do: mi). Listener C seemed to seek such a harmonic relation because he is a musician who uses the key of E-flat.

4.3.2 Pitch of "Complex Noise"

The purpose of this experiment using the complex noise is to determine that the amplitude of the first peak of the NACF, ϕ_1, influences pitch strength. The experimental method was the same as that of the experiment described in the previous section. The bandwidths of each partial noise, which consist of the band-pass noise with a cutoff slope of 1080 dB/octave, were changed. The center frequencies of the band-pass noise components were 600, 800, 1000, 1200, and 1400 Hz. The complex signal consisting of band-pass noises with different center frequencies is called here "complex noise." The bandwidths (Δf) of the four components were 40, 80, 120, and 160 Hz (Fig. 4.20). Their waveforms without any specific periodical envelopes are shown on the left side of Fig. 4.21. Measured results of the NACF for four conditions are shown on the right side of Fig. 4.21. The amplitude of the maximum peak (indicated by arrows in the figures) in the NACF is increased with decreasing Δf. Individual differences were also observed in the results obtained in tests with complex noises.

The probabilities of the matching data counted for each 1/12 octave band are shown in Fig. 4.22. All histograms show that there is a strong tendency to perceive a pitch of 200 Hz for any stimulus. This agrees with the prediction based on the value of τ_1. These results indicate that a stimulus with a narrow bandwidth gives a stronger pitch corresponding to 200 Hz than does a stimulus with a wide bandwidth. The standard deviation for the perceived pitches increased because the value of ϕ_1 decreased as Δf increased. The probability of pitch around 200 Hz being identified is plotted in Fig. 4.23 as a function of the ϕ_1. For narrow-band noise, the probability of a pitch of the fundamental frequency increases as the magnitude of the 5-ms peaks in the NACF increases. Thus, as the ϕ_1 increases, pitch strength also increases (r = 0.98). In this figure, the pitch-matching result from the previous section using the complex tones is also plotted at ϕ_1 = 1.0. Thus far, it was confirmed that the missing fundamental can be well described by using the ACF model.

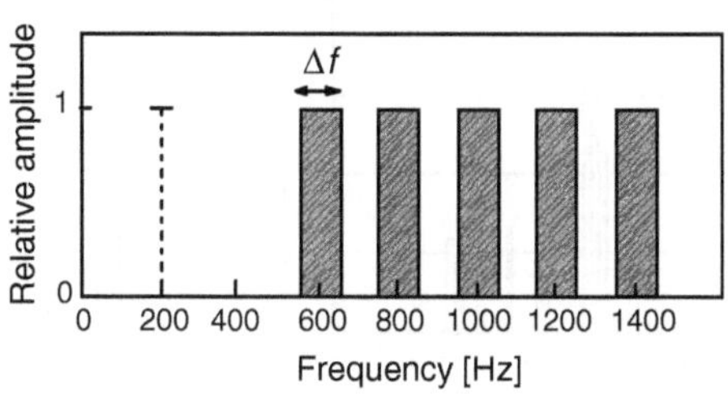

Fig. 4.20 "Complex noise" containing the center frequencies: 600, 800, 1000, 1200, and 1400 Hz used. Its fundamental frequency is centered on 200 Hz. The bandwidths of each partial noise consist of the band-pass noise filtered with a cutoff slope of 1080 dB/octave

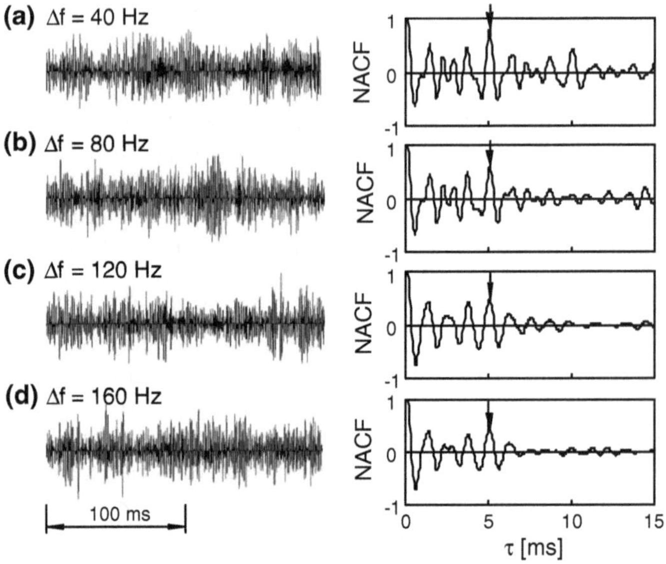

Fig. 4.21 Waveforms and the NACF of the four complex noises applied with Δf = **a** 40 Hz, **b** 80 Hz, **c** 120 Hz, and **d** 160 Hz

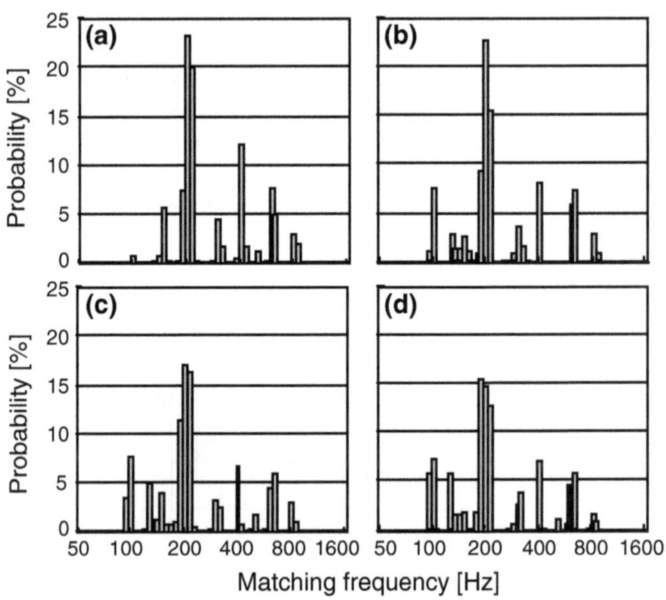

Fig. 4.22 Results of pitch-matching tests, with all five listeners. Δf: **a** 40 Hz, **b** 80 Hz, **c** 120 Hz, and **d** 160 Hz

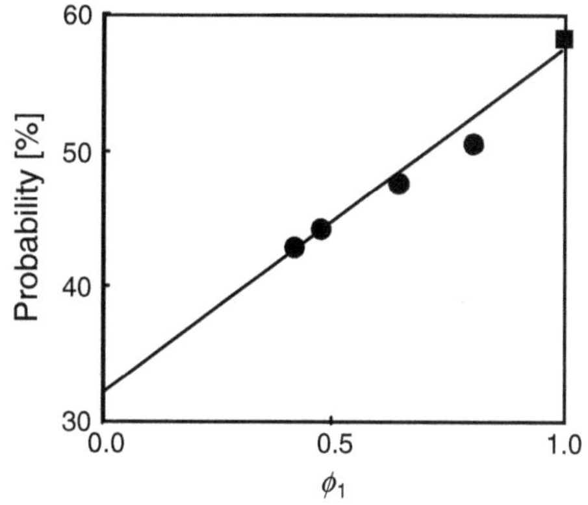

Fig. 4.23 Relationship between ϕ_1 and probability of the pitch being within 200 ± 16 Hz ($r = 0.98$, $p < 0.01$). For reference, the plot (*filled square*) at $\phi_1 = 1$ is the result with the pure tone

4.3.3 Frequency Limits of the ACF Model

In the former sections, the effectiveness of ACF factors, τ_1 and ϕ_1, for prediction of pitch of complex tone and complex noise was demonstrated. These ACF models are based on the time intervals between the spikes (interspike intervals) in the auditory nerve. The timings of spikes predominantly occur during one phase of the stimulus waveform. This is known as phase locking. Phase locking can exist up to approximately 1 kHz, dramatically declining as progressively higher frequencies are reached. To the extent that phase locking exists, interspike intervals reflect stimulus periodicities, that is, pitch (Cariani 1999). This suggests that there is a frequency range for pitch identification by the ACF model. The main purpose of the current experiment is to find the frequency range applicable to pitch identification by the ACF model.

For fundamental frequencies of 500, 1000, 1200, 1600, 2000, and 3000 Hz, stimuli consisting of two or three pure-tone components were used as stimuli (Inoue et al. 2001). The two-component stimuli consisted of the second and third harmonics of the fundamental frequency, and the three-component stimuli consisted of the second, third, and fourth harmonics. The starting phase of all components was adjusted to zero (in phase). The total SPL at the center of the listener's head was fixed at 74 dB. The NACF of all stimuli was calculated obtaining the peak τ_1 related to the fundamental frequency. The loudspeaker was placed in front of a listener in an anechoic chamber. The distance between the center of the listener's head and the loudspeaker was 0.8 m. Three 21- to 27-year-old musicians participated as listeners in the experiment. Pitch-matching tests were conducted using complex tones as test stimuli and a pure tone generated by a sinusoidal generator as a reference.

Results for all listeners are shown in Fig. 4.24. Whenever the fundamental frequency of the stimulus was 500, 1000, or 1200 Hz, more than 90 % of the responses obtained from all listeners under both conditions clustered around the fundamental frequency. When the fundamental frequencies of the stimuli were 1600, 2000, or 3000 Hz, however, the probability that the listeners adjusted the

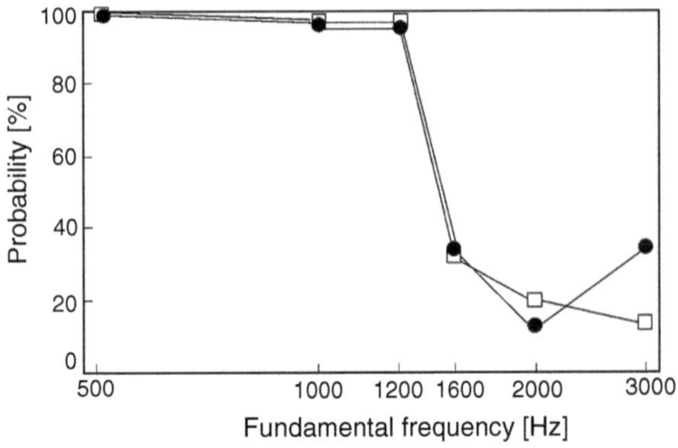

Fig. 4.24 Probability that three listeners adjusted a pure tone near the fundamental frequency of complex tones. *Filled circles* are results for two harmonics, and *open squares* are those for three harmonics (Inoue et al. 2001)

frequency of the pure tone to the calculated fundamental frequency was much lower. These results imply that the ACF model is applicable when the fundamental frequency of stimuli is below 1200 Hz.

The reasons for this upper limit are fairly straightforward. According to neuronal autocorrelation models, in order to evoke a "missing fundamental," one needs to satisfy at least one of two conditions that involve, respectively, either individual, cochlear-resolved harmonics or envelopes of unresolved, interacting harmonics (Cariani and Delgutte 1996a, b). In the first mode, interspike intervals associated with individual harmonics are produced and summed together across the auditory nerve. Here, one needs at least two resolved harmonics that are below the limit of significant phase-locked temporal information (~4000 Hz), such that interspike intervals associated with their common subharmonic, the fundamental, will predominate in the pooled ACF representation. In the second mode, pairs of unresolved adjacent harmonics beat together to produce interspike intervals associated with their beat period, which is the fundamental period. For several reasons, this mechanism that is based on interval representation of the stimulus envelope is less effective at producing intervals close to the fundamental period, and as a consequence, the pitches evoked are weaker than those associated with the first mechanism. In the current context, in order to represent a 1500-Hz missing fundamental using the envelope-based mechanism, one would need several pairs of unresolved harmonics, all at 9000 Hz or above ($n > 5$, for $F_0 = 1500$ Hz, $f_n > 9000$ Hz). Because there are relatively few auditory nerve fibers in humans that are responsive to such high frequencies, and intervals from all regions are pooled together, intervals associated with envelopes in these frequency regions are dwarfed by the spontaneous activity in the rest of the auditory nerve. The result is that the interval peaks associated with the F_0 envelope period are very shallow and do not rise above the signal/background threshold required for an audible low pitch.

On the low-frequency side of fundamental pitch perception, in psychophysical experiments, the lowest periodicities that produce pitches capable of supporting melodic recognition are approximately 30 Hz (Pressnitzer et al. 2001). This may be a consequence of a limitation in the longest interspike intervals that central auditory pitch processors analyze. Many current ACF models of pitch and consonance (e.g., Cariani 2001) therefore use a tapering interval weighting system that eliminates from consideration intervals longer than 33 ms.

It is worth noting that results of evoked brain magnetic response (N1 m latency) correspond to the fundamental frequency down to about 20 Hz (Yrttiaho et al. 2008; Soeta and Nakagawa 2008b). Thus, Eq. (4.9) could hold for the fundamental frequency, 20 Hz $< f_L(\tau_1) \leq$ 1200 Hz. Thus far, the results lead to the following two conclusions:

1. For the pitch, the ACF model is applicable when the fundamental frequency is below 1200 Hz and probably above 20 Hz.
2. Within this range, even if the complex tone has two frequency components, the pitch may be calculated by the delay time of the first peak in the NACF.

4.4 Timbre

Timbre is defined as an auditory perception of sound, which is neither pitch nor loudness (i.e., the quality of a musical note or sound or tone that distinguishes different types of sound production, such as voices, musical instruments, and so on). An attempt is made here to investigate the relationship between the temporal factor extracted from the ACF of an electric guitar sound and dissimilarity representing the difference of timbre with a difference of distortion. An electric guitar with "distortion" is a primary instrument in pop and rock music. Previously, Marui and Martens (2005) investigated timbre variations by the use of three types of nonlinear distortion processors with differing level of Zwicker sharpness (Zwicker and Fastl 1999). In this study, it was examined whether or not timbre is described by the temporal factor extracted from the ACF of the source signal that distinguishes notes of equal pitch and loudness, which are played at different distortion levels.

The sounds of an electric guitar were changed by the strength of distortion by the use of a computer or commercial effectors. The distortion of music signal $p(t)$ was processed by a computer program, such that when $|p(t)| \leq C$

$$p(t) = p(t), \tag{4.10a}$$

and when $|p(t)| \geq C$

$$p(t) = +C,\ p(t) \geq C; \quad p(t) = -C,\ p(t) \leq -C \tag{4.10b}$$

where C is the cutoff pressure amplitude, and its level is defined by

$$CL = 20\log 10(C/|p(t)|_{\max}) \tag{4.11}$$

and $|p(t)|_{max}$ is the maximum amplitude of the signal.

The value of CL was varied as 0–49 dB (7 dB step), so that eight stimuli were applied for test signals in Experiment 1. Nine stimuli were produced with three kinds of effect types (VINT, CHUTCH, and HARD) and three drive levels due to strength of distortion, 50, 70, and 90, by the effector Type MEAQ8 30 (Bose) in Experiment 2. The note (pitch) of the stimuli was 220 Hz by the use of third string and second fret. The SPL and duration were 80 dBA and 4.0 s in Experiment 1 and 70 dBA and 1.5 s in Experiment 2.

Listeners participated were 19 and 20 students (males and females 20 years of age) in Experiment 1 and Experiment 2, respectively. Listeners listened to three stimuli and judged dissimilarity. The number of combinations of the experiment was ${}_8C_3 = 56$ and ${}_9C_3 = 84$ triads in Experiment 1 and Experiment 2, respectively. The dissimilarity matrix was made according to the judgments giving the numbers 2 for the most different pair, 1 for the neutral pair, and 0 for the most similar pair. After the analysis of multidimensional scaling, the scale value was obtained. Contributions of the factors, such as ACF factors, $W_{\phi(0)}$ and ϕ_1, and the decay rate of SPL (dBA/s) to the scale value were analyzed.

It was found that the most significant factor contributing to the scale value was the $W_{\phi(0)}$. Certain correlations between the $W_{\phi(0)}$ and other factors were found, so that the $W_{\phi(0)}$ is considered as the representative. The scale value as a function of the $W_{\phi(0)}$ is shown in Fig. 4.25. The correlation between the scale value and the value of $W_{\phi(0)}$ is 0.98 ($p < 0.01$) and 0.92 ($p < 0.01$) in Experiment 1 and

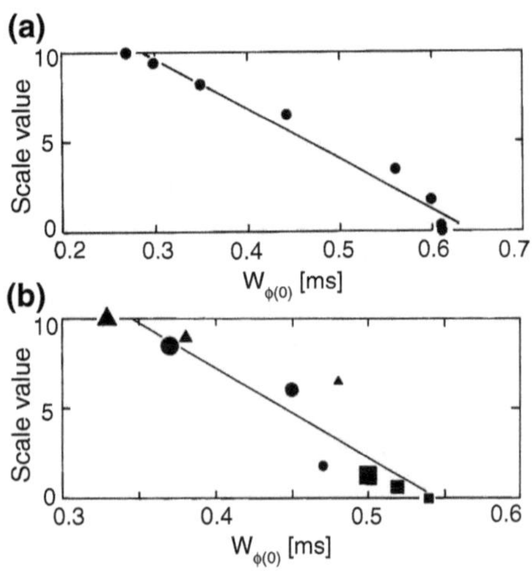

Fig. 4.25 Relationship between SV and the mean value of $W_{\phi(0)}$ in experiments **a** 1 and **b** 2

Experiment 2, respectively. $W_{\phi(0)}$ corresponds to the frequency component of the source signal. It is interesting that Ohgushi (1980) showed that the lowest and highest components govern primarily timbre.

4.5 Duration

The sensation of temporal duration is introduced here. Perceptually, the duration sensation depends on the physical signal duration, D, of course. In terms of internal auditory representations, temporal factors extracted from the ACF may influence the sensation of temporal duration. It was assumed that duration sensation is influenced by the time delay of the first major peak of the ACF, τ_1.

The perception of duration for complex tones was discussed (Saifuddin et al. 2002). An experimental study for pure and complex tones was performed by the paired comparison test. The SPL was fixed at 80 dBA throughout this investigation. Waveform amplitudes during stimulus onsets and offsets were ramped with rise/fall times of 1 ms for all stimuli tested, the time required to reach a threshold 3 dB below the steady level. The perceived durations of the two-component complex tones (3000 and 3500 Hz) having a fundamental at 500 Hz were compared with those evoked by a pure-tone stimuli at 500 and 3000 Hz. Pairs consisting of two stimuli were presented randomly to obtain scale values for duration sensation (DS). Three signal durations, including rise/fall segments, were used for each of the stimuli: $D = 140$, 150, and 160 ms. There were thus nine stimulus conditions and 36 pairwise stimulus combinations. The stimuli were presented in a darkened sound-proof chamber from a single loudspeaker at the horizontal distance of 74 (±1) cm from the center of the seated listener's head. Ten 22- to 36-year-old listeners with normal hearing levels participated in the experiment. Each pair of stimuli was presented five times randomly within every session for each listener.

Observed scale values for the perceived durations of the nine stimuli are shown in Fig. 4.26. Whereas signal duration and stimulus periodicity had major effects on perceived duration, the number of frequency components (1 vs. 2) did not. Perceived durations of tones with the same periodicity (f, $F_0 = 500$ Hz) were almost identical, whereas durations for pure tones of different frequencies (500 vs. 3000 Hz) differed significantly, by approximately 10 ms (judging from equivalent scale values, the 500-Hz pure tone appeared about 10 ms longer than the 3000-Hz tone). Thus, the DS of the higher-frequency pure tone (3000 Hz; $\tau_1 = 0.33$ ms) was found to be significantly shorter ($p < 0.01$) than that of either the pure tone (frequency: 500 Hz; $\tau_1 = 2$ ms) or the complex tone (fundamental frequency: 500 Hz; $\tau_1 = 2$ ms). Also, the scale values of DS between the two pure tones, $\tau_1 = 2$ ms (500 Hz) and 0.33 ms (3000 Hz), are almost parallel, so that the effects of periodicity (τ_1) and signal duration (D) on the apparent duration (DS) are independent and additive.

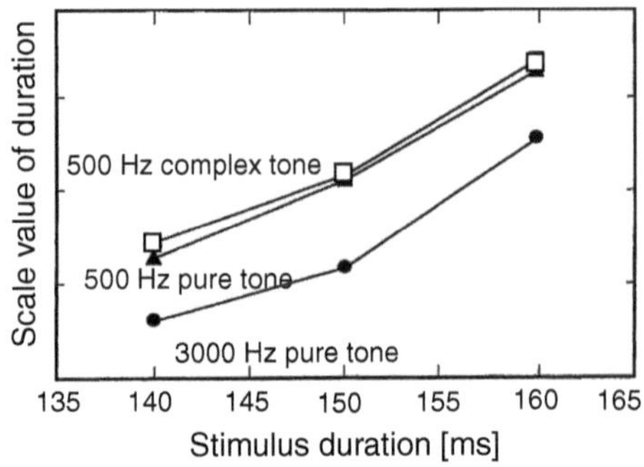

Fig. 4.26 Scale values of DS for (*open square*) complex tone ($F_0 = 500$ Hz) with 3000-Hz and 3500-Hz pure-tone components, (*filled triangle*) 500-Hz pure tone, and (*filled circle*) 3000-Hz pure tone

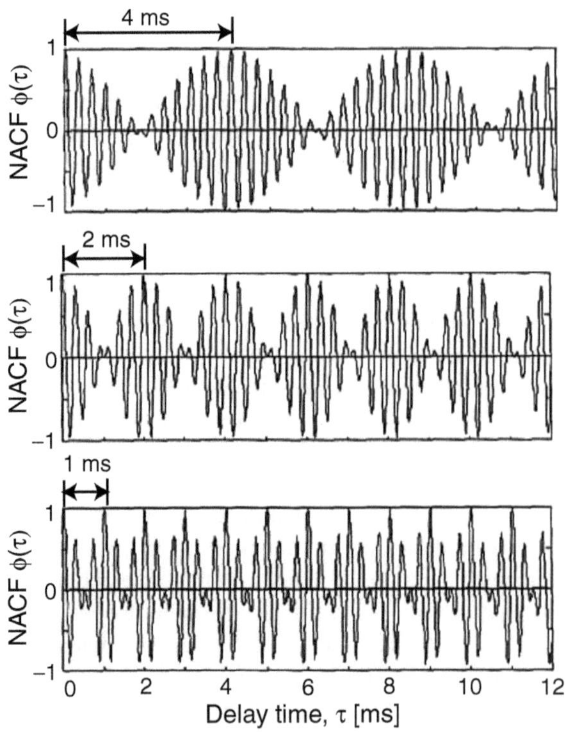

Fig. 4.27 Demonstrations of the NACF analyzed for the complex tone. **a** Complex tone with the components of 3000 and 3250 Hz ($F_0 = 250$ Hz). **b** Complex tone with the components of 3000 and 3500 Hz ($F_0 = 500$ Hz). **c** Complex tone with the components of 3000 and 4000 Hz ($F_0 = 1000$ Hz)

Therefore, for these experimental conditions, Eq. (4.12) may be reduced to

$$S = S_L = f_L(\tau_1, D) = f_L(\tau_1) + f_L D \tag{4.12}$$

where τ_1 is extracted from the stimulus ACF. Figure 4.27 shows the normalized stimulus ACF. Here, τ_1 corresponds to the missing fundamental, which is the pitch that can be heard for fundamental periodicities below roughly 1200 Hz (see Sect. 4.3.3).

The significant results of this experiment are summarized below.

1. Apparent stimulus duration DS depends primarily on the duration of the signal and secondarily on signal periodicity τ_1 (pure-tone frequency or complex-tone fundamental frequency).
2. Effects of the τ_1 extracted from the ACF on DS are almost the same on the scale value for the pure tone (τ_1 = 2 ms) and complex tone (τ_1 = 2 ms). The apparent duration DS of the pure-tone stimulus (τ_1 = 0.33 ms = 1/3000 Hz) with the higher pitch is significantly shorter than that of the pure-tone and complex-tone stimuli with the lower pitch (τ_1 = 2 ms = 1/500 Hz).
3. Apparent duration DS can be readily expressed as a function of D and τ_1 for both pure and complex tones.

References

Ando Y (1998) Architectural acoustics: blending sound sources, sound fields, and listeners. AIP Press, Springer, New York
Ando Y, Sato S, Sakai H (1999) Fundamental subjective attributes of sound fields based on the model of auditory brain system. In: Sendra JJ (ed) Computational acoustics in architecture. WIT Press, Southampton, pp 63–99
Cariani PA (1999) Temporal coding of periodicity pitch in the auditory system: an overview. Neural Plast 6:147–172
Cariani PA (2001) Neural timing nets. Neural Netw 14:737–753
Cariani PA, Delgutte B (1996a) Neural correlates of the pitch of complex tones. I. Pitch and pitch salience. J Neurophysiol 76:1698–1716
Cariani PA, Delgutte B (1996b) Neural correlates of the pitch of complex tones. II. Pitch shift, pitch ambiguity, phase invariance, pitch circularity, rate pitch, and the dominance region for pitch. J Neurophysiol 76:1717–1734
de Cheveigne A (1998) Cancellation model of pitch perception. J Acoust Soc Am 103:1261–1271
Dubrovskii NA, Chernyak RI (1969) Binaural loudness summation under varying degrees of noise correlation. Sov Phys Acoust 14:326–332
Fraisse P (1984) Perception and estimation of time. Annu Rev Psychol 35:1–36
Fujii K, Soeta Y, Ando Y (2001) Acoustical properties of aircraft noise measured by temporal and spatial factors. J Sound Vib 241:69–78
Fujii K, Atagi J, Ando Y (2002) Temporal and spatial factors of traffic noise and its annoyance. J Temporal Des Arch Environ 2:33–41
Goldstein JL (1973) An optimum processor theory for the central formation of the pitch of complex tones. J Acoust Soc Am 54:1496–1516
Greenwood DD (1961a) Auditory masking and the critical band. J Acoust Soc Am 33:484–502
Greenwood DD (1961b) Critical bandwidth and the frequency of the basilar membrane. J Acoust Soc Am 33:1344–1356
Inoue M, Ando Y, Taguti T (2001) The frequency range applicable to pitch identification based upon the auto-correlation function model. J Sound Vib 241:105–116
Katsuki Y, Sumi T, Uchiyama H, Watanabe T (1958) Electric responses of auditory neurons in cat to sound stimulation. J Neurophysiol 21:569–588

Kitamura T, Sato S, Shimokura R, Ando Y (2002) Measurement of temporal and spatial factors of a flushing toilet noise in a downstairs bedroom. J Temporal Des Arch Environ 2:13–19
Levitt H (1971) Transformed up-down procedures in psychophysics. J Acoust Soc Am 49:467–477
Licklider JCR (1951) A duplex theory of pitch perception. Experimenta 7:128–134
Lundeen C, Small AM (1984) The influence of temporal cue on the strength of periodicity pitches. J Acoust Soc Am 75:1578–1587
Marui A, Martens WL (2005) Timbre of nonlinear distortion effects: Perceptual attributes beyond sharpness? In: Proceedings of the conference on interdisciplinary musicology, Montreal, 2005
Mathews MV, Pfafflin SM (1965) Effect of filter type on energy-detection models for auditory signal detection. J Acoust Soc Am 38:1055–1056
Meddis R, Hewitt M (1991a) Virtual pitch and phase sensitivity of a computer model of the auditory periphery. I: pitch identification. J Acoust Soc Am 89:2866–2882
Meddis R, Hewitt M (1991b) Virtual pitch and phase sensitivity of a computer model of the auditory periphery. II: phase sensitivity. J Acoust Soc Am 89:2883–2894
Merthayasa IN, Hemmi H, Ando Y (1994) Loudness of a 1 kHz pure tone and sharply (1080 dB/Oct.) filtered noises centered on its frequency. Mem Grad School Sci Tech Kobe Univ 12A:147–156
Moore BCJ, Glasberg BR, Baer T (1997) A model for the prediction of thresholds, loudness, and partial loudness. J Audio Eng Soc 45:224–240
Moore BCJ, Vickers D, Baer T, Launer S (1999) Factors affecting the loudness of modulated sounds. J Acoust Soc Am 105:2757–2772
Ohgushi K (1978) On the role of spatial and temporal cues in the perception of the pitch of complex tones. J Acoust Soc Am 64:764–771
Ohgushi K (1980) Physical and psychological factors governing timbre of complex tones. J Acoust Soc Jpn 36:253–259
Pressnitzer D, Patterson RD, Krumbholz K (2001) The lower limit of melodic pitch. J Acoust Soc Am 109:2074–2084
Saifuddin K, Matsushima T, Ando Y (2002) Duration sensation when listening to pure tone and complex tone. J Temporal Des Arch Environ 2:42–47
Sakai H, Sato S, Prodi N, Pompoli R, Ando Y (2001) Measurement of regional environmental noise by use of a PC-based system. An application to the noise near the airport "G. Marconi" in Bologna. J Sound Vib 241:57–68
Sakai H, Hotehama T, Prodi N, Pompoli R, Ando Y (2002) Diagnostic system based on the human auditory-brain model for measuring environmental noise—an application to the railway noise-. J Sound Vib 250:9–21
Sato S, Kitamura T, Sakai H, Ando Y (2001) The loudness of "complex noise" in relation to the factors extracted from the autocorrelation function. J Sound Vib 241:97–103
Sato S, Kitamura T, Ando Y (2002) Loudness of sharply (2068 dB/Octave) filtered noises in relation to the factors extracted from the autocorrelation function. J Sound Vib 250:47–52
Soeta Y, Nakagawa S (2008a) Effect of the repetitive components of a noise on loudness. J Temporal Des Arch Environ 8:1–7
Soeta Y, Nakagawa S (2008b) The effect of pitch and pitch strength on an auditory-evoked N1 m. NeuroReport 19:783–787
Soeta Y, Maruo T, Ando Y (2004) Annoyance of bandpass filtered noises in relation to the factor extracted from autocorrelation function. J Acoust Soc Am 116:3275–3278
Soeta Y, Nakagawa S, Matsuoka K (2005) Effects of the critical band on auditory evoked magnetic fields. NeuroReport 16:1787–1790
Soeta Y, Yanai K, Nakagawa S, Kotani K, Horii K (2007) Loudness in relation to iterated rippled noise. J Sound Vib 304:415–419
Terhardt E (1974) Pitch, consonance, and harmony. J Acoust Soc Am 55:1061–1069
Thurstone LL (1927) A law of comparative judgement. Psychol Rev 34:273–289
van Noorden L (1982) Two channel pitch perception. In: Clynes M (ed) Music, mind and brain. Plenum, New York, pp 251–269
Wightman FL (1973a) Pitch and stimulus fine structure. J Acoust Soc Am 54:397–406

Wightman FL (1973b) The pattern-transformation model of pitch. J Acoust Soc Am 54:407–416
Yost WA (1996) Pitch strength of iterated rippled noise. J Acoust Soc Am 100:3329–3335
Yost WA, Hill R (1979) Models of the pitch and pitch strength of ripple noise. J Acoust Soc Am 66:400–410
Yost WA, Hill R, Perez-Falcon T (1978) Pitch and pitch discrimination of broadband signals with rippled power spectra. J Acoust Soc Am 63:1166–1173
Yost WA, Patterson R, Sheft S (1996) A time domain description for the pitch strength of iterated rippled noise. J Acoust Soc Am 99:1066–1078
Yrttiaho S, Tiitien H, May PJC, Leino S (2008) Cortical sensitivity to periodicity of speech sounds. J Acoust Soc Am 123:2191–2199
Zhang C, Zeng FG (1997) Loudness of dynamic stimuli in acoustic and electric hearing. J Acoust Soc Am 102:2925–2934
Zwicker E (1961) Subdivision of the audible frequency range into critical bands (Frequenzgruppen). J Acoust Soc Am 33:248
Zwicker E, Fastl H (eds) (1999) Psychoacoustics: facts and models. Springer, Berlin
Zwicker E, Flottorp G, Stevens SS (1957) Critical bandwidth in loudness summation. J Acoust Soc Am 29:548–557

Chapter 5
Spatial Primary Sensations of Noise

Abstract Spatial sensations include the localization, the subjective diffuseness (envelopment), and the apparent source width in the sound field. These are described by the multiple spatial factors extracted from the interaural cross-correlation function (IACF) for the signal arriving at two ears. As for the localization in the median plane, temporal factors extracted from the autocorrelation function (ACF) are also useful because the information arriving at two ears is same.

Keywords Spatial sensation · Localization · Subjective diffuseness · Apparent source width

5.1 Localization of Noise Source

It is considered that the perceived direction of a sound source in the horizontal plane is expressed in terms of the spatial factors extracted from the interaural cross-correlation function (IACF), such that

$$L_{\text{Horizontal}} = S_R = f_R\left(\sqrt{\Phi_{ll}(0)\Phi_{rr}(0)},\ \text{IACC},\ \tau_{\text{IACC}},\ W_{\text{IACC}}\right) \tag{5.1}$$

where $\Phi_{ll}(0)$ and $\Phi_{rr}(0)$ are the autocorrelation functions (ACFs) at $\tau = 0$ for the left and right ear entrances, and $\sqrt{\Phi_{ll}(0)\Phi_{rr}(0)}$ corresponds to SPL. It is well known that the most significant factor for the horizontal localization in the five spatial factors in Eq. (5.1) is the interaural delay time, τ_{IACC}, as well as sound energies at two ears including the interaural level difference. A well-defined direction is perceived when the normalized IACF has one sharp maximum with a large value of interaural correlation coefficients (IACC) and with a narrow value of width of the IACF, W_{IACC}, due to the high-frequency components above 2 kHz. On the other

Y. Soeta and Y. Ando, *Neurally Based Measurement and Evaluation of Environmental Noise*, Mathematics for Industry 20,
DOI 10.1007/978-4-431-55432-5_5

hand, subjective diffuseness or no spatial directional impression corresponds to a low value of IACC <0.15 (Damaske and Ando 1972) and a wide delay time of W_{IACC} due to the low-frequency components.

Apart from these five spatial factors of the localization in the horizontal plane given by Eq. (5.1), of particular interest is the localization in the median plane. In this plane, these spatial factors from the IACF are not significantly changed as a function of elevation (i.e., due to almost symmetrical shape of the head and pinnae), so that IACC is almost in unity, τ_{IACC} is zero, and $\Phi_{ll}(0) = \Phi_{rr}(0)$ = constant, and W_{IACC} = constant, which depend only on the spectrum of the source signal. It has been believed that the cues must be found in the spectrum of the sound signals. However, it is hard to find distinct cues in the head-related transfer function (HRTF) in the spectrum (Mehrgardt and Mellert 1977). In fact, the temporal factors extracted from the early delay range of the ACF of a sound signal arriving at the ear entrances might act as cues as discussed later (Sato et al. 2001).

We shall clearly show differences in the three temporal factors, τ_e, τ_1, and ϕ_1, extracted from the ACF as a function of the incident angle. According to the model described in Sect. 2.4, the sound localization in the median plane is assumed by a process associated with "left hemisphere specialization," because the related factors may be extracted from the ACF, such that

$$L_{\text{Median}} = S_L = f_L(\tau_e, \tau_1, \phi_1) \tag{5.2}$$

The amplitudes of the transfer functions for sound incident from the median plane to the ear entrances as measured by Mehrgardt and Mellert (1977) are transformed into the ACF. The following steps obtain the ACF:

1. Data sets were obtained from the figures from the Mehrgardt and Mellert paper using an optical image reader (scanner) with 300 data points each.
2. The amplitude as a function of frequency in the logarithmic scale was obtained.
3. The amplitude in decibel scale was converted to the real number, and the ACF was calculated by the inverse Fourier transform after passing through the A-weighting filter.

Examples of the normalized ACF (NACF) are shown in Fig. 5.1. There is a certain degree of correlation between both τ_n and τ_{n+i}, ϕ_n and ϕ_{n+i}, where τ_n and ϕ_n are the delay time and amplitude, respectively, of the *n*th peak of the NACF. Thus, τ_1 and ϕ_1 can be representatives for sets of τ_n and ϕ_n. Examples of plotting the amplitude of the ACF on a logarithmic scale as a function of the delay time are shown in Fig. 5.2. A straight line can fit the envelope of the decay of the ACF in the logarithmic scale, and τ_e was easily obtained from the delay at which the envelope drops below −10 dB. The value $\Phi(0)$ is not considered as a cue for sound localization here. Three factors extracted from the NACF are plotted in Fig. 5.3. The value of τ_1 for incident angles from 0° to 45° is almost the same, but τ_e for incident angle of 45° (τ_e = 3.1 ms) is much larger than that at 0° (τ_e = 2.1 ms). The value of τ_1 for incident angle of 180° is different from those for the above two angles;

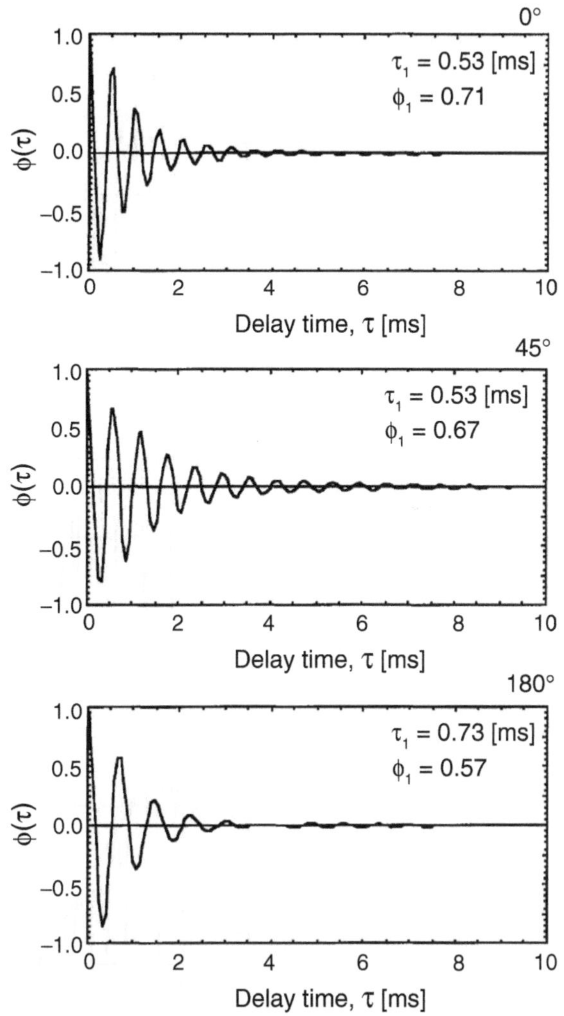

Fig. 5.1 Examples of the NACF of different incident angles in the median plane, 0°, 45°, and 180° (Sato et al. 2001)

however, the ϕ_1 is relatively small. Obviously, the angle in the median plane can be distinguished by three factors τ_1, ϕ_1, and τ_e. These factors, therefore, may have an important role in the perception of localization in the median plane.

5.2 Subjective Diffuseness

The subjective diffuseness or the spatial impression of sound fields is one of the important attributes used to describe good acoustics. Barron (1971) who investigated subjective effects of a single side reflection, and identified the effect of spatial

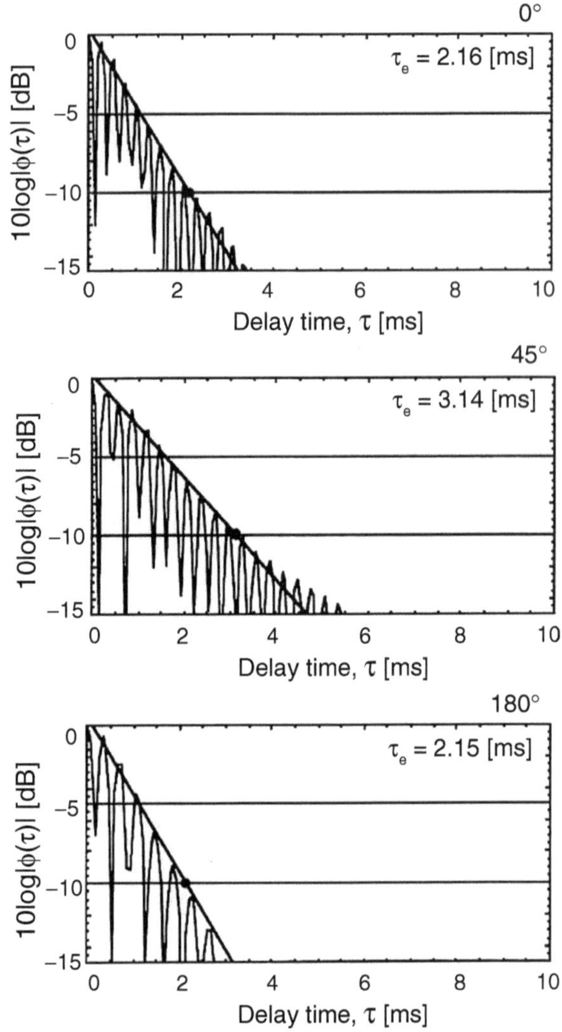

Fig. 5.2 Examples of the effective duration extracted from the NACF envelope (in logarithm) for different incident angles in the median plane, 0°, 45°, and 180° (Sato et al. 2001)

impression, concluded that the degree of spatial impression is probably related to the ratio of lateral to nonlateral sound. He also confirmed that a cross-correlation process is involved in subjective spatial impression. Barron and Marshall (1981) investigated the determinants of subjective effects created by early lateral reflections, that is, spatial impression and found the early lateral energy fraction is linearly related to perceived spatial impression. The purpose of this study was to determine which horizontal reflection angles are most effective in stimulating subjective diffuseness or a listener in a room.

To investigate effects of horizontal angles on the subjective diffuseness for each band-pass noise, two symmetrical reflections added to the direct sound were

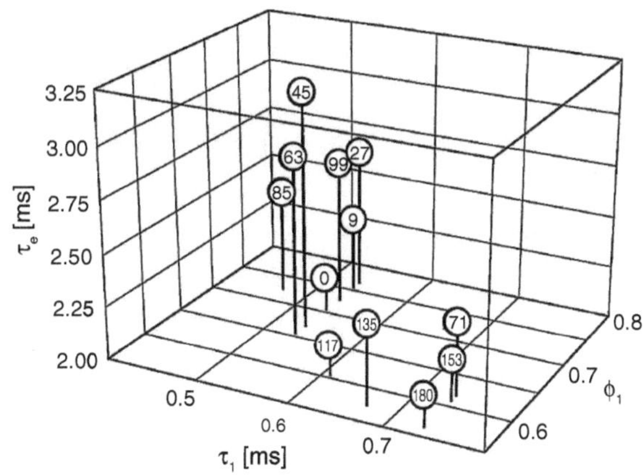

Fig. 5.3 Three-dimensional illustration plotted for three factors, ϕ_1, τ_1, and τ_e, which may distinguish different incident angles in the median plane. Numbers in the *circles* indicate incident angle in degree (Sato et al. 2001)

simulated and produced (Ando and Kurihara 1986; Singh et al. 1994). IACC was confirmed by corresponding measured value. As a source signal, the one-third octave band-pass noise, which roughly corresponds to critical bandwidth in the auditory system, was used. The center frequency of the noise was 0.25, 0.5, 1, 2, and 4 kHz. The time delays between the direct sound and two reflections were fixed at 40 and 60 ms. These delays were long enough to assure incoherence of the two reflections. The amplitude of the reflection was 3 dB below the level of the direct sound. The total SPL at the listener's ear was 65 dBA.

Paired comparison tests of nine sound fields for the changes in the horizontal angles were conducted. The signal duration of each stimulus was 3 s. Listeners (11 students 22 ± 1 year of age) were asked to judge in which of the two sound fields they perceived more diffuseness. The tests were performed for all combinations of pairs, i.e., 72 pairs for each band-pass noise, interchanging in order in each pair. Scale values of subjective diffuseness were obtained by applying the law of comparative judgment (Thurstone 1927).

Scale values of subjective diffuseness and the IACC as a function of the horizontal angle of incidence for a 1/3 octave band-pass noise with the center frequencies of 250 Hz, 500 Hz, 1 kHz, 2 kHz, and 4 kHz to a listener are shown in Fig. 5.4. A strong negative correlation between scale values and IACCs was found. The horizontal angle to the listener for reflections that contribute most to the subjective diffuseness was determined at the maximum scale value. For the frequency range below 500 Hz, the most effective angles were found in the range centered on 90°; for 1 kHz, it was 126°, which corresponds to 54° considering well-known front/back confusion; for 2 kHz, it was about 18°; and, for the 4-kHz range, it was 18°–54° in which the angle spread was due to individual differenced in judgments caused probably by interaural difference of time and amplitude, or individual IACCs.

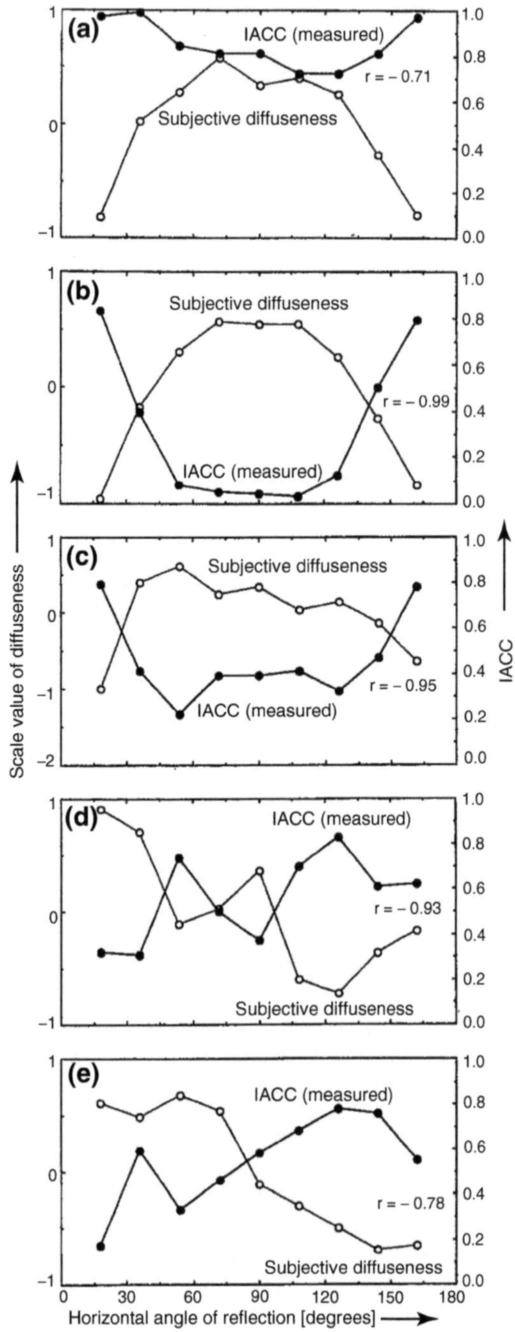

Fig. 5.4 Scale values of subjective diffuseness and the IACC as a function of the horizontal angle of incidence to a listener with 1/3 octave band-pass noise of center frequencies: **a** 250 Hz, **b** 500 Hz, **c** 1 kHz, **d** 2 kHz, **e** 4 kHz

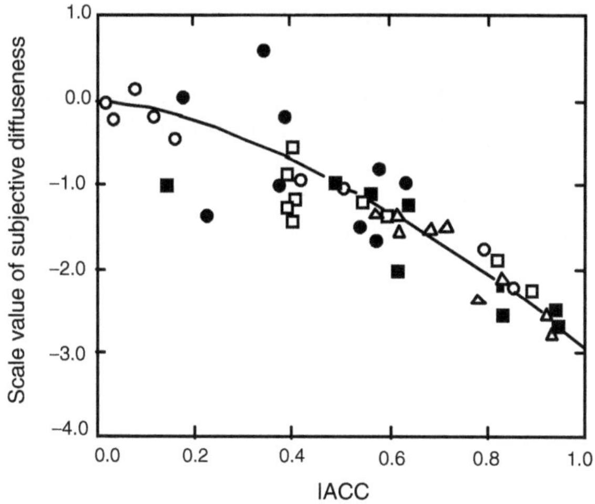

Fig. 5.5 Scale values of subjective diffuseness as a function of the IACC (calculated). Different symbols indicate different frequencies of the 1/3 octave band-pass noise: *open triangle* 250 Hz; *open circle* 500 Hz; *open square* 1 kHz; *filled circle* 2 kHz; *filled square* 4 kHz, *underline* Regression line by Eq. (5.3)

To find a relationship between the IACC and the scale value of subjective diffuseness, all data are plotted in Fig. 5.5, in which scale values of each band-pass noise are shifted to a certain value because of the relative nature of the scale value. This arrangement yielded almost the same results for all bands, and a single non-linear formulation of the scale value as a function of IACC. A remarkable finding is that the scale values of subjective diffuseness are inversely proportional to the IACC and may be formulated in terms of the 3/2 power of the IACC in a manner similar to that for the subjective preference values (Ando 1985, 1998), that is,

$$S = S_R \approx -\alpha(\text{IACC})^{\beta} \tag{5.3}$$

where $\alpha = 2.9$ and $\beta = 3/2$.

The results of the scale value from the paired comparison tests together with calculated values by Eq. (5.3) as a function of the IACC are shown in Fig. 5.5. There is a great variation of data in the range of the IACC <0.5; however, no essential difference may be found in results with different frequencies between 250 Hz and 4 kHz.

To control the subjective diffuseness, the most effective horizontal angles for early reflections to a listener have been found for each frequency band. The effective range of the angles from the median plane is around 55°, except for the high-frequency range above 2 kHz. These results are well described by minimizing the calculated IACC. The scale value of subjective diffuseness has been formulated in terms of the 3/2 power of the IACC.

5.3 Apparent Source Width (ASW)

Apparent source width (ASW) is an important subjective attribute of the sound field in a concert hall or an opera house. ASW is the acoustical width of the sound source as perceived by a listener. Generally, it depends on the spectral component of the source signal as well as the IACC. For example, even if the IACC is constant, the ASW increases as the center frequency of the octave band-pass noise decreases or as the low-frequency components increase (Morimoto and Maekawa 1988; Hidaka et al. 1995). The fact that wider ASW is perceived for a sound source with a predominately low frequency is reflected in the IACF. The peak of the function becomes broader as the center frequency of the band-pass noise decreases. Thus, the width of the IACF, W_{IACC}, which is defined as the interval of delay time at a value of 10 % below the IACC, has been introduced (Ando et al. 1999). SPL also affects ASW (Käsbach et al. 2014). It is assumed that when the τ_{IACC} is fixed at a constant value, the scale value of ASW, S, is expressed by

$$S = S_R \approx f_R\left(\sqrt{\Phi_{ll}(0)\Phi_{rr}(0)},\ \mathrm{IACC},\ W_{\mathrm{IACC}}\right) \tag{5.4}$$

where $\Phi_{ll}(0)$ and $\Phi_{rr}(0)$ are the ACFs at $\tau = 0$ for the left and right ear entrances, and $\sqrt{\Phi_{ll}(0)\Phi_{rr}(0)}$ corresponds to SPL.

5.3.1 ASW of the Band-Pass-Filtered Noise

On the basis of the auditory-brain model, the ASW may be described by the spatial factors extracted from the IACF as expressed in Eq. (5.4). It is assumed that this depends on the W_{IACC} instead of the spectral component of the source signal as well as the IACC and the SPL. It has been reported that even if the IACC is constant, the ASW increases as the low-frequency components increase (Morimoto and Maekawa 1988; Hidaka et al. 1995). The fact that the wider ASW is perceived for the sound source with a predominately low frequency might be reflected in the factor W_{IACC}.

For the sound field with a predominately low-frequency range, the IACF has no sharp peaks for the delay range of $|\tau| < 1$ ms, and W_{IACC} becomes wider (Fig. 5.6). The W_{IACC} of the band-pass noise may be calculated theoretically by the following equation (Ando 1998):

$$W_{\mathrm{IACC}}^{(\delta)} \approx \frac{4}{\Delta\omega_c}\cos^{-1}\left(1 - \frac{\delta}{\mathrm{IACC}}\right) \tag{5.5}$$

where $\Delta\omega_c = 2\pi(f_1 + f_2)$, and f_1 and f_2 are the lower and upper frequencies of an ideal filter. For the sake of simplicity, δ is defined by 0.1(IACC).

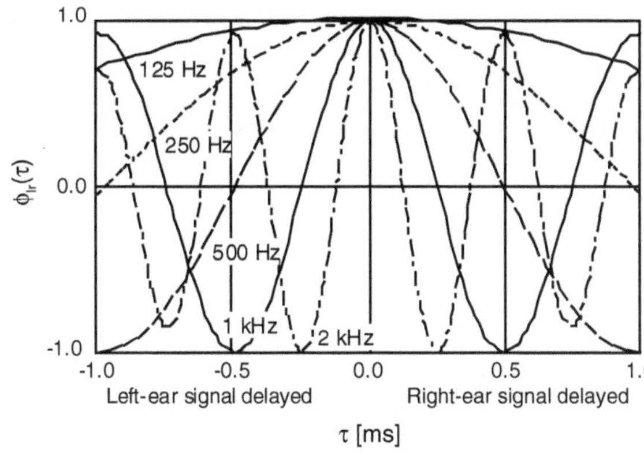

Fig. 5.6 Measured IACFs for 1/3 octave band-pass noises with center frequencies of 125 Hz–2 kHz

One-third octave band-pass noises were used as a source signal in Experiment 1 (Sato 1999). The center frequencies of the noise were 250 Hz, 500 Hz, 1 kHz, and 2 kHz. IACCs were set 0.90 and 0.70 for the center frequency of 250 Hz and 0.90, 0.70, and 0.40 for the center frequency of 500 Hz, 1 kHz, and 2 kHz. The IACC = 0.40 for the center frequency of 250 Hz cannot realize in this condition. The band-pass noises of 500 Hz center frequency were used as a source signal in Experiment 2 (Sato 1999). The bandwidths were one-third octave width, one octave width, two octave width, and three octave width.

Two symmetrical lateral reflections ($\xi = \pm 54°$) added to the frontal direct sound ($\xi = 0°$) were simulated in an anechoic chamber. The distances between the loudspeakers and the center on the listener's head were 0.90 ± 0.01 m. To produce incoherent conditions, the time delays between the direct sound and two reflections were fixed at 20 and 30 ms, respectively. The value of IACC was adjusted by controlling the SPL ratio between reflections ($\xi = \pm 54°$) and the direct sound ($\xi = 0°$). To avoid effects of the SPL on the ASW, the total SPL at the ear canal entrances of all sound fields was kept constant at a peak of 75 dBA.

To obtain a scale value for ASW, paired comparison tests were performed. The listeners were asked to judge which of two sound sources they perceived to be wider. The duration of each stimulus was 3 s, and the silent interval between stimuli was 1 s. Each pair of sound fields was separated by an interval of 4 s and the pairs were arranged in random order. The number of listeners was ten and four in Experiment 1 and 2. The test consisted of 55 ($N(N-1)/2$, $N = 11$) and 66 pairs ($N(N-1)/2$, $N = 12$) of stimuli, in total in Experiment 1 and 2, respectively. A scale value for ASW was obtained by applying the method, a modification of the Thurstone's method (Ando and Singh 1996).

Figure 5.7 shows the observed scale value of ASW for one-third octave band-pass noises as a function of the W_{IACC} and as a parameter of IACC. The scale value of ASW increased with increasing W_{IACC} and decreasing IACC. Results of the analysis of variance (ANOVA) for the scale value S_R indicate that both the factors IACC and W_{IACC} contribute to S_R independently ($p < 0.01$), which may be expressed by

$$S = S_R = f_R(\mathrm{IACC}) + f_R(W_{\mathrm{IACC}}) \approx \alpha(\mathrm{IACC})^{3/2} + \beta(W_{\mathrm{IACC}})^{1/2} \qquad (5.5)$$

where coefficients α and β are the coefficients to be evaluated. The values of the power, 3/2 and 1/2, for the terms of IACC and W_{IACC} in Eq. (5.5) were selected to give the best correlation between the scale values obtained by paired comparison tests and the scale values calculated by Eq. (5.5). The scale value of preference and subjective diffuseness are also expressed in terms of the 3/2 power of the IACC (Ando and Kurihara 1996; Singh et al. 1994). The curves in Fig. 5.7 confirm the calculated scale value of Eq. (5.5), $\alpha \approx -1.64$ and $\beta \approx 2.44$, which were obtained by the multiple regression analysis.

Figure 5.8 shows the observed scale value of ASW for the band-pass noises of 500 Hz center frequency as a function of the W_{IACC} and as a parameter of IACC. The scale value of ASW increased with increasing W_{IACC} and decreasing IACC. Results of the ANOVA for the scale value S_R indicate that both the factors IACC and W_{IACC} contribute to S_R independently ($p < 0.01$). The curves in Fig. 5.8 confirm the calculated scale value of Eq. (5.5), $\alpha \approx -3.26$ and $\beta \approx 3.55$, which were obtained by the multiple regression analysis.

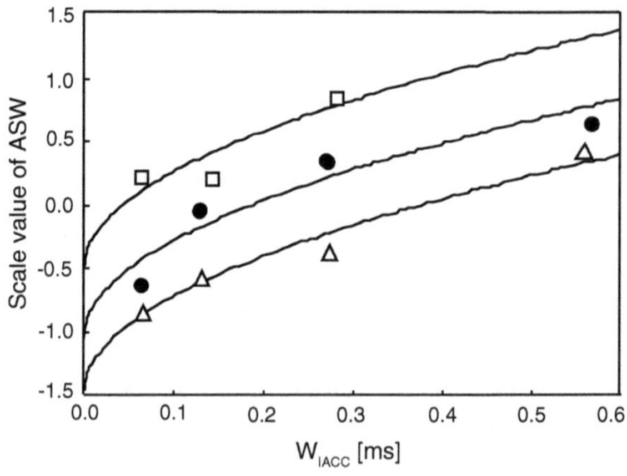

Fig. 5.7 Scale values of ASW for one-third octave band-pass noises as a function of W_{IACC} and as a parameter of IACC. *Open triangle* IACC = 0.90; *filled circle* IACC = 0.70; *open square* IACC = 0.40. The regression curve is expressed by Eq. (5.5), $\alpha \approx -1.64$ and $\beta \approx 2.44$

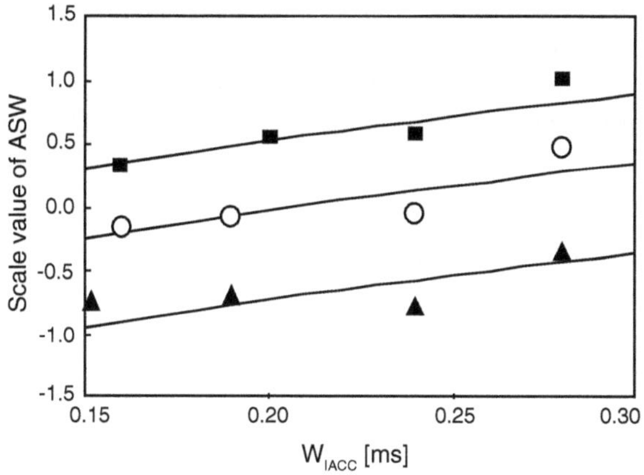

Fig. 5.8 Scale values of ASW for the band-pass noises of 500 Hz center frequency as a function of W_{IACC} and as a parameter of IACC. *Filled triangle* IACC = 0.82; *open circle* IACC = 0.65; *filled square* IACC = 0.51. The regression curve is expressed by Eq. (5.5), $\alpha \approx -3.26$ and $\beta \approx 3.55$

Coefficients α and β in the Eq. (5.5) obtained by the multiple regression analysis of the results from both experiments were −2.14 and 2.60, respectively. Negative value for α and positive value for β mean smaller IACC and larger W_{IACC} cause larger ASW. Figure 5.9 shows the relationship between the measured scale values of ASW and scale values of ASW calculated by Eq. (5.5). The correlation coefficient between the measured and calculated scale values is 0.92 ($p < 0.01$). The scale

Fig. 5.9 Relationship between the measured scale values of ASW and scale values of ASW calculated by Eq. (5.5) for (*open circle*) one-third octave band-pass noises and (*filled circle*) band-pass noises of 500 Hz center frequency. The correlation coefficient is $r = 0.92$ ($p < 0.01$)

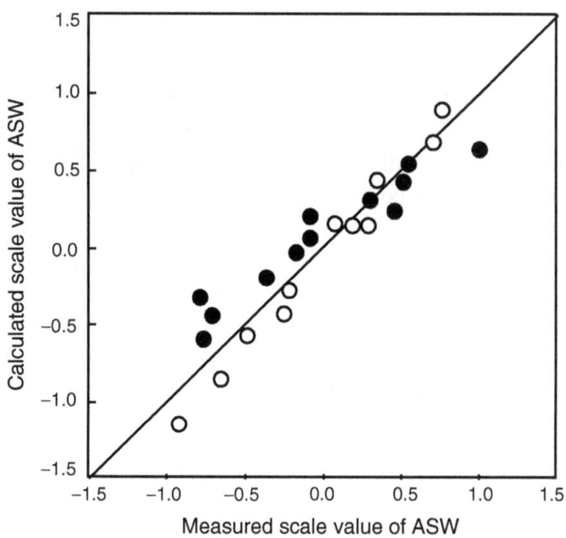

value of ASW for one-third octave band-pass noises and the band-pass noises of 500 Hz center frequency obtained by paired comparison tests can be described by two-dimensional factors: IACC and W_{IACC}. The ASW may be formulated by superposition of two terms the 3/2 power of IACC and the 1/2 power of W_{IACC}.

5.3.2 ASW of "Complex Noise"

The ASW of the complex noise, which consists of the band-pass noises whose center frequencies are the harmonics of the fundamental frequency, was examined. The question is whether or not pitch (the fundamental frequency) describes the ASW (Sato and Ando 2002). The scale value of ASW for complex noises was compared with that for band-pass noises. If the ASW of complex noises is judged from the frequency component, ASW of the complex noise can be described by W_{IACC} of the source signal. On the other hand, if ASW of complex noises is judged from the fundamental frequency, ASW can be described by the τ_1 value of the source signal. In latter case, ASW of the complex noise is equal to that of the band-pass noise, whose center frequency is the same as the fundamental frequency of the complex noise.

Band-pass noises with center frequencies of 200, 400, and 800 Hz and complex noises with fundamental frequencies of 200, 400, and 800 Hz were used as the source signals. The complex noises consisted of three band-pass noise components, and the center frequencies of lowest frequency components were fixed at 1600 Hz. All partial components had the same SPL. The bandwidth of the band-pass noises and that of the components of the complex noise were 80 Hz with a cutoff slope of 2068 dB/octave, which was obtained by the combination of two filters.

Two symmetrical lateral reflections ($\xi = \pm 54°$) added to the frontal direct sound ($\xi = 0°$) were simulated in an anechoic chamber. To produce incoherent conditions, the time delays of reflections were fixed at 20 and 40 ms, respectively. To reconfirm the effects of SPL on ASW, the SPL at the listener's head position was changed from 70 to 75 dB. The values of the IACC of all sound fields were adjusted to 0.90 ± 0.01 by controlling the identical amplitude of the reflections. The IACF was measured with two 1/2-inch condenser-type microphones, placed at the ear entrance of a dummy head, and the IACC was confirmed.

The paired comparison tests of 12 sound fields (6×2) were performed on five listeners with normal hearing ability. Listeners were seated in an anechoic chamber and asked to judge which of two stimuli they perceived to be wider. The duration of each sound stimulus was 3 s, the rise and fall times were 50 ms, and the silent interval between stimuli was 1 s. Each pair of stimuli was separated by an interval of 4 s, and the pairs were presented in random order. Twenty-five responses (five listeners × five repeats) to each stimulus were obtained. The scale value of the ASW was obtained by applying the law of comparative judgment as usually.

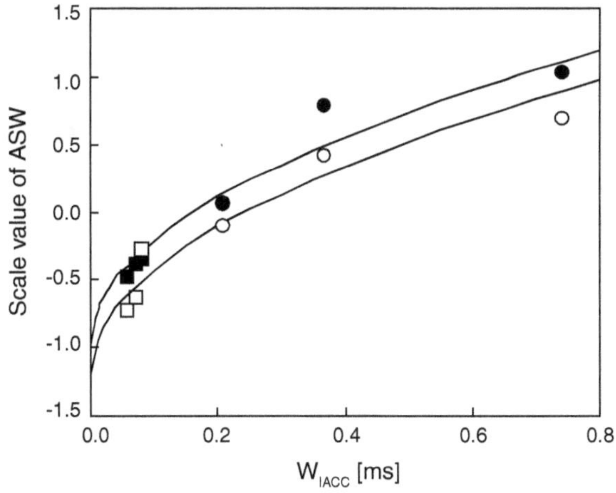

Fig. 5.10 Scale values of ASW as a function of W_{IACC} and as a parameter of SPL. *Filled circle* band-pass noise, SPL = 75 dB; *open circle* band-pass noise, SPL = 70 dB; *filled square* complex noise, SPL = 75 dB; *open square*: complex noise; SPL = 70 dB. The regression curve is expressed by Eq. (5.6) with α = 2.40 and β = 0.003

The relationship between the scale value of the ASW and W_{IACC} of the source signal is shown in Fig. 5.10. There is a significant difference between the scale value of the ASW of the single band-pass noise and that of the complex noise ($p < 0.01$) even when pitch is the same. However, the factor W_{IACC} explains these differences. The scale value of ASW increased with increasing W_{IACC}. Results of the ANOVA for scale values of the ASW revealed that the explanatory factors W_{IACC} and SPL are significant ($p < 0.01$) and independent, so that

$$S = S_R = f_R(\mathrm{W_{IACC}}) + f_R(\mathrm{SPL}) \approx a(\mathrm{W_{IACC}})^{1/2} + b(\mathrm{SPL})^{3/2} \qquad (5.6)$$

where α and β are coefficients. The powers 1/2 and 3/2 for the terms of W_{IACC} and SPL in Eq. (5.6), respectively, were determined to obtain the best correlation between the scale value measured and its calculated value. The curves in Fig. 5.10 confirm the calculated scale values by Eq. (5.6) with coefficients $\alpha \approx 2.40$ and $\beta \approx 0.003$. Positive values for α and β mean larger W_{IACC} and SPL cause larger ASW. The correlation coefficient between the measured and calculated scale values is 0.97 ($p < 0.01$). Figure 5.11 shows the relationship between the measured scale value of the ASW and that of the ASW calculated from Eq. (5.6).

It is noteworthy that the weighting coefficients of $(W_{IACC})^{1/2}$ in Eqs. (5.5) and (5.6) are apparently similar, so that it might read to a common formula

$$S = S_R = f(\mathrm{IACC}) + f(W_{IACC}) + f(\mathrm{SPL}) \approx \alpha(\mathrm{IACC})^{3/2} + \beta(W_{IACC})^{1/2} + \gamma(\mathrm{SPL})^{3/2} \qquad (5.7)$$

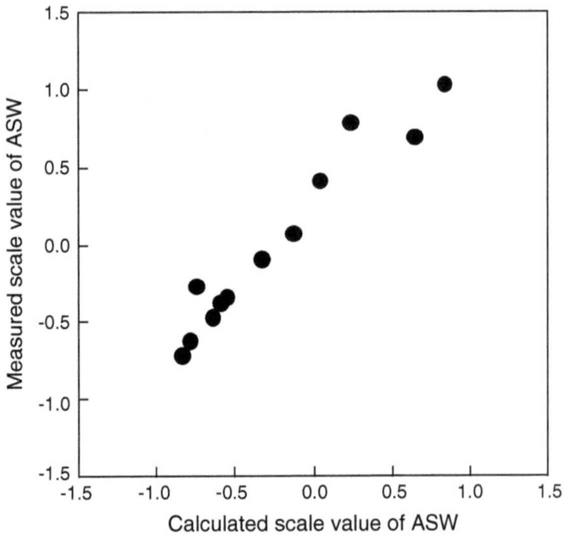

Fig. 5.11 Relationship between the measured scale values of ASW and the scale values of ASW calculated by Eq. (5.6) with $\alpha = 2.40$ and $\beta = 0.003$. The correlation coefficient is $r = 0.97$ $(p < 0.01)$

where $\alpha \approx -1.64$, $\beta \approx 2.42$, and $\gamma \approx 0.005$. The factor W_{IACC} is determined by the frequency component of the source signal; thus, the pitch or the fundamental frequency represented by the temporal factor τ_1 is not necessary to describe the ASW.

References

Ando Y (1985) Concert hall acoustics. Springer, Heidelberg

Ando Y (1998) Architectural acoustics: blending sound sources, sound fields, and listeners. AIP Press Springer, New York

Ando Y, Kurihara Y (1986) Nonlinear response in evaluating the subjective diffuseness of sound fields. J Acoust Soc Am 80:833–836

Ando Y, Singh PK (1996) A simple method of calculating individual subjective responses by paired comparison tests. Mem Grad School Sci Technol Kobe Univ 14-A:57–66

Ando Y, Sato S, Sakai H (1999) Fundamental subjective attributes of sound fields based on the model of auditory brain system. In: Sendra JJ (ed) Computational acoustics in architecture. WIT Press, Southampton, pp 63–99

Barron M (1971) The subjective effects of first reflections in concert halls—the need for lateral reflections. J Sound Vib 15:475–494

Barron M, Marshall AH (1981) Spatial impression due to early reflections in concert halls: the derivation of a physical measures. J Sound Vib 77:211–232

Damaske P, Ando Y (1972) Interaural crosscorrelation for multichannel loudspeaker reproduction. Acustica 27:232–238

Hidaka T, Beranek LL, Okano T (1995) Interaural cross-correlation, lateral fraction, and low- and high-frequency sound levels as measures of acoustical quality in concert halls. J Acoust Soc Am 98:988–1007

Kasbach J, May T, Le Goff N, Dau T (2014) The importance of binaural cues for the perception of apparent source width at different sound pressure levels. In: Proceedings of DAGA 2014

Mehrgardt S, Mellert V (1977) Transformation characteristics of the external human ear. J Acoust Soc Am 61:1567–1576

Morimoto M, Maekawa Z (1988) Effects of low frequency components on auditory spaciousness. Acustica 66:190–196

Sato S (1999) On the subjective responses based on the auditory-brain model in relation to the factors extracted from the interaural cross-correlation mechanism and the auto-correlation mechanism of sound fields. PhD Dissertation, Kobe University

Sato S, Ando Y (2002) Apparent source width (ASW) of complex noises in relation to the interaural cross-correlation function. J Temporal Des Arch Environ 2:29–32

Sato S, Kitamura T, Sakai H, Ando Y (2001) The loudness of "complex noise" in relation to the factors extracted from the autocorrelation function. J Sound Vib 241:97–103

Singh PK, Ando Y, Kurihara Y (1994) Individual subjective diffuseness responses of filtered noise sound fields. Acustica 80:471–477

Thurstone LL (1927) A law of comparative judgement. Psychol Rev 34:273–289

Chapter 6
Noise Measurements

Abstract Environmental noises are a fact of our life. Aircraft, traffic, railway, industry, machinery, and community noises are commonplace. These environmental noises have been evaluated according to sound pressure level (SPL) and frequency characteristics (e.g., Namba and Kuwano in J Acoust Soc Jpn 5:135–148, 1984; Kuwano et al. in Noise Control Eng J 33:107–115, 1989; Ayr et al. in Appl Acoust 64:129–145, 2003). Noise criterion (NC) curves, preferred noise criterion (PNC) curves, and balanced noise criterion (NCB) curves were developed to measure the SPL and its frequency characteristics (Beranek in Noise Control 3:19–27, 1957; Beranek in J Acoust Soc Am 86:650–664, 1989; Beranek et al. in J Acoust Soc Am 50:1223–1228, 1971). For evaluating fluctuation noise such as traffic noise and industry noise, the equivalent sound level (*L*eq) has been used widely. These NCs are not considered sufficiently as perceived acoustical properties. It has been reported that environmental noises not only disturb our sleep and conversation, but also affect the growth of unborn babies, infants, and children (Ando and Hattori in J Acoust Soc Am 47:1128–1130, 1970; Ando and Hattori in J Sound Vib 27:101–110, 1973); Ando and Hattori in J Acoust Soc Am 62:199–204, 1977a; Ando and Hattori in Br J Obstet Gynaecol 84:115–118, 1977b; Ando in J Sound Vib 55:600–603, 1977). In noise measurement, it is important to use parameters related to psychological attributes such as annoyance. In this chapter, we try to characterize some kinds of noise by the autocorrelation function (ACF) and interaural cross-correlation function (IACF) analyses. It has been found that information about perceived pitch and its strength (i.e., tonality) of complex sounds is extracted from the maximum peak in the ACF, and loudness of narrowband noise is related to the decay rate of the ACF (Merthayasa et al. in Mem Grad School Sci Tech Kobe Univ 12A:147–156, 1994; Sato et al. in J Sound Vib 250:47–52, 2002; Soeta et al. in J Acoust Soc Am 116:3275–3278, 2004). In addition, information of the source direction, diffuseness, and source width can be evaluated by the peak value of IACF. Continuous measurement of these factors can evaluate the noise characteristics. These ACF and IACF factors could be possible measures for describing the various parameters relating to annoyance.

Y. Soeta and Y. Ando, *Neurally Based Measurement and Evaluation of Environmental Noise*, Mathematics for Industry 20,
DOI 10.1007/978-4-431-55432-5_6

Keywords Noise evaluation · Aircraft noise · Road traffic noise · Railway noise · HVAC noise

6.1 Aircraft Noise

Aircraft noise disturbs peoples' daily lives and sometimes causes serious problems such as hearing loss or has an adverse impact on the growth of unborn babies, infants, and children (Ando and Hattori 1970, 1973, 1977a, b; Ando 1977). A lot of effort has been put on noise research and noise reduction technologies (Stephens and Cazier 1996). Significant progress has been made on reducing noise level, but a big problem remains. Noise has only been evaluated by statistic sound pressure level (SPL), but perceived acoustical properties have not been considered sufficiently. In particular, the relationship between physical properties and psychological effects is not clear. For example, a sound may exist that has a SPL below standards such as effective perceived noise level (EPNL) or weighted equivalent continuous perceived noise level (WECPNL), but that is perceived to be noisy in a given situation. Such an annoyance may be explained by ACF and IACF factors based on the mechanisms in the human auditory-brain system.

Measurements were taken outdoors near the flight course of Kansai International Airport on January 12, 2000, and near Osaka International Airport on December 13, 1999. Measurement locations are illustrated in Fig. 6.1 (Fujii et al. 2001). For measurement of noise along the flight course of Kansai Airport, a dummy head was set near coast. This location is 20 km southwest from the airport, and the flight course for landing is about 1.0 km from the shore. The altitude of the plane used in the measurement was about 1.0 km above sea level, according to the flight data from the airport. It was cloudy and windless on the ground level during the measurement. The average temperature for the day was 12 °C. Ambient noise level in this area was 43 ± 2 dB. At the Osaka Airport, two locations were chosen close to a runway to measure the noise from aircraft landing and taking off. The distances between the runway and each measuring point were about 100 m. Ambient noise level in this area was higher because of road traffic (60 ± 2 dB). It was cloudy and windless on the ground level. Temperature was about 10 °C during the measurement. Noise signals were received by two half-inch condenser microphones set at both ear positions of a sphere representing a human head. This dummy head is made of 20-mm-thick styrofoam with a diameter of 200 mm. Microphones were set at 1.5 m above the ground.

An aircraft flying overhead produces a noise on the ground, which rises above the ambient level, reaches a maximum when the aircraft is approximately overhead, and then decreases again below the ambient level. The properties of the aircraft noise vary throughout the flight. The typical case is one in which the noise is

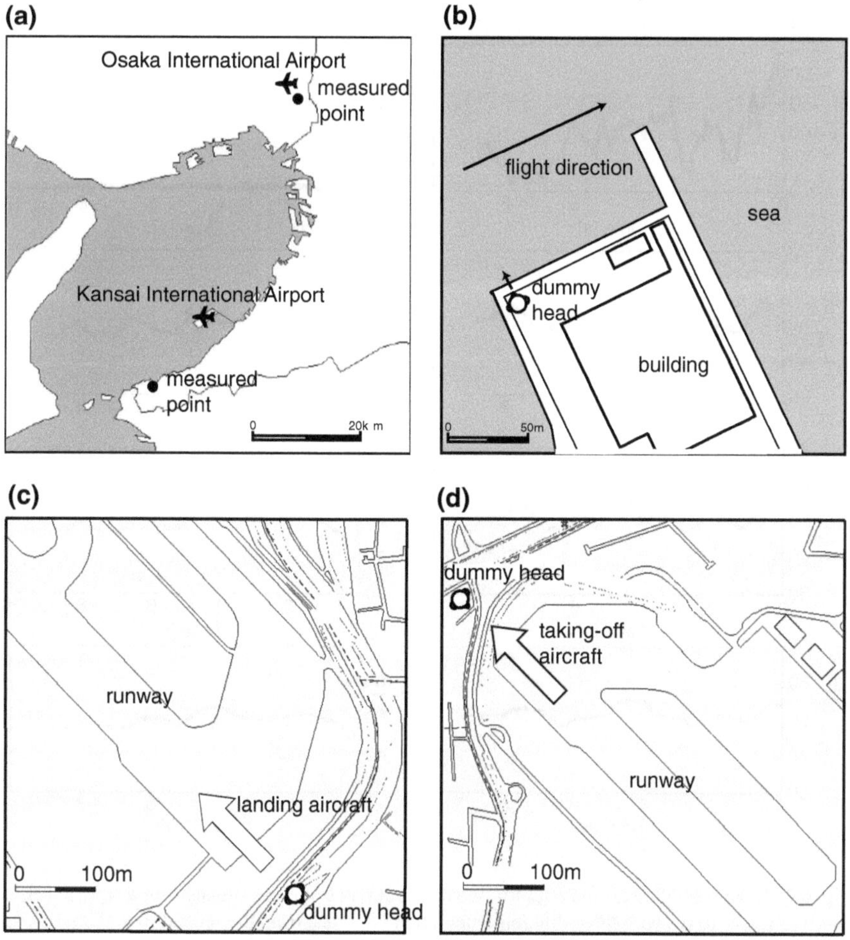

Fig. 6.1 Location of two airports and the measurement points (Fujii et al. 2001)

predominantly of high frequency while the aircraft is approaching and is predominantly of low frequency after the aircraft has passed over and is receding. Such characteristics are clearly represented by the factors from the ACF as shown in Fig. 6.2a–c for landing condition.

Measured A-weighted equivalent continuous SPL (L_{Aeq}) is shown in Fig. 6.3a as a function of time; at $t = 5.0$ s, the aircraft was directly overhead. The duration above the ambient level was about 10 s. The delay time and the amplitude of the first peak in ACF, τ_1 and ϕ_1, represent the perceived pitch and its strength. The reciprocal of τ_1 corresponds to the perceived pitch. Results indicate that the perceived pitch varied throughout the flight. As the aircraft approached, τ_1 was about 1 ms with the value of ϕ_1 increasing. The strongest pitch of 3300 Hz was perceived when the aircraft passed overhead, at which the value of τ_1 was 0.3 ms. Such a strong tonal component is

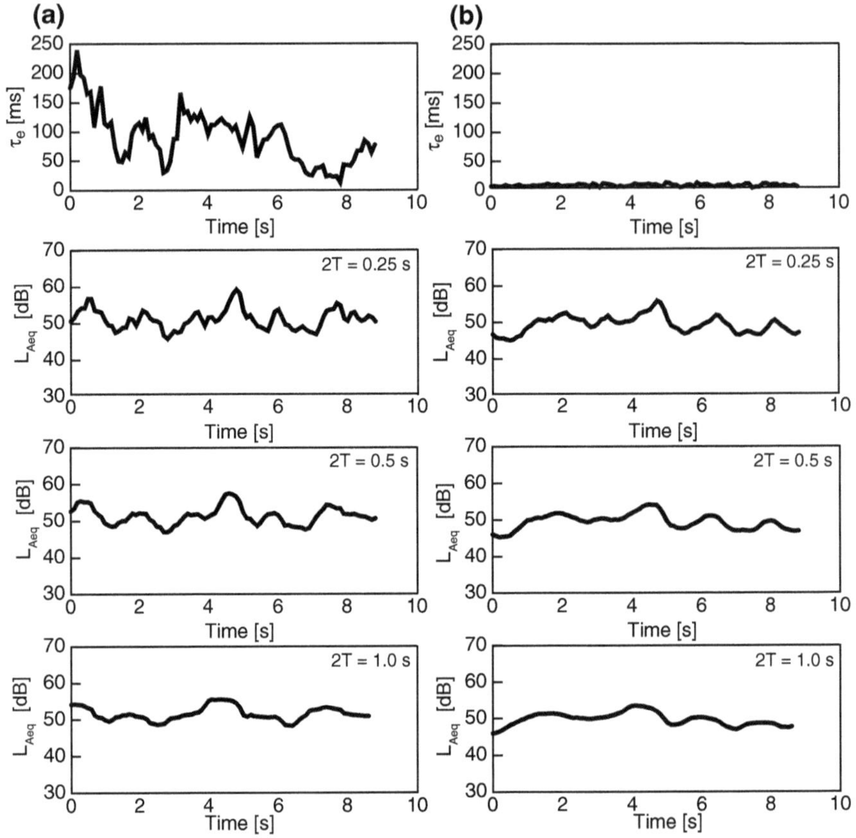

Fig. 6.2 Examples of measured L_{Aeq} for two different types of noise signals with **a** $(\tau_e)_{\min} = 20$ ms and **b** $(\tau_e)_{\min} = 10$ ms with three different integration intervals, from the second row, 0.25, 0.5, and 1.0 s (Fujii et al. 2001)

emitted from fan exhaust. After the aircraft passed over, τ_1 value increased and ϕ_1 value decreased simultaneously, indicating that the noise was dominated by the lower frequency components produced by jet exhaust.

Power spectra and the ACF measured at $t = 1.0$, 5.0, and 7.0 s are illustrated in Fig. 6.3b, c. They show that τ_1 and ϕ_1 represent the properties of aircraft noise clearly; at $t = 1.0$ s, there is a small peak around 1000 Hz, which is perceived as a noise with a weak pitch; at $t = 5.0$ s, there is a high-frequency component at 3300 Hz perceived as a tonal sound; and at $t = 7.0$ s, the strong peak disappears and the lower frequency components increase below 500 Hz, which is perceived as white noise. There are three major components in ground-perceived aircraft noise: some tonal components, broadband noise, and an interference pattern, although latter is not really noise source but rather the result of interference between direct and reflected noises (Berckmansa et al. 2008). Tonal and broadband noise components were observed in this measurement.

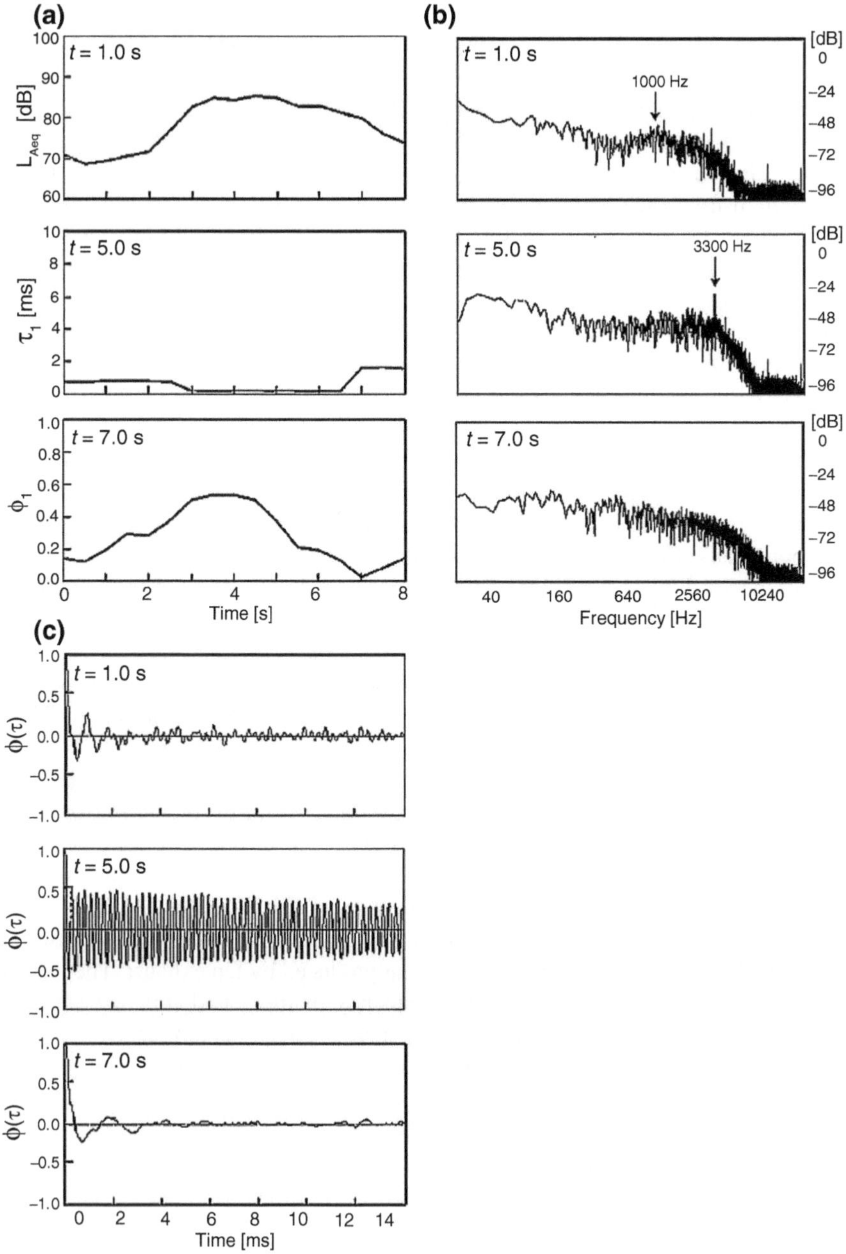

Fig. 6.3 Results for **a** measured L_{Aeq} (*top*), τ_1 (*middle*), and ϕ_1 (*bottom*) for landing aircraft as a function of time, **b** power spectra, and **c** normalized autocorrelation functions at t = 1.0, 5.0, and 7.0 s (Fujii et al. 2001)

For the same type of aircraft during taking off, the duration above ambient level was approximately 20 s, longer than that of landing condition. The acoustical properties of taking-off aircraft were somewhat different from those of landing aircraft. The ϕ_1 value was always below 0.2, which means the high-frequency tonal components were lessened and low-frequency components were pronounced. This is possibly because the aircraft is at a higher altitude and the engine power is higher. Thus, high-frequency components attenuate, and more jet noise is produced.

Figure 6.4a shows the measured factors for the aircraft during level flying overhead at an altitude of about 1 km, and measured power spectra and normalized ACF are shown in Fig. 6.4b, c. The noise for level flying aircraft was classified into two typical cases. The L_{Aeq} throughout the flight fluctuated in the same manner, but the values of τ_1 and ϕ_1 were extremely different for each case. The value of τ_1 for the two cases was almost the same (mean value 3.06 and 2.45 ms), but the ϕ_1 value for one case varied dramatically throughout the flight. At times with high ϕ_1, a tonal sound was heard and its pitch strength fluctuated in relation to $\mathrm{L_{Aeq}}$. In other words, when the noise contained a strong tonal component, the total $\mathrm{L_{Aeq}}$ increased. This phenomenon may be related to diffusion, air absorption, or the scattering reflection of sound caused by air conditions such as wind or clouds.

For the level flying condition, aircraft noise was classified as either tonal noise or untonal noise with low-frequency components. For the tonal noise, the value of τ_e becomes longer because of the repetitive component of the signal. It has been reported that loudness increases with increasing the value of τ_e (Merthayasa et al. 1994; Sato et al. 2002; Soeta et al. 2004). It is possible that the aircraft noise including tonal component is perceived louder than untonal noise.

The normalized IACF is shown in Fig. 6.5a for landing, taking-off, and level flying conditions. The values of IACC, τ_{IACC}, and W_{IACC} were found from them. Measured IACC is shown in Fig. 6.5b as a function of time. For the landing and taking-off conditions, the IACF had a strong peak at $\tau \approx 0$, meaning that the direction of the noise source is perceived clearly. The value of IACC decreased when the aircraft passed overhead for landing, possibly because the noise was dominated by the high-frequency component produced by fan exhaust. The value of W_{IACC} is dependent on the dominant frequency of the sound. For the taking-off condition, W_{IACC} was large because of the low-frequency components.

The value of IACC was generally small for the level flying condition. In this case, subjective diffuseness became high or no spatial impression was perceived. For flying aircraft at high altitude, the sound signals may come from various directions because of diffusion or the scattering reflections by clouds. It was also found that the value of IACC increased when the noise was dominated by tonal components. It is possible that such a tonal component reached the ground from the aircraft directly.

The value of τ_{IACC} for landing and taking-off aircraft was always close to zero, which means that the sound source is perceived for a frontal direction. On the contrary, for level flying condition, the value of τ_{IACC} could not be calculated in many cases because the peak of the IACF shifted over 1 ms. The value of W_{IACC} was also larger for level flying than for landing or takeoff conditions. It was found

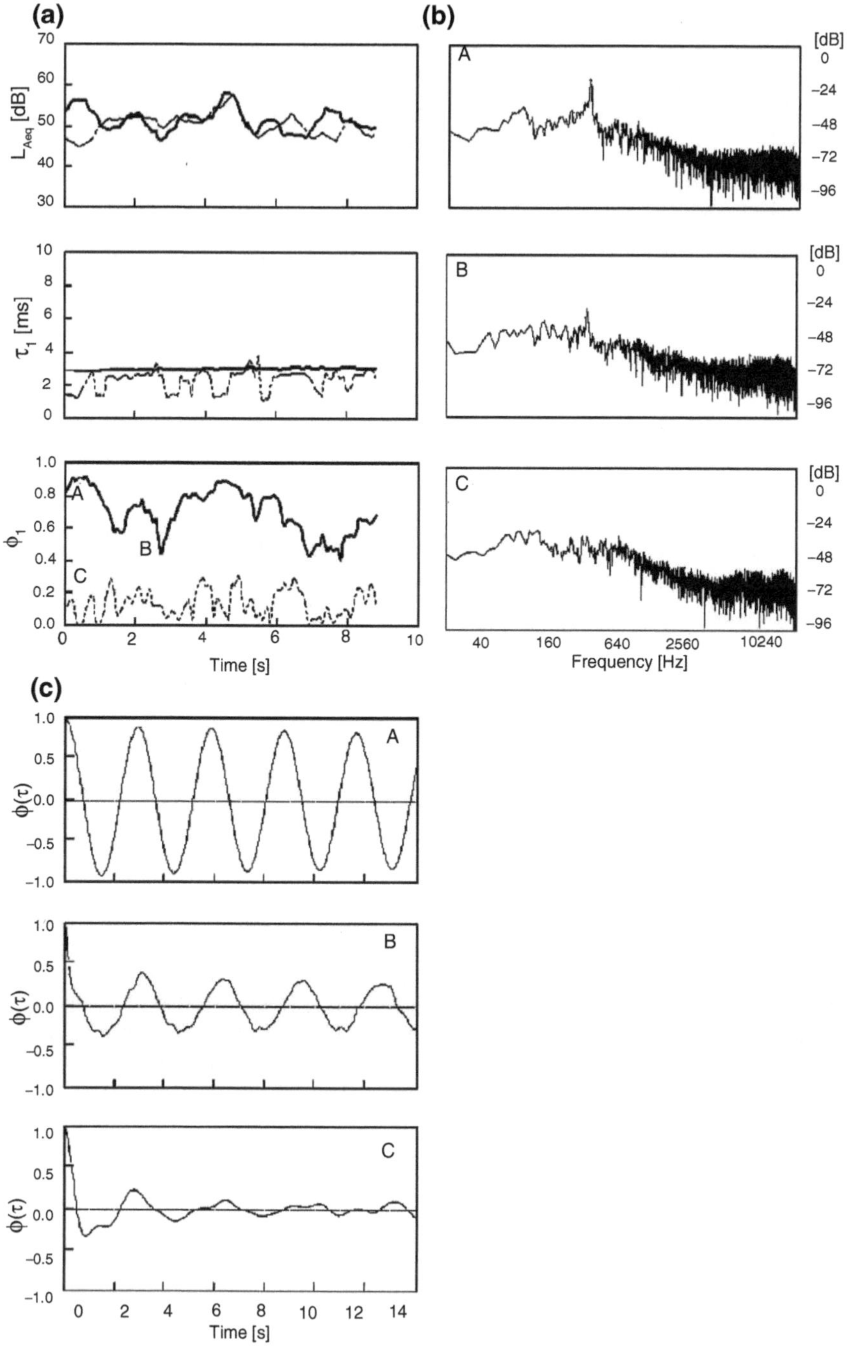

Fig. 6.4 **a** Measured L_{Aeq}, τ_1, and ϕ_1 for level flying aircraft. *underline* represents tonal noise and *dotted lines* untonal noise. **b** Examples of spectrum. **c** Normalized ACF, measured at *A*, *B*, and *C* shown in Fig. 6.4a (Fujii et al. 2001)

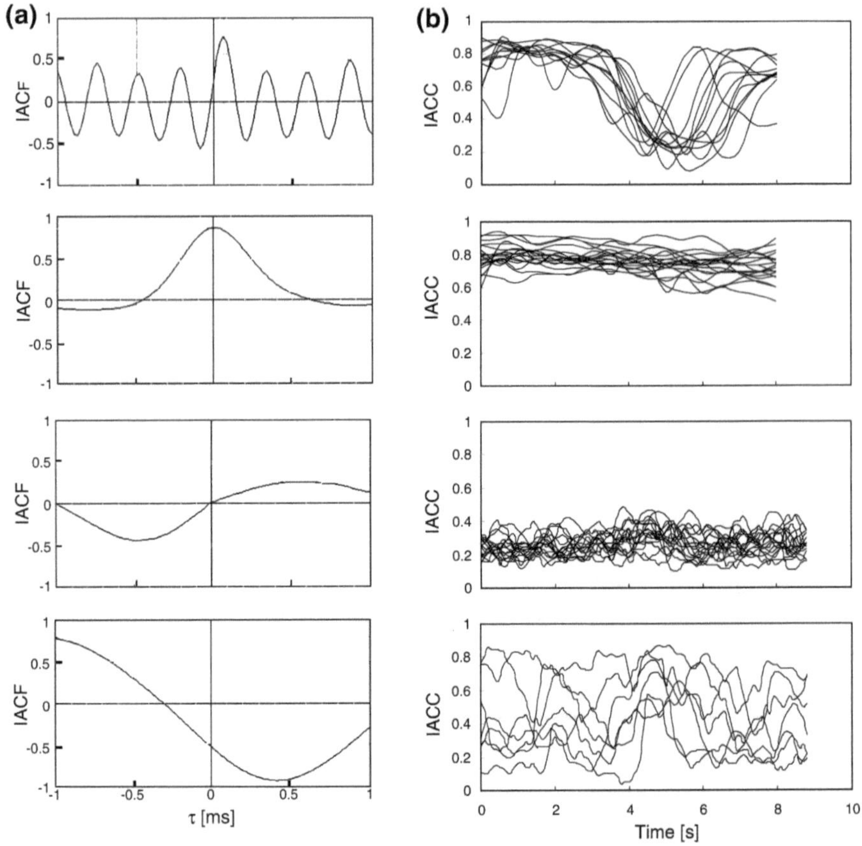

Fig. 6.5 **a** Normalized IACF and **b** measured IACC as a function of time for the conditions of landing (*top*), takeoff (*second row*), level flying 1 (*third row*), and level flying 2 (*bottom*) (Fujii et al. 2001)

that the value of IACC decreases and that of W_{IACC} increases for the level flying condition. This phenomenon may be related to the scattering reflection by clouds in the sky. The aircraft noise for level flying condition may cause higher subjective diffuseness and wider apparent source width (ASW).

6.2 Road Traffic Noise

In the previous literature, there is a sizable study describing acoustic parameters relating to annoyance. They include, for example, the frequency distribution, tonality, temporal fluctuation, and impulsivity. As for tonal color, or timbre of sound, previous studies were intended to quantify sound qualities in terms of their spectral shapes. For example, Versfeld and Vos (1997) and Cermak and Cornillon

(1976) proposed several measures for describing the shapes of the frequency spectra, such as the location and the level difference between the high-frequency (500–2000 Hz) and low-frequency (50–100 Hz) peaks in the spectrum. Also, as for the frequency distribution of noise, it is known that both of the high- and the low-frequency parts of sound have effects on annoyance. The high-frequency components of a sound make it sound sharp, and sharpness of sound increases annoyance (Zwicker and Fastl 1999). In contrast, some studies have demonstrated the more effects of low-frequency noise on annoyance (Berglund et al. 1996).

As for tonality of sound, a number of tone corrections are proposed for the evaluation of the perceived noise level. This is based on the finding that the tone-to-noise ratio in the spectrum increases annoyance of noise. Generally, a corrected value for extracted tonal component is added to the "perceived noise level" (PNL) to give the "tone-corrected perceived noise level" (PNLT). However, the calculation for this correction is lengthy, and their accuracy is not well established yet (May 1978). As for fluctuation of SPL, Hiramatsu et al. (1978, 1983) investigated annoyance of temporally fluctuating white noise (duration of 50 s) and found that standard deviation, equivalent SPL (L_{eq}), and L10 were good measures for describing the annoyance of fluctuating sound. Also, it was found that the fluctuation in the sound location and the diffuseness also much affect subjective annoyance (Jeon 2001; Kitamura et al. 2002; Sato et al. 2004). Such an annoyance may be explained by ACF and IACF factors based on the mechanisms in the human auditory-brain system.

Sound recordings were made of civil road traffic, such as a passenger car, a bus, a truck, and a motorbike (Fujii et al. 2002). The measurement point was 5 m from the center of a road, along a line perpendicular to the road. Sounds were received by two 1/2-inch condenser microphones set at the ear positions of a sphere representing a human head. This dummy head was made of 20-mm-thick styrofoam with a diameter of 200 mm. The ear positions were set at 1.5 m above the ground. Received sounds were recorded on a digital audio tape recorder at a sampling rate of 48 kHz and simultaneously stored on a hard disk of an analyzing computer at a sampling rate of 44.1 kHz for the following analysis.

The measured L_{Aeq} and three ACF factors are represented in Fig. 6.6 as a time function. Thick lines (a motorbike) and thin lines (a passenger car) show the two extremes of measured sounds: One has a clear pitch sensation, and the other has a weak pitch. Here, we call these two sounds tonal noise and untonal noise. As shown in Fig. 6.6a, the SPL throughout the period varies in the same manner. When the vehicles pass through the receiver, the SPL rises above the ambient level, reaches a maximum, and decreases again. Rise and fall time, and fluctuation of the level depend on the vehicle type and the operating conditions. As is shown in Fig. 6.6b–d, the ACF factors also vary throughout the measurement period.

The value of τ_1 varied between 1 and 10 ms, meaning that perceived pitch varied between 1000 and 100 Hz for both noises. The strength of perceived pitch increases with increasing ϕ_1. For a tonal noise, the ϕ_1 value reaches maximum around 0.6 and 0.7. At this time, a strong tonal sound is heard, having a pitch of τ_1. When the τ_1 value varies with a high ϕ_1 value, we can perceive the variation in the pitch.

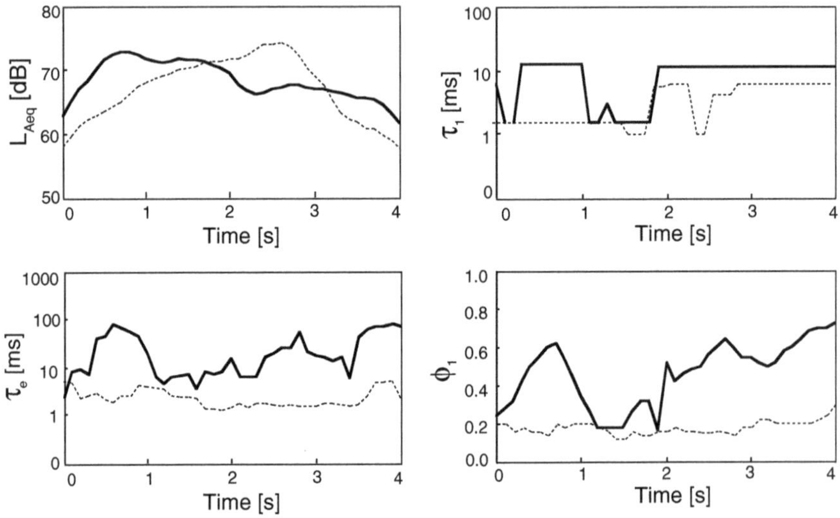

Fig. 6.6 Two extreme examples of measured L_{Aeq} and three ACF factors as time functions. *Thick lines* show factors of tonal noise (motorbike), and *thin lines* show factors of untonal noise (passenger car) (Fujii et al. 2002)

However, the ϕ_1 value for an untonal noise remained constant around 0.2, despite the variation in τ_1. The perceived pitch for an untonal noise is therefore very weak, and it is hard to discriminate pitch fluctuation.

The calculated power spectrum and ACF for tonal and untonal noises are shown in Fig. 6.7. For the "tonal noise", there are several peaks in the spectrum. Generally, the spectrum consists of harmonic components (discrete part) and noise components (continuous part), but it is difficult to identify which peak is a fundamental

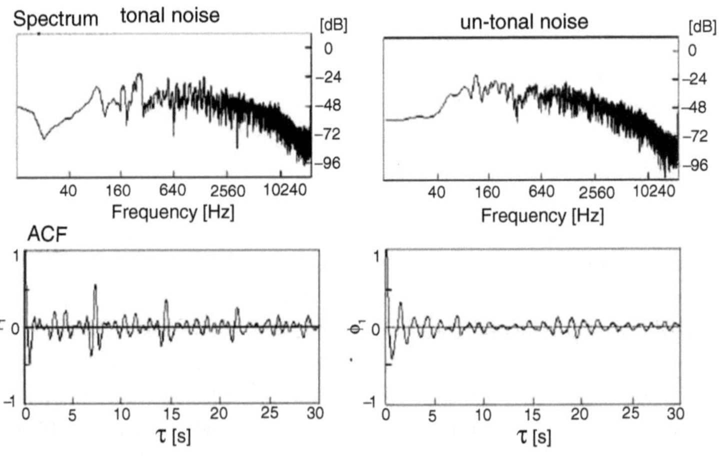

Fig. 6.7 Calculated power spectrum and ACF of tonal and untonal noises (Fujii et al. 2002)

frequency in the spectrum for a complex sound. When the same sound is analyzed by the ACF, its harmonic structure is easily extracted. Strong periodical peaks in the ACF show that a periodicity corresponding to the pitch is present in the sound. Minor peaks within a period of the ACF give information about the higher frequency components or timbre of the sound (Meddis and Hewitt 1991; Cariani and Delgutte 1996). For the "untonal noise," there is no particular peak in the spectrum. This means that the sound has no particular periodicity perceived as pitch. In this case, the ACF decreases to zero without strong periodical peaks. Because the envelope of the ACF is related to the value of ϕ_1, the value of τ_e is a good measure of the periodical structure of the sound signal. Previous study indicated that spectra of road traffic noises showed a peak at 1 kHz, which is mainly caused by the contact of tires on the road (Can et al. 2010; Buratti et al. 2014). This is consistent with present findings.

The perceived direction of a sound source is represented by the maximum peak of the IACF, τ_{IACC}, because it corresponds to an interaural time difference. As a sound source moves from left to right, the value of τ_{IACC} varies from a minus to plus value, as illustrated in Fig. 6.8. The measured values of τ_{IACC} show that the vehicles passed through the receiver from left (right) to right (left). It is found recently that the temporal fluctuation in spatial sensations such as localization and diffuseness affects subjective annoyance (Jeon 2001; Kitamura et al. 2002; Sato et al. 2004).

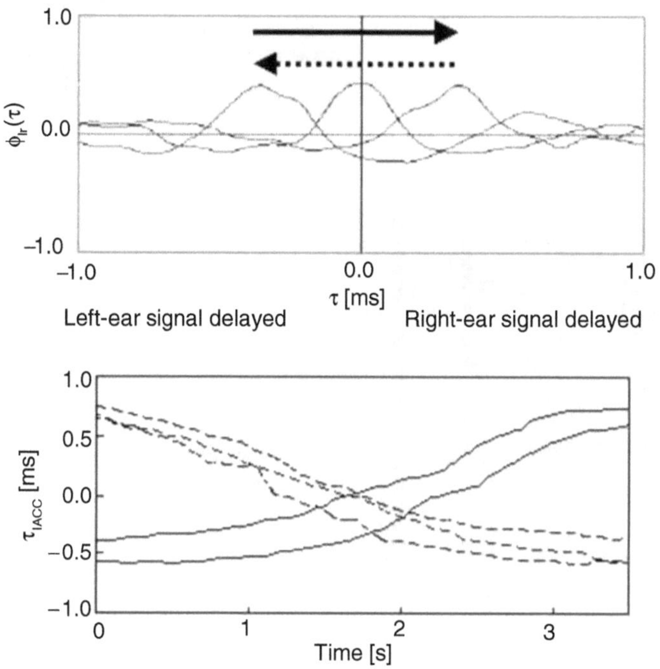

Fig. 6.8 Examples of IACF and the values of τ_{IACC} (Fujii et al. 2002)

6.3 Flushing Toilet Noise

In a three-floored apartment, located in a quiet living area of Kobe, one resident was very annoyed by the flushing noise of an upstairs toilet that spreads to the downstairs bedroom. The resident accused the construction company of improper construction, although the L_{Aeq} was only about 35 dBA. The construction company attempted to improve noise reduction in the bedroom. As a result, the SPL for the flushing toilet noise was improved by about 5 dBA. However, the resident still found the noise annoying. This fact implies that psychological effects cannot be fully evaluated by only the SPL. To clarify why the resident was annoyed by a flushing noise of an upstairs toilet despite a low L_{Aeq}, the noise was analyzed by ACF and IACF factors (Kitamura et al. 2002).

Binaural noise measurements were conducted to record the flushing noise of an upstairs toilet. The location of the noise recording was bedside (Fig. 6.9), because a resident was annoyed in sleeping time. Noise signals were received by two

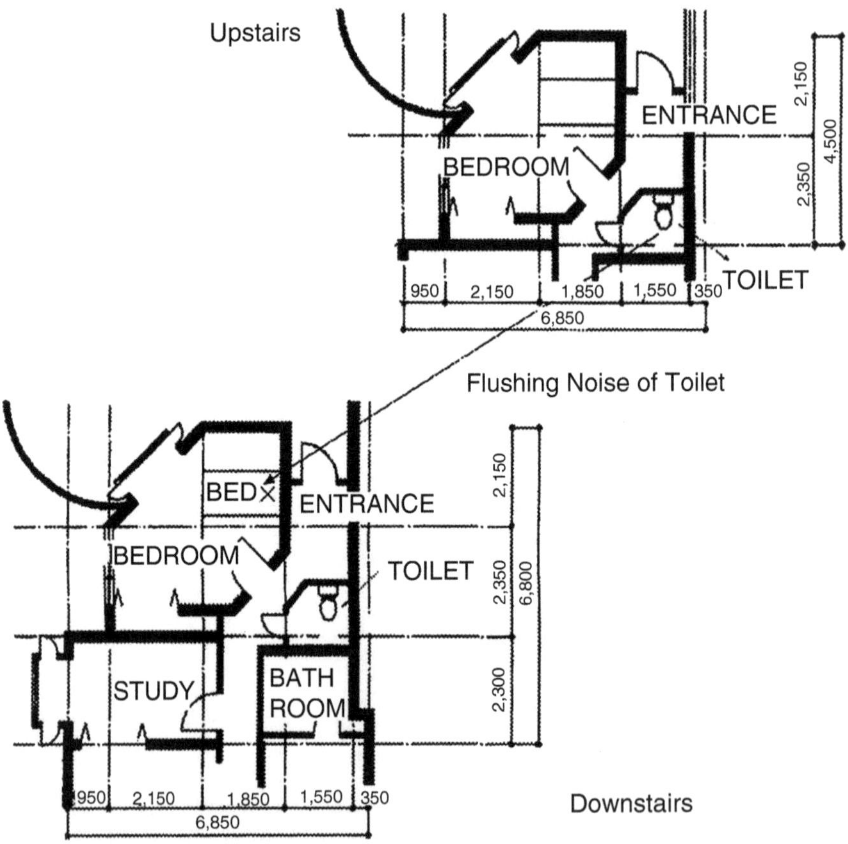

Fig. 6.9 Plan of the rooms in the apartment. The flushing noise of an upstairs toilet was recorded at bedside (Kitamura et al. 2002)

half-inch condenser microphones set at both ear positions of a sphere representing a human head. This dummy head is made of 20-mm-thick styrofoam with a diameter of 200 mm. In addition to recording the flushing toilet noise, background noise was recorded in order to compare it with the acoustic properties for a flushing toilet noise. The recording was conducted from 9 p.m. to 1 a.m. on the first day, and 6 a.m. to 8 a.m. and 9 p.m. to 12 p.m. on the second day. During the measurement, all windows and the bedroom door were closed and an air conditioner was turned off.

The measured factors extracted from the ACF are shown in Fig. 6.10a–d. Solid lines indicate values for a typical example of the flushing toilet noise, and dotted lines indicate the background noise. The measurement time was 5 s. The values of all factors were obtained every 100 ms with an integration interval of 0.5 s.

As shown in Fig. 6.10a, the $\Phi_{ll}(0)$ values, which corresponds to L_{Aeq}, for the flushing toilet noise are distributed between 30 and 35 dBA, but the background noise is about 23 dBA. The signal-to-noise ratio was 12 dBA at maximum. If the

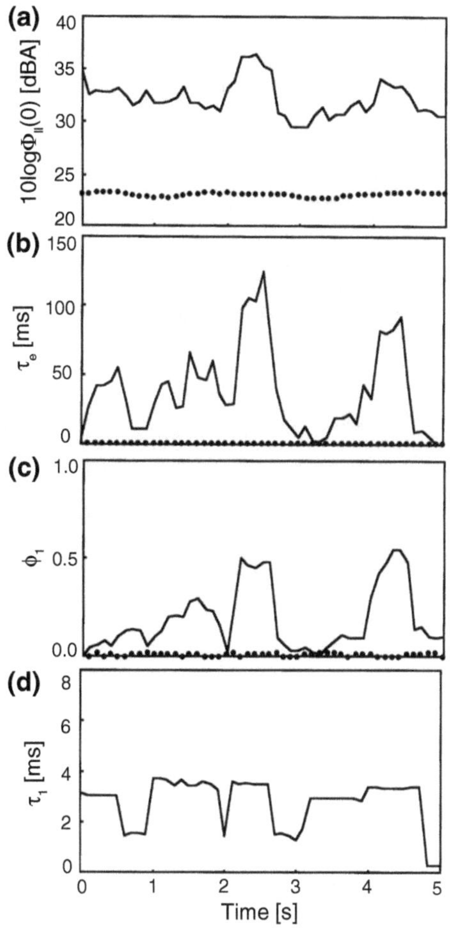

Fig. 6.10 Measured factors extracted from ACF. *Solid line* indicates values for the flushing toilet noise, and *dotted line* indicates the background noise. **a** $\Phi_{ll}(0)$, **b** τ_e, **c** τ_1; and **d** ϕ_1. Measurement time was 5 s. The values of all factors were obtained every 100 ms with an integration interval of 0.5 s (Kitamura et al. 2002)

difference between the background noise level and the noise signal level is greater than 10 dB, the background noise does not affect significantly the noise signal measurement (Beranek 1971). Thus, it seems that the resident was able to perceive easily the flushing toilet noise as a noise.

As shown in Fig. 6.10b, the τ_e values for the flushing toilet noise sometimes exceed 50 ms. On the other hand, the τ_e value for the background noise was below 0.1 ms in all measurement times. Thus, the flushing toilet noise had many more repetitive features than the background noise. Although the loudness is not influenced by τ_e when τ_e is below 10 ms or over 100 ms, the loudness increases as τ_e increases when τ_e is between 10 and 100 ms (Soeta and Nakagawa 2008). The τ_e values for the flushing toilet noise became larger near the $\Phi_{ll}(0)$ peak, as shown in Fig. 6.10a. Judging from these results, we may say that the value of τ_e increased the loudness perception of the flushing toilet noise although their SPL was low.

As shown in Fig. 6.10c, the ϕ_1 values for the flushing toilet noise were much higher than those for the background noise, which were below 0.1 in all measurement times. The background noise did not have clear pitch and tonal components as white noise components. In such cases, the τ_1 values for the background noise do not correspond to a specific pitch. This is clearly represented in Fig. 6.11a, which shows the example of a measured normalized ACF for the background noise

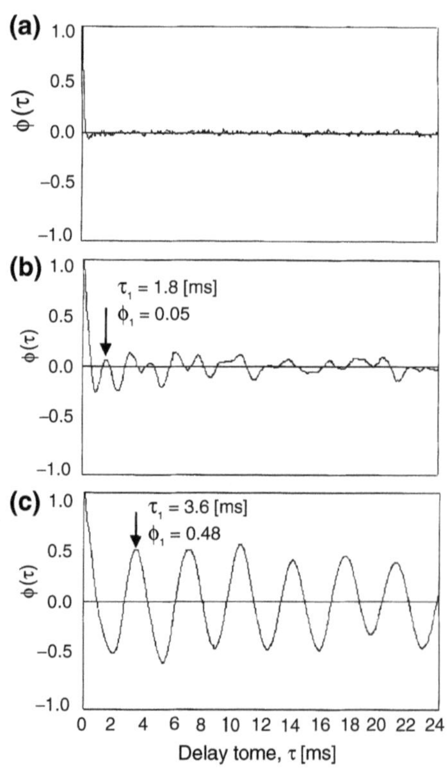

Fig. 6.11 Example of normalized ACF for the background noise and the flushing toilet noise. **a** Normalized ACF for the background noise measured at t = 0.1 s. **b** and **c** Normalized ACF for the flushing toilet noise measured at t = 0.9 and 2.5 s, respectively (Kitamura et al. 2002)

measured at $t = 0.1$ s. The amplitude of the first peak was very small. In addition, as shown in Fig. 6.12a, which shows the power spectrum for the background noise measured at $t = 0.1$ s, there were no specific frequency components. Thus, the τ_1 values for the background noise are not shown in Fig. 6.10d. On the contrary, the ϕ_1 values for the flushing toilet noise had strong peaks, as shown in Fig. 6.11b, c, which expresses an example of the measured normalized ACF for the flushing toilet noise measured at $t = 0.9$ and 2.5 s, respectively. In such cases, the tonal components increased and the perceived pitch became stronger at the dominant pitch corresponding to its τ_1 values.

As shown in Fig. 6.10d, the τ_1 values for the flushing toilet noise are distributed mostly at about 1.8 and 3.6 ms, which means that the perceived pitches of the flushing toilet noise varied about 550 and 275 Hz. This is clearly represented in Fig. 6.12b, c, which shows the power spectra for the flushing toilet noise measured at $t = 0.9$ and 2.5 s. The flushing toilet noise had peak frequency components at 550 and 275 Hz. These frequency components correspond to the values of τ_1, expressing the perceived pitch, as shown in Fig. 6.11b, c. As shown in Fig. 6.12b, the peak of the frequency components at 550 Hz was weak, because the amplitude of the first ACF peak, ϕ_1, was very small, as shown in Fig. 6.11b.

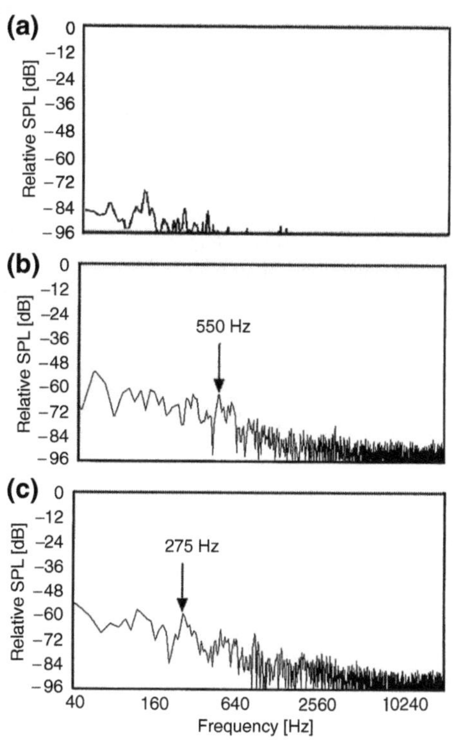

Fig. 6.12 Power spectra of noise sources. **a** Power spectrum for the background noise measured at $t = 0.1$ s and **b** and **c** power spectra for the flushing toilet noise measured at $t = 0.9$ and 2.5 s, respectively (Kitamura et al. 2002)

Both τ_1 and ϕ_1 values for the flushing toilet noise changed extremely as a function of time, as shown in Fig. 6.10c, d. This is also represented in Fig. 6.11b, c. This indicates that the perceived pitch and its strength were changing extremely as a function of time.

The measured factors extracted from the IACF are shown in Fig. 6.13a–c. As shown in Fig. 6.13a, the IACC values for the flushing toilet noise were much higher than those for the background noise. The IACC values for the background noise were below 0.05 in all measurement times. Thus, the background noise was considered to perceive no specific directions and had little spatial information. In this case, the effects of the τ_{IACC} and W_{IACC} values for the background noise are minor.

The IACF peak changed dramatically as a function of time, corresponding to the behavior of the τ_{IACC} values for the flushing toilet noise, as shown in Fig. 6.13b. In such cases, the receiver perceives the noise sources from various directions. This fact corresponds to the resident's comments. Further investigation must focus on the relationship between the characteristics of the room shape and the directional information of the flushing noise of an upstairs toilet. The values of the W_{IACC} also changed dramatically as a function of time, as shown in Fig. 6.13c. As a result, the receiver felt various ASW (Sato and Ando 2002).

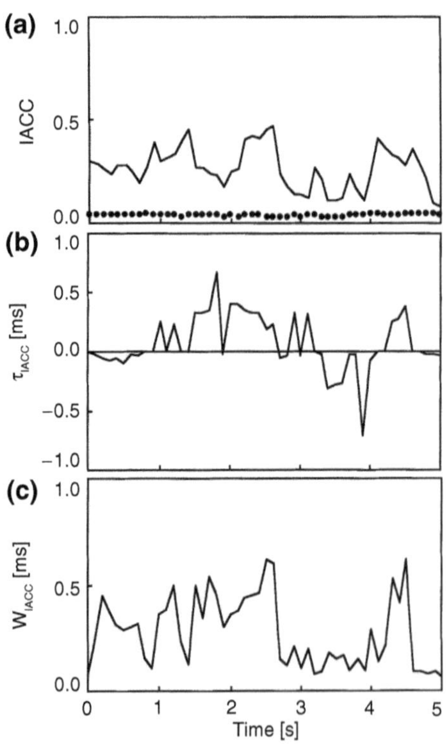

Fig. 6.13 Measured factors extracted from IACF. *Solid line* indicates values for the flushing toilet noise, and *dotted line* indicates the background noise. **a** IACC, **b** τ_{IACC}, and **c** W_{IACC}. Measurement time was 5 s. The values of all factors were obtained every 100 ms with an integration interval of 0.5 s (Kitamura et al. 2002)

It was found that the temporal and spatial factors extracted from the ACF and IACF for the flushing toilet noise had specific characteristics. Temporal and spatial factors for the flushing toilet noise changed dramatically as a function of time. According to the human auditory-brain system, temporal information is mainly processed in the left hemisphere and spatial information is mainly processed in the right hemisphere (Ando 1998). Judging from the above results, the flushing noise of an upstairs toilet stimulated dramatically both the left and right hemispheres of the resident. This is why the resident felt that the flushing noise of an upstairs toilet was annoying despite its low SPL.

6.4 Railway Noise in a Train Station

The train noise in a station is identified by a series of train noises occurring while a train arrives, stops, and leaves a station. Train noise can annoy passengers and reduces the speech intelligibility of public address systems in stations (Bhattachaya 1994; Endoh et al. 2007; Isumi and Fujii 2007). Thus, the control of train noise is important for the comfort, convenience and safety of passengers, transit workers, and operators. For example, a recent pilot survey showed that the mean of the L_{Aeq} in 17 US subway stations was 85.7 dB (Gershon et al. 2005). However, when studying the environmental noise, it is necessary to take into account not only noise levels but also sound quality. To evaluate the train noise quantitatively and qualitatively, the noise level (e.g., the L_{Aeq}) and factors extracted from the ACF and IACF were calculated for train stations.

The architectural conditions of stations are diverse, and their sound fields differ according to types of stations. In terms of location, an above-ground station covered by only a roof is similar to a free sound field, while a completely covered underground station is a reverberant sound field (Fig. 6.14). In terms of platform style, stations can be principally divided into the side (two platforms at the side and the rail tracks in the center) and island (one platform at the center and the rail tracks on the sides) platforms (Fig. 6.14). In an above-ground station, the side platform is semiclosed with a wide roof and lateral walls, while the island platform is an almost-free sound field. In an underground station, the train runs in the center of station with the side platforms, while it runs against the lateral wall of station with the island platform. With these different architectural elements, the noise propagation and sound qualities may also differ.

The train noise was recorded in 16 above-ground and underground stations on 8 railway lines by dummy head microphones (KU100; Neumann) or binaural microphone (Type 4101; B&K). The architectural data of stations were listed in previous studies (Shimokura and Soeta 2011; Soeta and Shimokura 2012, 2013a). The dummy head microphone is an artificial model of a human head with two microphones inserted in the ear canal. They were located at the three receiver positions. An experimenter positioned the binaural microphones (Type 4101; B&K) next to his ears and stood at the receiver positions. The height of the ears of the

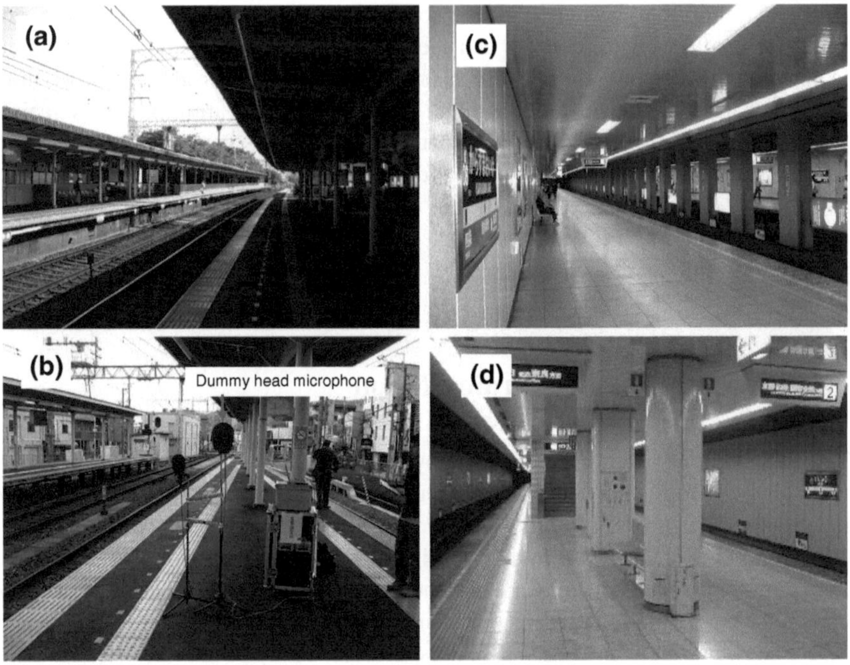

Fig. 6.14 Views of the above-ground stations with **a** side and **b** island platforms and underground stations with **c** side and **d** island platforms (Shimokura and Soeta 2011)

dummy head microphone or the experimenter was 1.6 m above the ground. Although the different head-related transfer functions of the two microphones may affect the calculation of ACF and IACF factors, the preliminary measurements verified that the difference in the microphones did not affect the results (Shimokura and Soeta 2011). Simulating a passenger waiting for a train, the dummy head microphone or the experimenter was always facing and perpendicular to the railway track. In each station, three receiver positions were set up, the first at the entrance end ($r1$), the second in the middle ($r2$), and the third at the exit end ($r3$) of the platform. For all measurements, train noises were saved on a personal computer (Let's note, Panasonic) via an analog-to-digital converter (48 kHz sampling rate, 24 bit sampling size). The noise measurements were taken once at three receiver positions in each station during the day (10:00–15:00) to avoid the rush hours. The duration of the measurement at each receiver position was between 60 and 220 s.

Considering the moving noise source, the train noise was analyzed for a number of train conditions. The train condition is defined as either "come," "stop," or "go" (Fig. 6.15). The interval "come" started when the L_{Aeq} of the train noise exceeded the background noise level by 3 dB. The interval "stop" was the duration the door was open. The interval "go" started when the door was closed and ended when the L_{Aeq} decreased to 3 dB above the background noise level. The values of the L_{Aeq}, τ_1, ϕ_1, $W_{\phi(0)}$, and IACC were calculated for each interval.

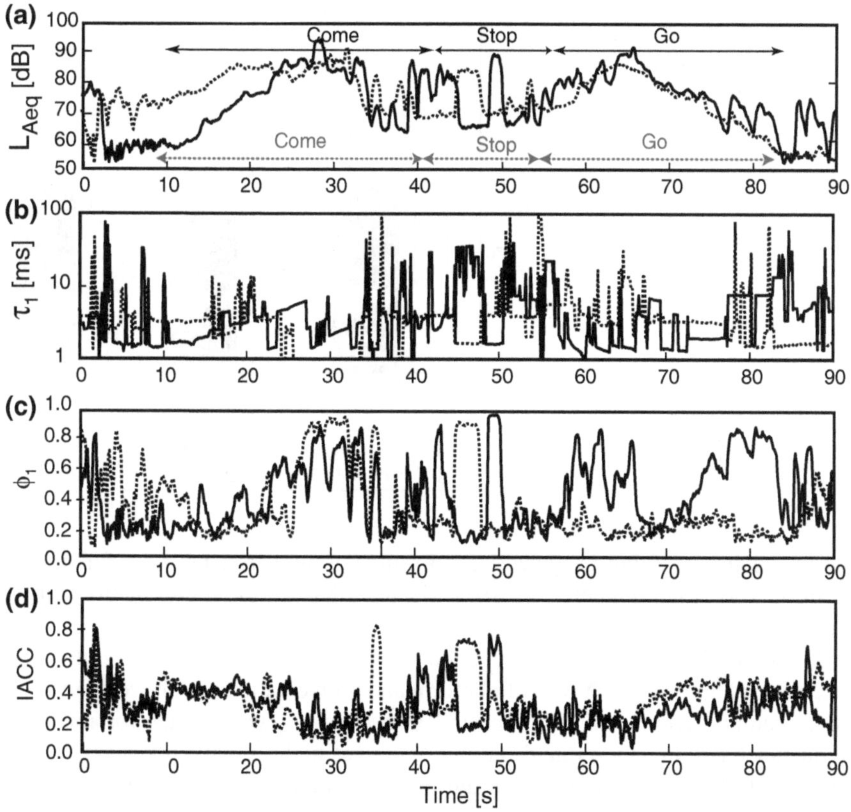

Fig. 6.15 **a** L_{Aeq}, **b** τ_1 and **c** ϕ_1, **d** $W_{\phi(0)}$, and **e** IACC as a function of time in the underground stations. *Straight* and *dashed lines* indicate side and island platforms, respectively. The receiver position was *r*2 (Soeta and Shimokura 2013a)

Figure 6.16 shows the noise level (L_{Aeq}) averaged across the three receiver positions for each type of platform and train condition in above-ground and underground stations. L_{Aeq} in the underground station was approximately 1–8 dB higher than that in above-ground stations. The ballast track can absorb running train noise (Stüber 1975), and this might explain why L_{Aeq} in the above-ground stations is lower. The other explanation is reverberation from the tunnel during the train coming and going intervals and reflections in the station during the train stopping interval.

In the case of above-ground stations, L_{Aeq} for the side platform was approximately 1–5 dB higher than that for the island platform. While an island platform is similar to a free sound field, a side platform is partly surrounded by a wide roof and lateral walls in above-ground stations (Fig. 6.14). Since there were ballast tracks in the above-ground stations, the higher L_{Aeq} for the side platforms may be due to reflections from the boundaries. On the island platforms, only direct sound reaches

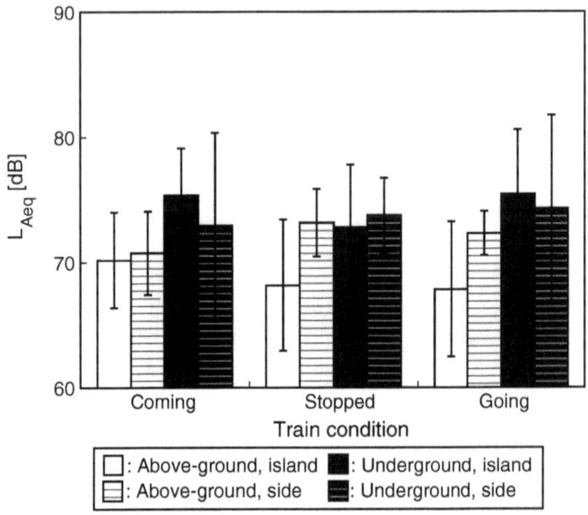

Fig. 6.16 L_{Aeq} in each location, platform style, and time interval (Soeta and Shimokura 2013a)

the passenger, resulting in lower L_{Aeq}. In underground stations, L_{Aeq} for the side platform was approximately 1–2 dB lower than that for the island platform in the coming and going intervals.

Figure 6.17 shows the octave band power level averaged across the three receiver positions for each type of platform and train condition in above-ground and underground stations. In above-ground stations, L_{Aeq} was higher for the lower frequency band. In underground stations, L_{Aeq} was highest in the frequency band of 250 or 500 Hz. The platform style did not have a clear effect on the octave band power level.

The pitch of the train noise, τ_1, was not affected by the type of platform in above-ground and underground stations (Fig. 6.18). Figure 6.19 shows ϕ_1 averaged across the three receiver positions for each type of platform and train condition. ϕ_1 in underground stations was greater than that in above-ground stations except for coming intervals.

Figure 6.20 shows $W_{\phi(0)}$ averaged across the three receiver positions for each type of platform and train condition. In the coming and going intervals, $W_{\phi(0)}$ in the underground stations was larger than that in the above-ground stations. The larger $W_{\phi(0)}$ means a lower spectral centroid of the noise. This suggests that underground stations have low-frequency components.

Figure 6.21 shows the IACC averaged across the three receiver positions for each type of platform and train condition. In the case of above-ground stations, the IACC for the island platforms was higher than that for the side platforms. The higher IACC for the island platforms in the coming and going intervals suggests that noise from the train car was directional. On the island platforms, only direct sound reaches the passenger, resulting in lower L_{Aeq}.

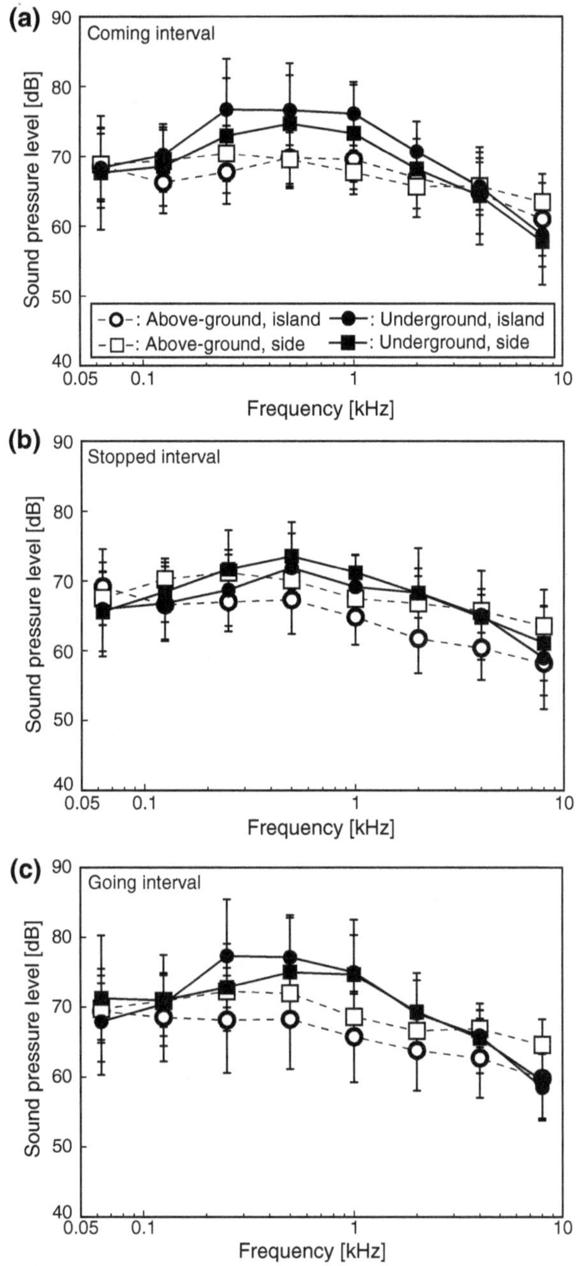

Fig. 6.17 Measured SPL as a function of the 1/1 octave band center frequency in each location, platform style, and time interval (Soeta and Shimokura 2013a)

Platform screen doors (PSDs) at above-ground or underground stations screen the platform from the train. They are a relatively new addition to stations around the world and widely used in Asian and European metro systems. PSDs help to prevent

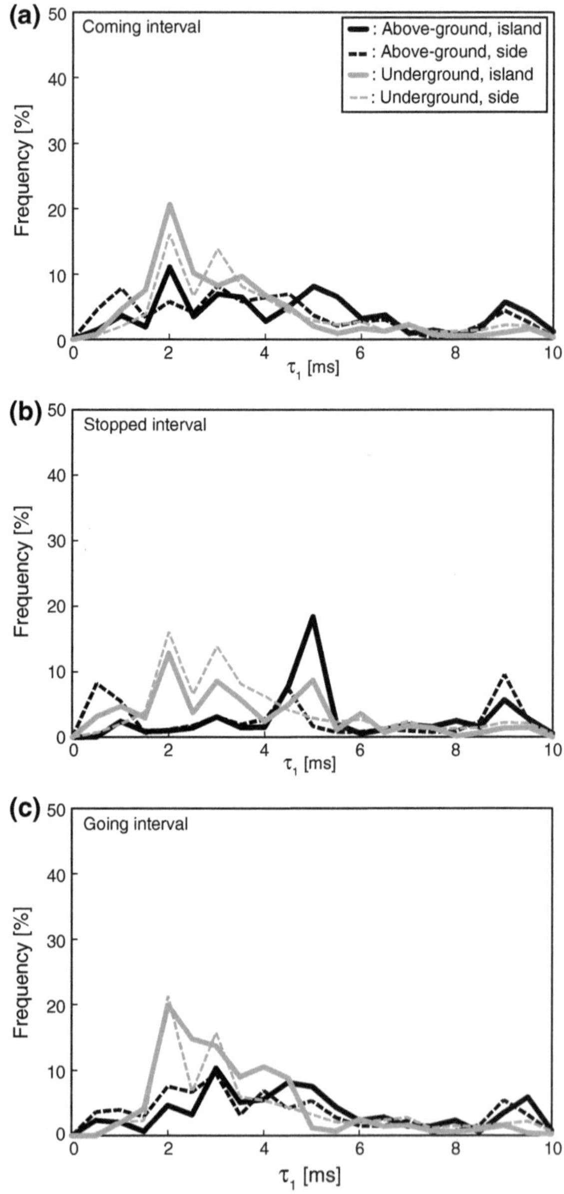

Fig. 6.18 Frequency distributions of τ_1 obtained from short-time moving analyses for each location, platform style, and time interval (Soeta and Shimokura 2013a)

falls from the platform onto the track area, to reduce the risk of accidents from service trains passing through the station at high speeds, and to improve climate control, e.g., heating, ventilation, and air conditioning (HVAC), within the station. PSDs have an impact on safety in train stations, and they might also have an effect on train noises in stations since they are a kind of wall which blocks the noise

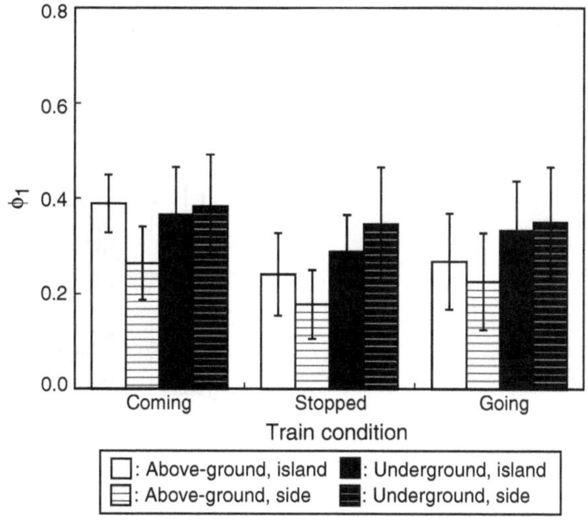

Fig. 6.19 ϕ_1 in each location, platform style, and time interval (Soeta and Shimokura 2013a)

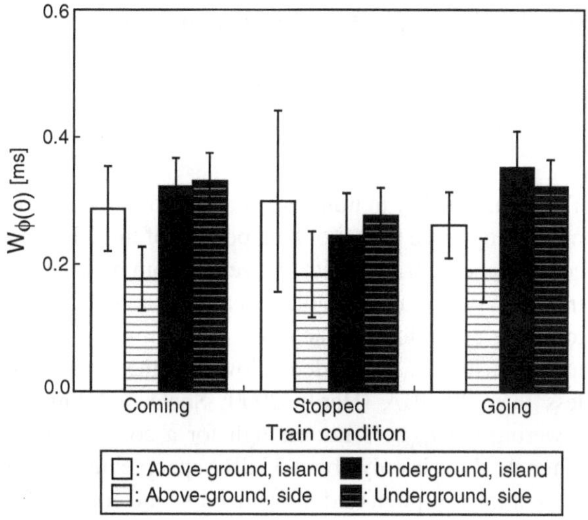

Fig. 6.20 $W_{\phi(0)}$ in each location, platform style, and time interval (Soeta and Shimokura 2013a)

generated by a train. The effects of PSDs on acoustical characteristics have been investigated (Soeta and Shimokura 2012, 2013a). Noise level emitted by trains was reduced by PSDs in both above-ground and underground stations. IACC was decreased by PSDs in both above-ground and underground stations, suggesting that PSDs made train noises more diffused. $W_{\phi(0)}$ was decreased by PSDs in both above-ground and underground stations, which means that the train noises in station with PSDs have higher spectral centroid, suggesting that the PSDs blocked the lower frequency components of train noises.

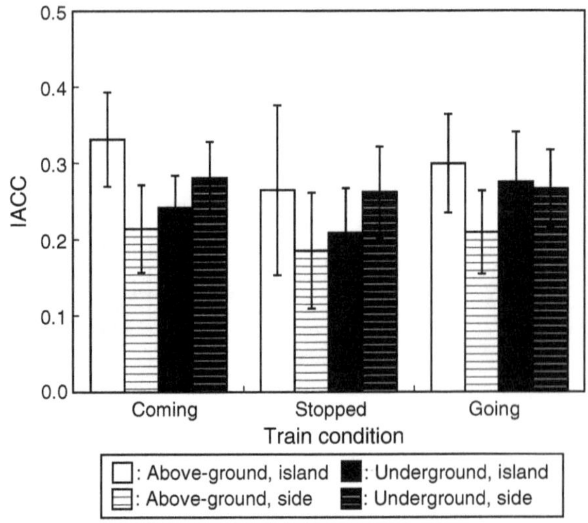

Fig. 6.21 IACC in each location, platform style, and time interval (Soeta and Shimokura 2013a)

6.5 Railway Noise in a Train Car

6.5.1 Effect of Train Type and Noise Caused by Wheel–Rail Interaction

The sound quality in train cars is an important concern for the railway industry since many people use trains for long periods of time. For example, approximately 40 % of commuters use trains, with an average commuting time of more than 140 min (round trip), in a typical metropolitan area of Japan such as Tokyo. The fundamental issue is to create a comfortable acoustic environment in a train car for the passenger. The train industry has succeeded in lowering the equivalent L_{Aeq} in train cars to a mean of less than 70 dBA (Hardy 2000; Soeta and Shimokura 2009, 2011). However, lowering the L_{Aeq} is not enough for a comfortable acoustical environment. Even when the L_{Aeq} is only about 35 dBA, people may feel annoyed by conditions such as fluctuations of pitch and a localized sound source (Kitamura et al. 2002).

People hear various types of noise in train cars. Researchers have studied mechanisms for generating different types of noise such as rolling, impact, curve squeal (Remington 1987; Van Ruiten 1988; Thompson and Jones 2000, 2006; Vincent et al. 2006), motor (fans) (Mellet et al. 2006; Schulte-Werning et al. 2006), and aerodynamic noises (Cleon and Williame 2000; Nagakura 2006). Some studies have modeled the noise in train cars to help predict and reduce it (Talotte et al. 2003; Dittrich and Zhang 2006; Thron and Hecht 2010; Forssen et al. 2012). Although some studies have quantitatively (Yoshihisa et al. 1968; Chang and Hermann 1974; Eade and Hardy 1977; Kono et al. 1982; Hardy 2000) investigated the noise characteristics inside running train cars, the effects of noise sources on the noise characteristics have not been evaluated qualitatively. To evaluate the noise characteristics inside running train cars quantitatively and qualitatively, the noise level (e.g., the L_{Aeq}) and factors extracted from the ACF and IACF were calculated (Soeta and Shimokura 2009, 2011). In particular, the effects of the rolling, impact, curve squeal, and motor (fans) noises inside train cars were focused on. Acoustic treatment (i.e., sound absorption, insulation, and active noise control) is an effective solution for improving the sound environment inside train cars. To consider an appropriate acoustic treatment, it is necessary to clarify the characteristics of the noise inside train cars.

Noises in a train car were measured continually traveling from one terminal station to another terminal station on 16 lines. Different types of train car were run on each line. Active noise control is not installed in the train car. Trains are classified according to location (above-ground and underground), main motor (linear or three-phase induction motors), and speed (normal and high speed). Trains usually run on a slab track underground or on a ballast track above the ground. Only lines *G* and *H* are classified as high-speed trains. Line I is a monorail with straddle-beam systems, and lines *J* and *K* are automated guideway transits with rubber tires.

The three types of noise caused by wheel–rail interaction are rolling, impact, and curve squeal. The rolling type was defined as noises in running train cars that include no impact or curve squeal noises. The recorded sounds were listened to identify the impact and curve squeal noises. The noises were cut out from the data, and the durations were at least 10 s. It was difficult to determine beforehand that measurement was taken in motor or trailer cars because various types of trains usually run on the line. It was only on lines *L*, *M*, *N*, and *O* that we could record the noise in motor and trailer cars distinctly because the cars used were limited. On other lines, data from motor and trailer cars were mixed.

A dummy head microphone (KU100, Nuemann) or a binaural microphone (Type 4101, B&K) was used to record noise inside a train car. The microphone was placed in the area for a wheelchair user. The microphone always faced the inside of the train car and was perpendicular to the direction of travel except for lines *G* and *H*, which had no available area for wheelchair users. The height of the dummy head and binaural microphones was 1.6 m. For all measurements, noise was recorded on a personal computer (Let's Note, Panasonic) via an AD/DA converter (AudioFire8, Echo Digital Audio) at a sampling rate of 48 kHz and sampling resolution of 24 bits. The measurements were taken from 10:00 to 15:00 to avoid rush-hour commuting.

Figure 6.22 shows the averaged L_{Aeq} and octave band levels for rolling, impact, and curve squeal noises in running above-ground and underground trains. Impact noises had larger components at lower frequencies (less than 500 Hz) compared with rolling noises in both above-ground and underground trains. Curve squeal noises had larger components at frequencies higher than 125 Hz in underground trains and at frequencies between 125 and 500 Hz s in above-ground trains compared with rolling noise. Underground trains often run sharp curves because of geographical constraints, while above-ground trains rarely run sharp curves. High-frequency components are enhanced by reflections in tunnels. This might affect the difference in higher frequency components in above-ground and underground trains.

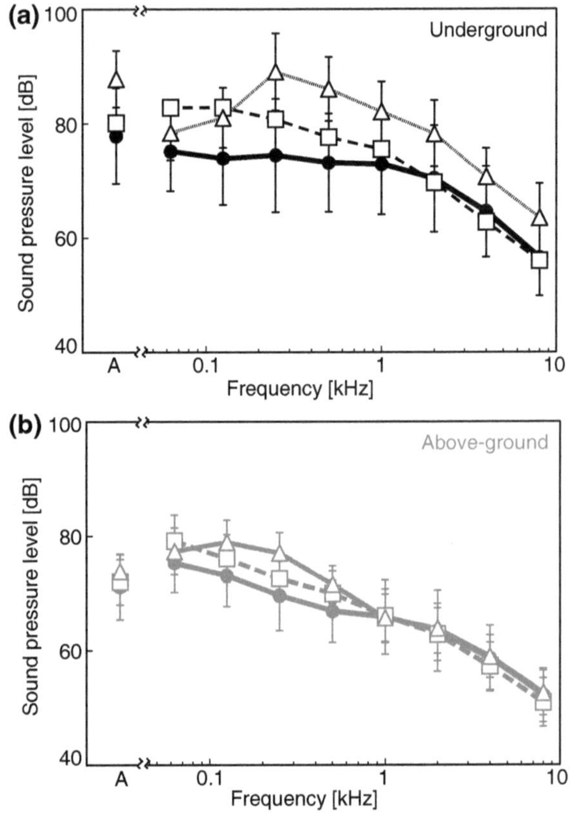

Fig. 6.22 Measured SPL as a function of the 1/1 octave band center frequency and L_{Aeq} for (*filled circle*) rolling, (*open square*) impact, and (*open triangle*) curve squeal noises in **a** underground and **b** above-ground trains (Soeta and Shimokura 2013b)

Figures 6.23 and 6.24 show the frequency distributions and averaged ACF and IACF factors for rolling, impact, and curve squeal noises in above-ground and underground trains. The τ_1 values concentrated around 1.5, 4.0, and 5.5 ms. Curve squeal noises had longer τ_1 values compared to rolling and impact noises in both above-ground and underground trains. Curve squeal noise may be perceived louder than rolling and impact noises when the L_{Aeq} is the same because longer τ_1 causes louder sensation in the τ_1 range between 0.5 and 8.0 ms (Soeta and Nakagawa 2008). Curve squeal noises had the largest ϕ_1 values in underground trains, suggesting the effect of both wheel–rail friction and reflections in tunnels. Curve squeal noises showed the largest $W_{\phi(0)}$ and IACC values in above-ground trains, suggesting an increase in lower frequency components and binaural coherence. Curve squeal noises showed the largest variation in IACC values in above-ground trains, which may cause higher annoyance (Sato et al. 2004).

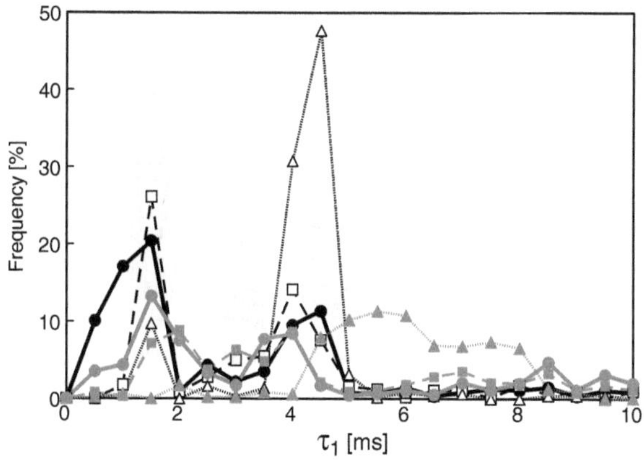

Fig. 6.23 Frequency distribution of τ_1 values for (*filled circle*) rolling, (*open square*) impact, and (*open triangle*) curve squeal noises in underground trains and (*filled circle*) rolling, (*open square*) impact, and (*open triangle*) curve squeal noises in the above-ground trains (Soeta and Shimokura 2013b)

Figure 6.25 shows the averaged L_{Aeq} and octave band levels for different types of train. Noise levels increase as the speed of the train increases (Moritoh et al. 1996; Mellet et al. 2006). The noise level for high-speed trains with linear motors was 6–9 dB larger than that for high-speed trains with three-phase induction motors, because high-speed trains with linear motors run faster. There were no clear differences in frequency characteristics between high-speed trains with linear or three-phase induction motors. This was also true of underground trains. High-speed trains had strong frequency components that were lower than 250 Hz compared with normal speed trains. This confirmed the effects of countermeasures on aerodynamic noises. These countermeasures included the louvres of air inlets and outlets equipped on car surfaces and insulators on car roofs, which are necessary for bus cables to extend to different cars, car surfaces near the front nose, where the speed of the air flow is very high, and gaps between any two adjoining cars (Moritoh et al. 1996; Nagakura 2004; Kitagawa 2008). The difference between the noise levels of underground and above-ground trains with normal speed was large at frequencies around 250 and 500 Hz, agreeing with previous results (Chang and Hermann 1974; Soeta and Shimokura 2011). Noise levels in automated guideway transits with rubber tires were approximately 2 dB lower at lower frequency (63 and 125 Hz) than those in normal speed trains and monorails with straddle-beam systems, suggesting the effects of rubber tires. Noise levels in normal speed trains were 2–3 dB lower at higher frequency (more than 1000 Hz) than those in automated guideway transits with rubber tires and monorails with straddle-beam systems.

Figures 6.26 and 6.27 show the ACF and IACF factors for different types of train. High-speed trains had longer τ_1 values, confirming the effects of countermeasures on aerodynamic noises. This means that the fundamental frequency of the noise was lower and suggests that the noise pitch in high-speed train cars was low.

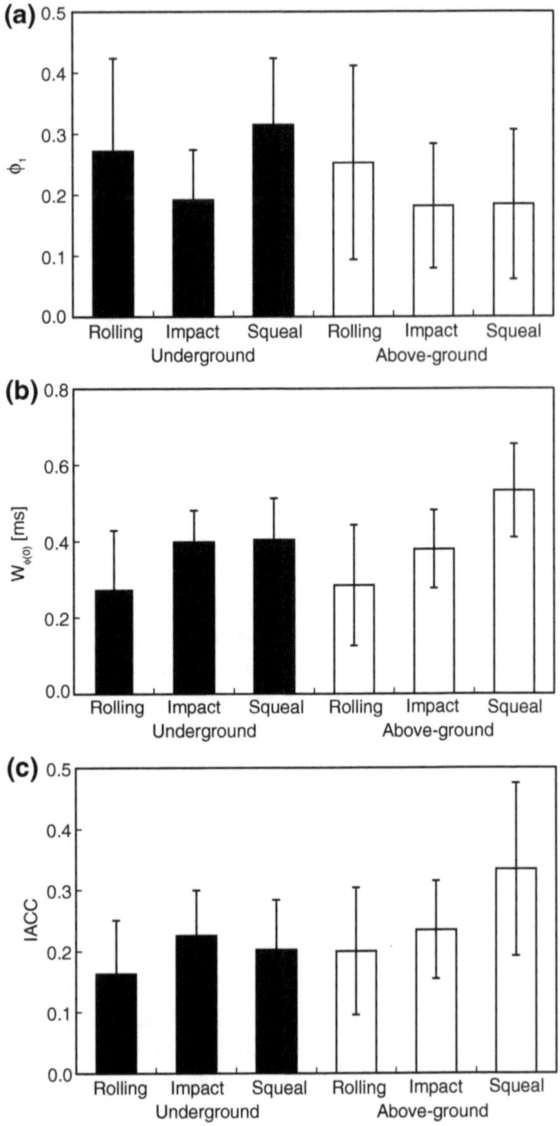

Fig. 6.24 Averaged **a** ϕ_1, **b** $W_{\phi(0)}$, and **c** IACC values for rolling, impact, and curve squeal noises in underground and above-ground trains. Error bars are standard deviations (Soeta and Shimokura 2013b)

Underground trains had slightly shorter τ_1 values than did above-ground trains, meaning the major frequency component shifts from lower to higher frequency ranges by reflections in tunnels and agreeing with previous results (Chang and Hermann 1974; Soeta and Shimokura 2011). High-speed trains with linear motors had larger ϕ_1 values than those with three-phase induction motors, suggesting better noise reduction measures for those with three-phase induction motors. This suggests that high-speed trains with linear motors cause higher annoyance since sound

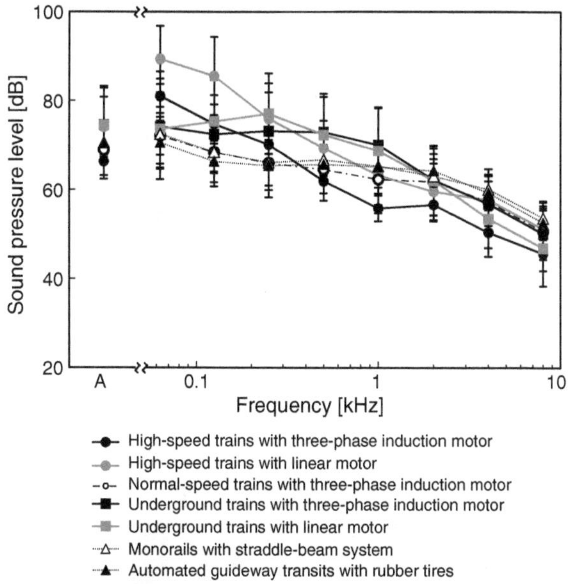

Fig. 6.25 Measured SPL as a function of the 1/1 octave band center frequency and L_{Aeq} different types of train (Soeta and Shimokura 2013b)

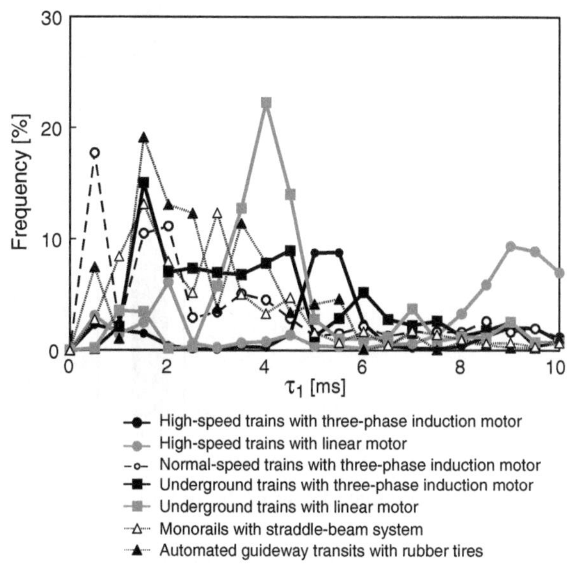

Fig. 6.26 Frequency distribution of τ_1 values for different types of train (Soeta and Shimokura 2013b)

with larger ϕ_1 is perceived more annoying (Sato et al. 2002, Soeta et al. 2004). High-speed trains with three-phase induction motors had smaller ϕ_1 values than did normal speed trains, because high-speed trains usually have more noise reduction measures than normal speed trains do (Moritoh et al. 1996; Nagakura 2004). High-speed trains had the largest $W_{\phi(0)}$ values. Underground trains had larger $W_{\phi(0)}$ values than did above-ground normal speed trains, because low-frequency

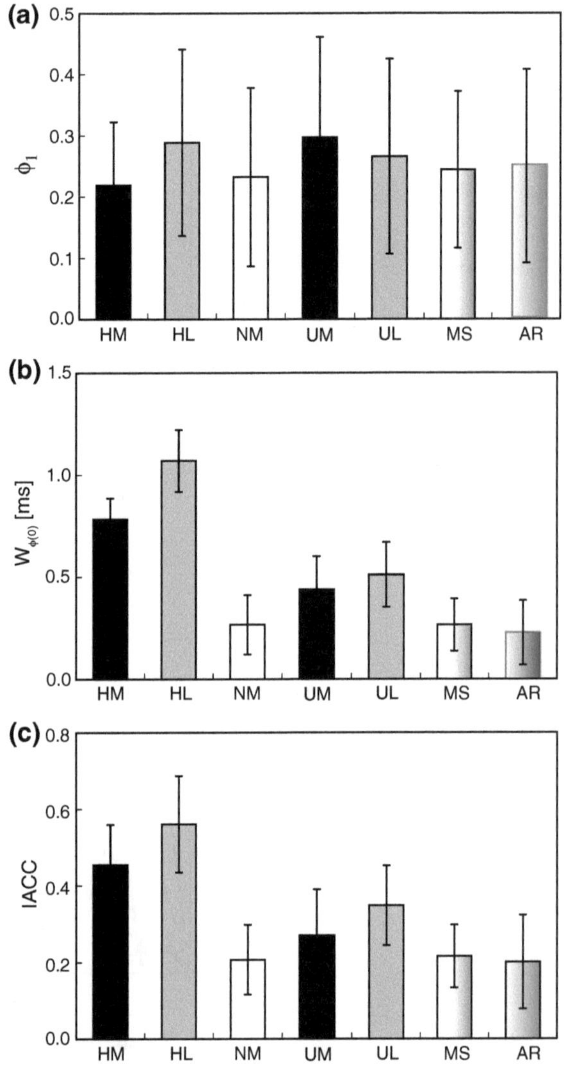

Fig. 6.27 Averaged **a** ϕ_1, **b** $W_{\phi(0)}$, and **c** IACC values for high-speed trains with three-phase induction motors (HM), high-speed trains with linear motors (HL), normal speed above-ground trains with three-phase induction motors (NM), underground trains with three-phase induction motors (UM), underground trains with linear motors (UL), monorails with straddle-beam systems (MS), and automated guideway transits with rubber tires (AR) (Soeta and Shimokura 2013b)

components attenuated less in the tunnel (Imaizumi et al. 1997, 2000). High-speed trains had larger IACC because of the lower frequency components. This suggests that the perceived annoyance is lower since annoyance of noise in train stations decreases with increasing IACC of the noise source (Soeta and Shimokura 2013a). Underground trains with linear motors had larger IACC values than underground trains with three-phase induction motors due to tunnel effects. Underground trains with linear motors, which have smaller cars, can often run in tunnels excavated via the boring machine method (BMM), which usually have smaller volume and concentrated sounds, causing a larger IACC (Soeta and Shimokura 2011).

Regarding the difference between motor and trail cars, noise levels in motor cars were 2–3 dB larger than those in trailer cars. There were no discernible effects of motor or trailer cars on ACF and IACF factors.

The effects of the rolling, impact, curve squeal, and motor (fans) noises on noise inside train cars were analyzed. The results show the following: (1) Impact noises had larger components at lower frequencies (less than 500 Hz) compared with rolling noises. (2) Curve squeal noises had larger components at frequencies above 125 Hz, which have stronger pitch, in underground trains and at frequencies between 125 and 500 Hz in surface trains compared with rolling noises. (3) Curve squeal noises had longer τ_1 values compared with rolling and impact noises, meaning that tonality expressed by longer τ_1 values increases loudness. (4) High-speed trains had strong frequency components that were lower than 250 Hz compared with normal speed trains. (5) High-speed trains had lower frequency centroid and higher binaural coherences, meaning that lower frequency is dominant and annoyance is lower in high-speed trains. (6) Motors, tires, and differences between motor and trailer cars showed no discernible effects.

Technological advances in designing trains and tracks (e.g., wheel absorbers, damping devices, variable-frequency drives, and long rails without joints) have reduced train noise. The train industry has succeeded in lowering the L_{Aeq} in train cars. However, the results of this study show that active noise control in high-speed trains and countermeasures for aerodynamic noises in normal speed trains can be effective in improving the acoustic characteristics in train cars. This is because high-speed trains have more lower frequency components, and this study has confirmed the effects of countermeasures on aerodynamic noises in high-speed trains. Environmental quality standards for noise in Japan determine the highest daytime standard value to be 70 dB or less. This means that noise should be reduced by a few dB for above-ground trains and approximately 10 dB for underground trains. In addition, ACF/IACF factors such as τ_1, ϕ_1, and IACC should be introduced for evaluating various types of noises because the factors affect loudness and annoyance even if the L_{Aeq} values are equal and because they change according to types of noise sources and the surroundings.

6.5.2 *Effects of External Environments*

The sound environment inside the train car is determined not only by the train noise, but also by the outside environment. It has been shown that the noise level inside underground train cars is higher than that inside above-ground train cars (Kono et al. 1982), because of reflections from the tunnel walls. Train noise, such as the noise of rolling wheels and the engine, for above-ground railways is emitted outside the train and does not reflect back into the train car, while train noise for underground railways is also emitted outside the train, but is reflected off tunnel walls back into the train car.

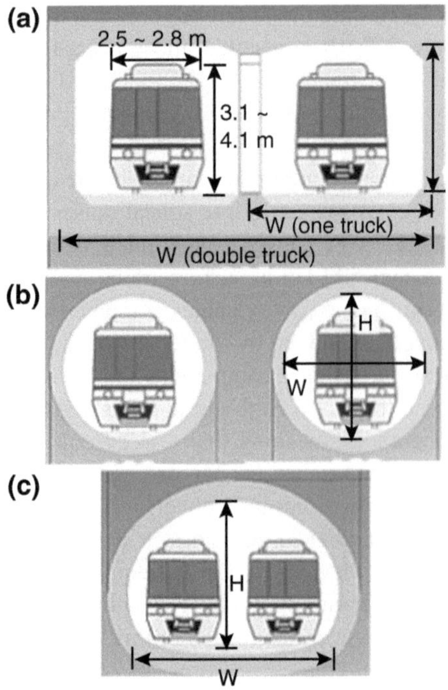

Fig. 6.28 Cross sections of **a** CCM, **b** BMM, and **c** NATM tunnels (Soeta and Shimokura 2011)

The characteristics of underground railway tunnels vary according to the excavation method. The cut-and-cover method (CCM) excavates a tunnel from ground level downward, constructs the tunnel roof and sides, and buries the tunnel. Since the flat side walls and straight capping beams shape the configuration of tunnel, the cross section of the tunnel is rectangular and relatively large (Fig. 6.28a). The BMM drills tunnels using a boring machine and lines the inside with concrete, steel, or ductile iron segments. Since the boring machine has a circular cutter head at the front, the cross section of the tunnel is circular and relatively small (Fig. 6.28b). The new Austrian tunneling method (NATM) excavates a tunnel that is maintained using retentive characteristics of the natural ground, the surface of which is quickly strengthened with shotcrete, and the insertion of reinforcing locking bolts. The characteristics of the sound fields in these three tunnel types are more than likely to differ.

Noise was measured continually in a train car traveling from one terminal station to another terminal station on five lines. Different types of train car run on each line. Intervals between stations were selected for analysis according to the outside environment of the train car, that is, CCM (single and double tracks), BMM, and NATM tunnels, and the above-ground environment. The numbers of intervals in which measurements were made was 4, 15, 13, 5, and 20 for the CCM (single track), CCM (double track), BMM, and NATM tunnels, and above the ground, respectively. All tunnels were covered by concrete except for a small part of the

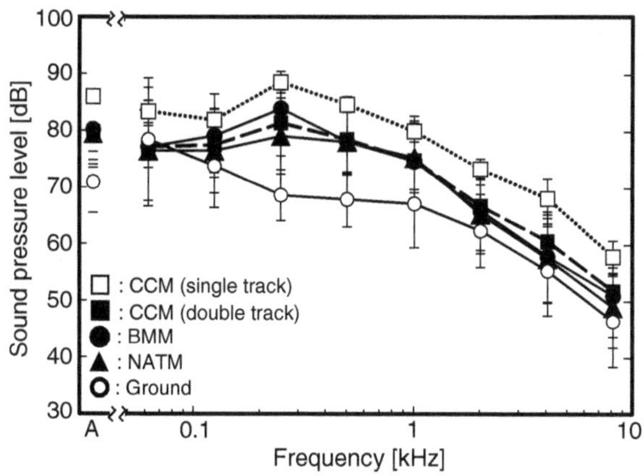

Fig. 6.29 Measured SPL as a function of the 1/1 octave band center frequency and L_{Aeq} in (*open square*) CCM (single track), (*filled square*) CCM (double track), (*filled circle*) BMM, (*filled triangle*) NATM tunnels, and (*open circle*) above the ground (Soeta and Shimokura 2011)

BMM tunnels, which was covered by steel segments. Trains run on a slab track in the tunnels and on a ballast track above the ground.

A dummy head microphone (KU100, Nuemann) was used to record noise inside a train car. The microphone was located in the area for a wheelchair user. The microphones always faced the inside of the train cars and were perpendicular to the direction of travel. The height of the dummy head was 1.6 m. For all measurements, noise was recorded on a personal computer (Let's Note, Panasonic) via an AD/DA converter (AudioFire8, Echo Digital Audio) at a sampling rate of 48 kHz and sampling resolution of 24 bits. The measurements were taken from 10:00 to 15:00 to avoid rush-hour commuting.

Figure 6.29 shows the L_{Aeq} and octave band power levels. The L_{Aeq} in the tunnels was 8–15 dB higher than that above the ground. Trains usually run on tracks mounted on concrete slabs in the tunnels and on ballasted track above the ground. Tracks mounted on concrete slabs are generally found to be noisier than ballasted tracks, typically by 3–5 dB (Stüber 1975; Thompson and Jones 2006), suggesting that the reflections in the tunnels increased L_{Aeq} by 3–12 dB. In all types of tunnel, the power level had a peak around 250 Hz. In contrast, the power level dipped around 250 Hz above the ground. The difference between the power levels in tunnels and above the ground was large in frequency bands around 250 and 500 Hz. These results suggest the effect of reflections on tunnels can be mainly observed around 250 and 500 Hz, and the source could be rolling, locomotive exhaust, and cooling fan noise (Lotz 1977; Thompson and Jones 2000; Talotte et al. 2003; Schulte-Werning et al. 2006; Kitagawa 2008). When a train enters a tunnel, the major frequency component shifts from lower to higher frequency ranges, agreeing with the present results. The sectional area of the CCM tunnel has clear effects on noise levels. The smaller CCM tunnels had larger noise levels.

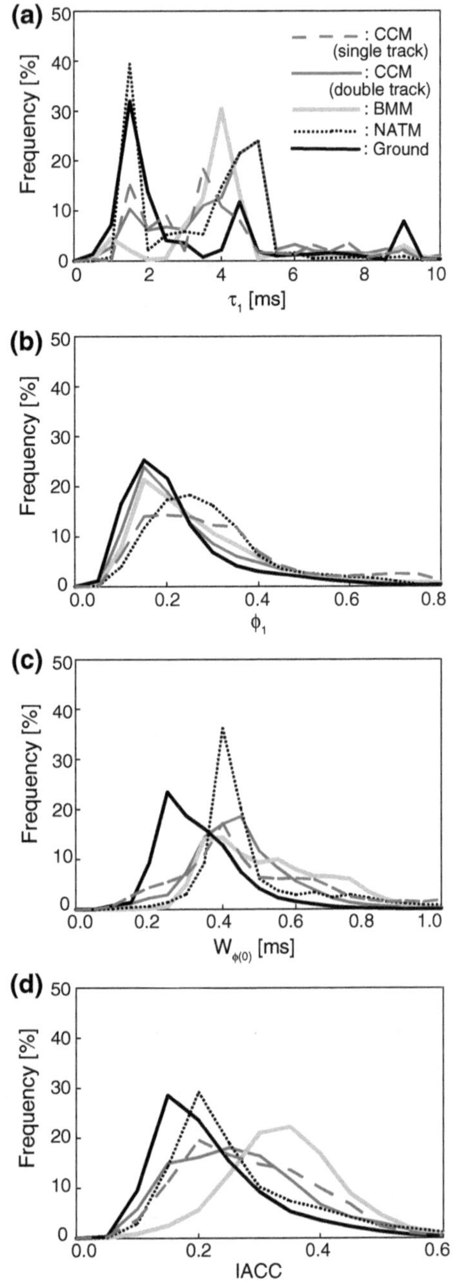

Fig. 6.30 Frequency distributions of **a** τ_1, **b** ϕ_1, **c** $W_{\phi(0)}$, and **d** IACC obtained from short-time moving analyses for each outside environment. The *dashed gray*, *gray*, *light gray*, *dashed black*, and *black line* indicate results for CCM (*single track*), CCM (*double track*), BMM, and NATM tunnels and above the ground, respectively (Soeta and Shimokura 2011)

On the basis of a running analysis, the frequency distributions of the τ_1, ϕ_1, $W_{\phi(0)}$, and IACC for each outside environment were obtained (Fig. 6.30). The values of τ_1 were concentrated at 1 ms, which correspond to a frequency of 1000 Hz, for the NATM tunnel and above the ground. The noise at 1000 Hz could be rolling noise emitted from the wheels and rails (Stüber 1975; Thompson and Jones 2006), locomotive exhaust and cooling fan noise (Lotz 1977; Talotte et al. 2003; Schulte-Werning et al. 2006; Kitagawa 2008). The values of τ_1 were concentrated at 4 and 5 ms, which correspond to a frequency of 250 and 200 Hz, for the CCM, BMM, and NATM tunnels. The frequency of 250 Hz corresponded to the dominant frequency component in the BMM tunnel (Fig. 6.30a). The noise at 200 and 250 Hz could be locomotive exhaust and cooling fan noise (Lotz 1977; Schulte-Werning et al. 2006). ϕ_1 was largest for the NATM tunnel, smaller for the BMM and CCM tunnels, and smallest for above the ground. The reflections in tunnels could increase ϕ_1, and the perceived pitch strength of noise inside the train cars might be stronger in tunnels and cause more annoyance (Sato et al. 2002, 2004). $W_{\phi(0)}$ was largest for the BMM tunnel, smaller for the CCM and NATM tunnels, and smallest for above the ground. This suggests that the tunnel lowers the spectral centroid of noise inside the train cars. The IACC was largest for the BMM tunnel, smaller for the CCM and NATM tunnels, and smallest for above the ground. More reflections were evident inside the train cars in the BMM tunnels because of the circular cross section. In contrast, fewer reflections were evident inside the train cars in the NATM tunnels because of the complex cross section. The sectional area of the CCM tunnel has no clear effect on IACF and ACF parameters.

To clarify the impact of the external environment on noise inside a train car, the noise inside a train car was evaluated using the octave band power levels, L_{Aeq}, τ_1, ϕ_1, $W_{\phi(0)}$, and IACC. The results indicate that the noise inside train cars traveling through the tunnels had salient sound energy around 250 Hz and the reflections in the tunnels increased L_{Aeq} by 3–12 dB. This suggests the effectiveness of acoustic treatment for the low frequency. The geometry of the tunnel affected the IACC. The IACC was largest for the BMM tunnel. More reflections might enter the train cars in BMM tunnels because of their circular cross section. The dimensions of the tunnel affected the L_{Aeq}, with smaller CCM tunnels having larger noise levels.

This work focuses on the effects of the outside environment on noise inside train cars, such as the effects of the above-ground or underground environment and the cross section of a tunnel. The acoustic treatment (i.e., sound absorption, insulation, and active noise control) of tunnels is an effective solution for improving the sound environment inside train cars.

6.6 Floor Impact Sound

In apartment buildings, the predominate dwelling in Korea, floor impact sounds are mainly caused by walking, running, and jumping and are regarded as the most annoying type of noise in these buildings. To simulate and evaluate the sound of humans walking and running, heavyweight floor impact sources, such as the bang machine and impact ball, have been used (JIS A 1418, 2000; KS F 2810-2 2001; ISO 140 2005). Floor impact sound is evaluated according to SPLs and their frequency characteristics. Although the noise level is one of the most important parameters for the subjective evaluation of noise, and many studies try to find effective parameters to express human perception of the noises (Tachibana et al. 1988; Parmanen 1995), objective parameters other than noise level and frequency characteristics, such as ACF and IACF factors, may also affect the perception of the noise source. Thus, floor impact sounds have been analyzed by the ACF factors.

The heavyweight impact sounds were generated with the bang machine and the impact ball at the central position of each room. According to ISO standards for evaluating floor impact noise levels (ISO 140 2005), microphones were positioned either at fixed locations including the periphery of the room, or at varying locations using a moving microphone. Sounds were generated by a bang machine and an impact ball and were measured in various rooms in apartments with different sound insulation treatments. The details of the measurement, the sound insulation treatments for the floors, and so on are explained in the previous study (Jeon and Sato 2008). The heavyweight impact sounds were recorded binaurally through a dummy head (B&K 4100) positioned in the center of the downstairs room representing the typical listening location of a tenant. For the ACF analysis, integration interval was chosen to be 0.5 s and the running step was 10 ms.

The relationship between the ACF factors of the bang machine and the impact ball is shown in Fig. 6.31. As shown in Fig. 6.31a, $\Phi(0)$ was smaller than the inverse A-weighted impact SPL, $L_{i,F\max,\mathrm{AW}}$. Here, i refers to impact, $F_{\max}$ is the maximum SPL measured by the sound level meter with the time constant Fast, and AW is a consideration of the inverse A-weighting curve (JIS A 1419-2 2000). This was because $\Phi(0)$ included the tail (attenuation part) of the impact sound as well as the initial part, while $L_{i,F\max,\mathrm{AW}}$ was calculated only from the maximum sound level. The impact sounds were measured in the units whose floor of the above room with viscoelastic damping materials (Visco_f) had a smaller $\Phi(0)$ for both the bang machine and the impact ball. $\Phi(0)$ for the bang machine was always greater than that for the impact ball. The floor impact sound levels of the bang machine always had a longer response time than those of the impact ball due to the higher impact force level regardless of the floor insulation treatment.

The difference between sound insulation treatments may be described by τ_1 regardless of the type of impact source, as shown in Fig. 6.31b. The inverse of τ_1 corresponds to the frequency of the pitch. In units with installed resilient isolators (RI), a longer τ_1 (429 ms), that is, a lower perceived pitch, was observed for the sound insulation treatment that included both the floor and the walls. In contrast, the Visco_f units produced impact sounds with shorter τ_1 values (higher pitch sensation). In cases where concrete slabs and upper layers are separated by a resilient isolator, they act independently in a rigid body mode. In contrast, concrete slabs and upper layers connected by an adhesive viscoelastic damping material act as a single body, and thus, the impact energy was absorbed in the damping layer. Moreover, increasing the resonance frequency of a floor structure effectively reduced sound

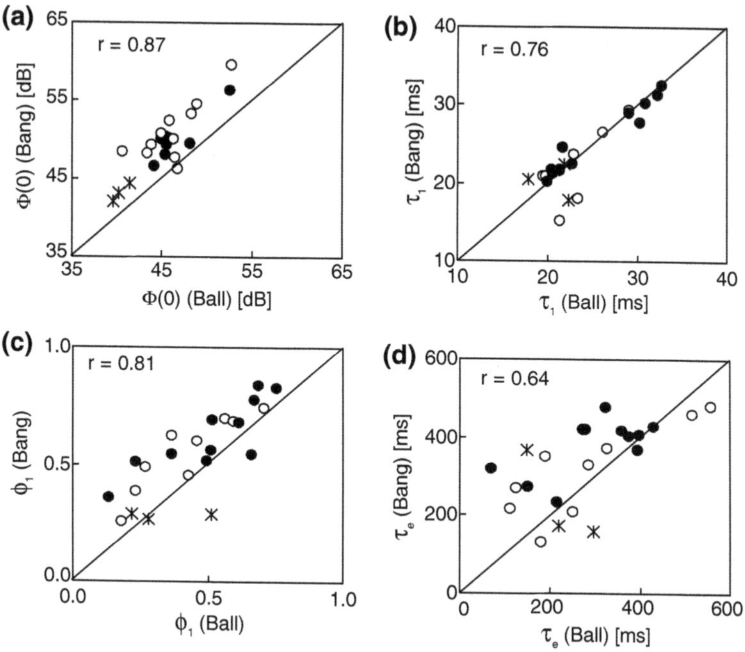

Fig. 6.31 Relationship between the ACF parameters for the bang machine and impact ball. **a** $\Phi(0)$, **b** τ_1, **c** ϕ_1, and **d** τ_e. *Filled circle* with noise isolators, *open circle* without noise isolators, and *division times* viscoelastic damping material (Jeon and Sato 2008)

levels from heavyweight impact sources. The inverse of τ_1 indicates the dominant frequency, such as the resonance frequency, of a sound. Therefore, a decreasing τ_1 was also related to the level of reduction of the heavyweight floor impact sound. Floors with viscoelastic damping material reduced sound levels, thus decreasing annoyance of the floor impact sound. Conversely, floors with resilient isolators do not always decrease annoyance.

As shown in Fig. 6.31c, d, the values of ϕ_1 and τ_e for the bang machine were larger than those for the impact ball. Floors of the Visco$_f$ units produced impact sounds with smaller ϕ_1 and τ_e for both the bang machine and the impact ball, suggesting that viscoelastic damping material effectively reduces the lower frequency component, causing the higher frequency component to become alternatively dominant. Table 6.1 compares the resonance frequencies calculated from the ACF of vibration signals and the frequencies calculated by the τ_1 of the impact sounds. Both the resonance frequency and the inverse of τ_1 for the floors of the Visco units are higher than those of RI because viscoelastic damping materials affect resonance and its center frequency and half-width (Fig. 6.32).

The relationship of the variance of the ACF factors for the bang machine and the impact ball is shown in Fig. 6.33. The correlation between the variance of $\Phi(0)$ (VAR_$\Phi(0)$ for the bang machine and the impact ball is 0.65, and the VAR_$\Phi(0)$

Table 6.1 Resonance frequencies of the unit structures calculated from the ACF of vibration signals and dominant frequencies calculated from τ_1 for units with installed viscoelastic damping materials (Visco) and units with installed resilient isolators (RI) (Jeon and Sato 2008)

Room	Source	Resonance frequency (Hz)	$1/\tau_1$ (Hz)
102 (Visco)	Bang	24	44
	Ball	25	45
103 (Visco)	Bang	29	41
	Ball	29	46
109 (Visco)	Bang	33	46
	Ball	33	46
110 (RI)	Bang	20	36
	Ball	20	35

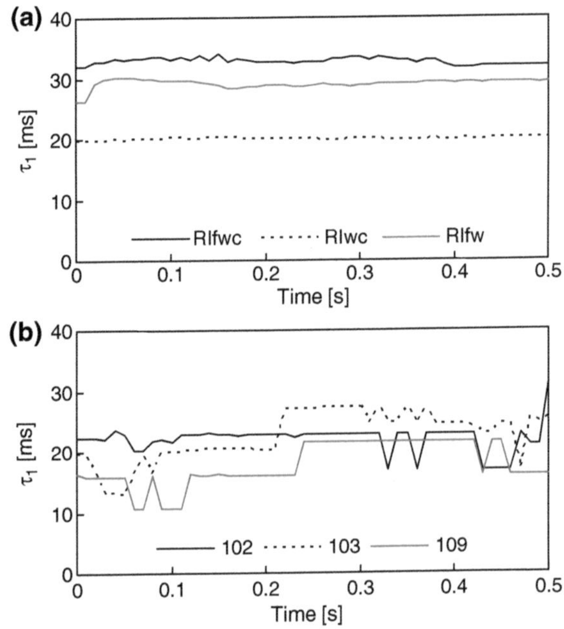

Fig. 6.32 Examples of the running τ_1 of impact sound generated by the impact ball: **a** τ_1 for floors with different resilient isolators (RI) and **b** τ_1 for floors with viscoelastic damping material in each room (Jeon and Sato 2008)

for the floors with viscoelastic damping material is smaller than that for those with isolators. With respect to VAR_τ_1, the values for the floor with viscoelastic damping material are greater than those for the floor with isolators. The correlation between VAR_τ_1 for the bang machine and the impact ball is 0.83. Lastly, with respect to VAR_ϕ_1 and VAR_τ_e, the correlation between the values for the bang machine and the impact ball is not significant.

The ACF indicates how similar a signal is to a time-delayed version of itself and is useful for detecting the periodicity (regularity). The ACF analysis can describe

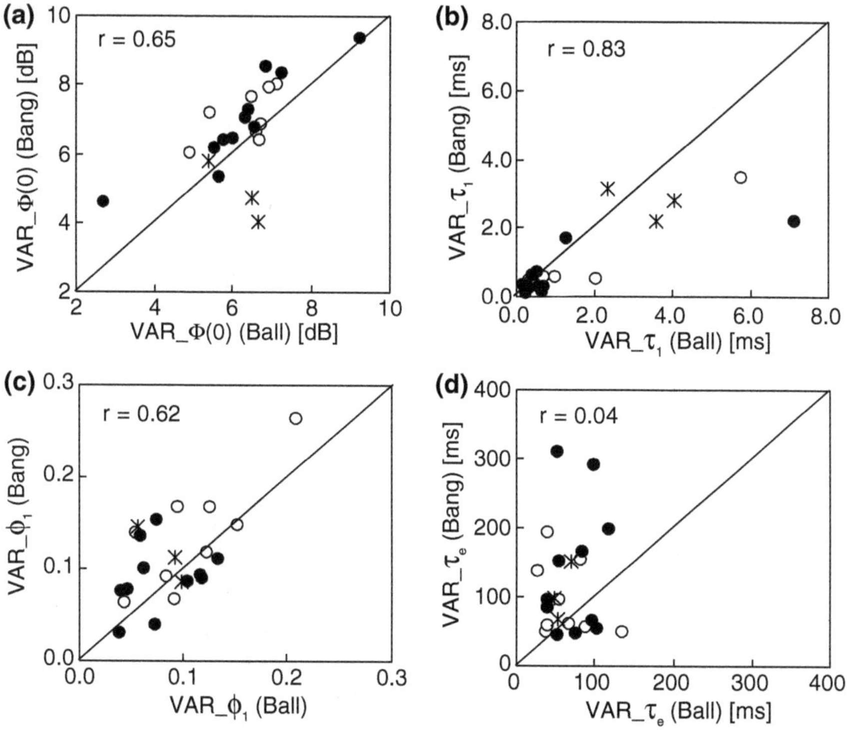

Fig. 6.33 Relationship between the variance of ACF parameters for the bang machine and the impact ball: **a** variance of Φ(0), **b** variance of τ_1, **c** variance of ϕ_1, and **d** variance of τ_e. *Filled circle* with noise isolators; *open circle* without noise isolators; and *division times* viscoelastic damping material (Jeon and Sato 2008)

not only a particular frequency such as the resonance frequency, but also the relationship between the resonance frequency and its harmonics. Figure 6.34 shows the relationship between ϕ_1 and the difference of levels at the maximum peak and the second or the third peak. The floor impact sounds have their maximum peaks between 30 and 60 Hz. The second and the third maximum peaks appear between 60 and 100 Hz and 100 and 200 Hz, respectively. The larger level difference corresponds to a larger ϕ_1. This means that if the frequency component concentrates at a particular frequency, ϕ_1 becomes large.

The ACF analysis of heavyweight floor impact sounds has shown the following results. τ_1 can describe the dominant frequency, such as the resonance frequency, of floor impact sounds. Sound insulation using floor and wall sound isolators resulted in a longer τ_1. Viscoelastic damping materials absorb the energy of heavyweight impact sounds and also reduce τ_1. Viscoelastic damping material also reduced ϕ_1, which is related to the dominance of the resonance frequency of the floor slab structure and its harmonics.

Fig. 6.34 Relationship between ϕ_1 and the level difference. **a** Definition of L1 and L2, **b** L1, and **c** L2 (*filled circle* ball and *open circle* bang)

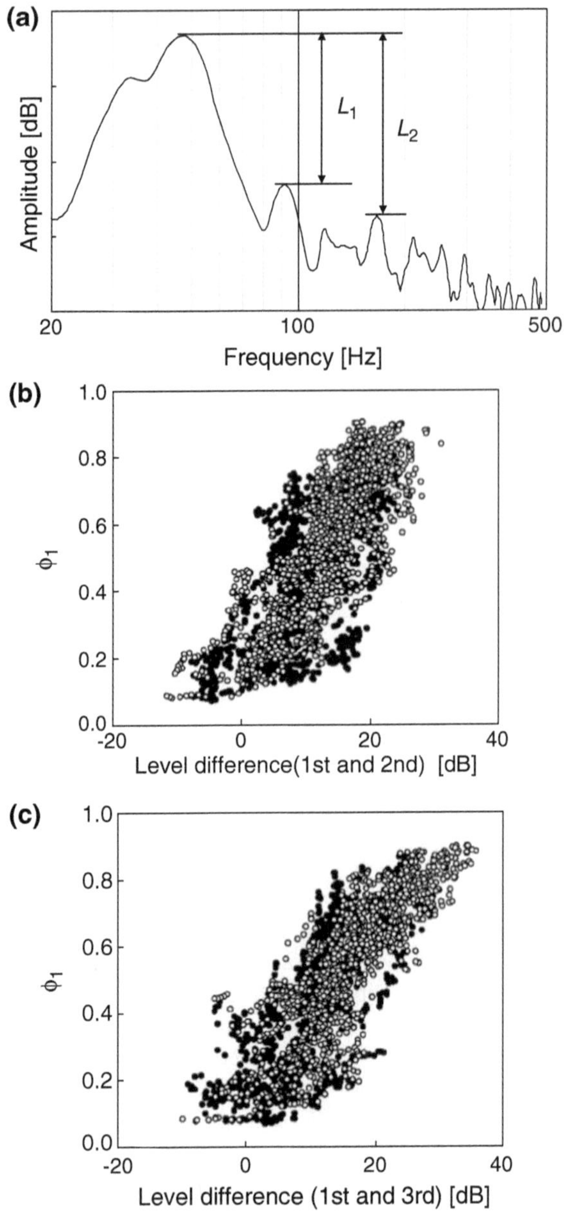

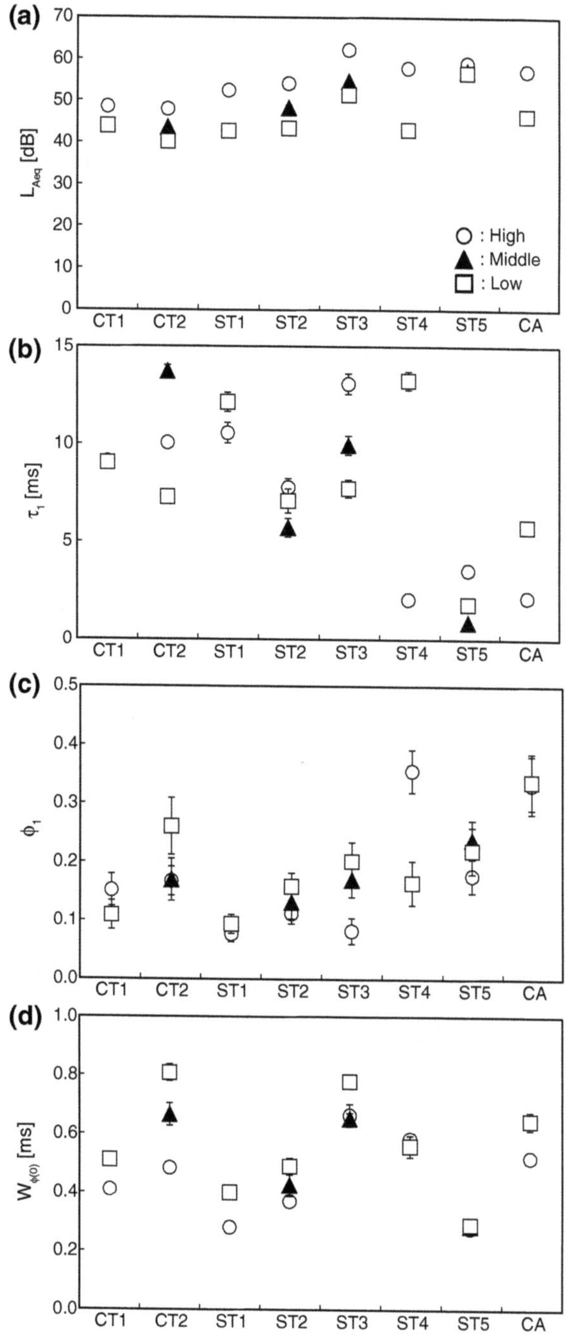

Fig. 6.35 Mean **a** L_{Aeq}, **b** τ_1, **c** ϕ_1, and **d** $W_{\phi(0)}$ values (±SDs) of noises generated by each air conditioner (CT1, CT2, ST1, ST2, ST3, ST4, ST5, and CA) at each level: (*open circle*) high, (*filled triangle*) middle, and (*open square*) low

6.7 Heating, Ventilating, and Air Conditioning (HVAC) Noise

Major noise sources of indoor environments are regarded as electrical appliances such as air conditioner, refrigerator, vacuum cleaner, and washing machine. Because air conditioners are widely used at classroom, office, and residence for a long time, many efforts have been put to reduce the sound level, and the sound level of air conditioners becomes low (Tang and Wong 2004; Ayr et al. 2003). Therefore, sound quality of air conditioners becomes more important.

Noises generated from two cassette-type air conditioners (CT1 and CT2), five split-type air conditioners (ST1, ST2, ST3, ST4, and ST5), and one central air conditioning system (CA) at two or three levels (high, middle, or low) were measured by omnidirectional microphone. The microphone was placed just below the split-type air conditioners, in front of split-type air conditioners, and the place where the noise from the central air conditioning system was clearly heard. For all measurements, noise was recorded on a personal computer at a sampling rate of 44.1 kHz and sampling resolution of 24 bits. ACF factors, L_{Aeq}, τ_1, ϕ_1, and $W_{\phi(0)}$ were calculated as a function of time to evaluate the noises quantitatively and qualitatively. For all calculations, the integration interval was $2T = 0.5$ s, and the running step was $s = 0.1$ s in all calculations.

Figure 6.35 shows the averaged L_{Aeq} and ACF factors (τ_1, ϕ_1, and $W_{\phi(0)}$) of noises generated by each air conditioner. L_{Aeq} values were between 40 and 60 dBA. τ_1 values show large difference between air conditioners. Most ϕ_1 values were less than 0.2, which are lower than those measured in train stations (Soeta and Shimokura 2011) and cars (Soeta and Shimokura 2013b) and show that the noises generate no clear pitch. Most $W_{\phi(0)}$ values were larger than 0.4, which are larger than those measured in train stations (Soeta and Shimokura 2013a) and cars (Soeta and Shimokura 2013b) and indicate that the noise includes lower frequency components.

References

Ando Y (1977) Effects of noise on duration experience. J Sound Vib 55:600–603

Ando Y (1998) Architectural acoustics: blending sound sources, sound fields, and listeners. AIP Press Springer-Verlag, New York

Ando Y, Hattori H (1970) Effect of noise during fetal life upon postnatal adaptability (statistical study of the reaction of babies to air-craft noise). J Acoust Soc Am 47:1128–1130

Ando Y, Hattori H (1973) Statistical studies on the effect of intense noise during human fetal life. J Sound Vib 27:101–110

Ando Y, Hattori H (1977a) Effects of noise on sleep of babies. J Acoust Soc Am 62:199–204

Ando Y, Hattori H (1977b) Effects of noise on human placental lactogen (HPL) levels in maternal plasma. Br J Obstet Gynaecol 84:115–118

Ayr U, Cirillo E, Fato I, Martellotta F (2003) A new approach to assessing the performance of noise indices in buildings. Appl Acoust 64:129–145

Beranek LL (1957) Revised criteria for noise in buildings. Noise Control 3:19–27

Beranek LL (1971) Noise and vibration control. McGraw-Hill, New York

Beranek LL (1989) Balanced noise-criterion (NCB) curves. J Acoust Soc Am 86:650–664

Beranek LL, Blazier WE, Figwer JJ (1971) Preferred noise criterion (PNC) curves and their application to rooms. J Acoust Soc Am 50:1223–1228

Berckmansa D, Janssensb K, Van der Auweraerb H, Sasa P, Desmeta W (2008) Model-based synthesis of aircraft noise to quantify human perception of sound quality and annoyance. J Sound Vib 311:1175–1195

Berglund B, Hassmen P, Job RFS (1996) Sources and effects of low-frequency noise. J Acoust Soc Am 99:2985–3002

Bhattachaya SK (1994) Assessment of noise environment in a major railway station in India. Ind Health 32:187–192

Buratti C, Belloni E, Moretti E (2014) Facade noise abatement prediction: new spectrum adaptation terms measured in field in different road and rail-way traffic conditions. Appl Acoust 76:238–248

Can A, Leclercq L, Lelong J, Botteldooren D (2010) Traffic noise spectrum analysis: dynamic modeling vs. experimental observations. Appl Acoust 71:764–770

Cariani PA, Delgutte B (1996) Neural correlates of the pitch of complex tones. I. Pitch and pitch salience. J Neurophysiol 76:1698–1716

Cermak GW, Cornillon PC (1976) Multidimensional analysis of judgments about traffic noise. J Acoust Soc Am 59:1412–1420

Chang HC, Hermann ER (1974) Acoustical study of a rapid train system. Am Ind Hyg Assoc J 35:640–653

Cleon LM, Williame A (2000) Aero-acoustic optimization of the fans and cooling circuit on SNCF's X 72500 railcar. J Sound Vib 231:925–933

Dittrich MG, Zhang X (2006) The harmonoise/IMAGINE model for traction noise of powered railway vehicles. J Sound Vib 293:986–994

Eade PW, Hardy AEJ (1977) Railway vehicle internal noise. J Sound Vib 57:403–415

Endoh H, Suzuki H, Izumi Y, Hio Y (2007) Appropriate sound level of broadcasting on station platforms under noisy environment. RTRI REPORT 21:41–46 (In Japanese)

Forssen J, Tober S, Corakci AC, Frid A, Kropp W (2012) Modelling the interior sound field of a railway vehicle using statistical energy analysis. Appl Acoust 73:307–311

Fujii K, Soeta Y, Ando Y (2001) Acoustical properties of aircraft noise measured by temporal and spatial factors. J Sound Vib 241:69–78

Fujii K, Atagi J, Ando Y (2002) Temporal and spatial factors of traffic noise and its annoyance. J Temporal Des Arch Environ 2:33–41

Gershon RRM, Qureshi KA, Barrera MA, Erwin MJ, Goldsmith F (2005) Health and safety hazards associated with subways: a review. J Urban Health 82:10–20

Hardy AE (2000) Measurement and assessment of noise within passenger trains. J Sound Vib 231:819–829

Hiramatsu K, Yamanaka K, Takagi K, Yamamoto T (1978) Annoyance of fluctuating noise (effects of standard deviation and fluctuation frequency of sound level). J Acoust Soc Jpn 34:376–386 (in Japanese)

Hiramatsu K, Takagi K, Yamamoto T (1983) Experimental investigation on the effect of some temporal factors of nonsteady noise on annoyance. J Acoust Soc Am 74:1782–1793

Imaizumi H, Kinoshita M, Kunimatsu S, Isei T (1997) Sound propagation and speech hearing in a curved reverberant tunnel. J Acoust Soc Jpn (E) 18:129–137

Imaizumi H, Kunimatsu S, Isei T (2000) Sound propagation and speech transmission in a branching underground tunnel. J Acoust Soc Am 108:632–641

International Standard Organization ISO (140:2005) Acoustics-measurement of sound insulation in buildings and of building elements—part 11: laboratory measurements of the reduction of transmitted impact sound by floor coverings on lightweight reference floors

Isumi Y, Fujii K (2007) An investigation into the actual condition and subjective evaluation test of acoustical environment in railway station. RTRI REPORT 21:4147–5246 (In Japanese)

Japanese Industrial Standard JIS A 1418, Part 2 (2000) Acoustics measurement of floor impact sound insulation of buildings. Part 2: method using standard heavy impact sources

Japanese Industrial Standard JIS A 1419, Part 2 (2000) Acoustics rating of sound insulation in buildings and of building elements, Part 2: floor impact sound insulation

Jeon JY (2001) Subjective evaluation of floor impact noise based on the model of ACF/IACF. J Sound Vib 241:147–155

Jeon JY, Sato S (2008) Annoyance caused by heavy-weight floor impact sounds in relation to the autocorrelation function and sound quality metrics. J Sound Vib 311:767–785

Kitagawa T (2008) Noise and vibration control of railway vehicles. J Acoust Soc Jpn 64:629–634 (in Japanese)

Kitamura T, Sato S, Shimokura R, Ando Y (2002) Measurement of temporal and spatial factors of a flushing toilet noise in a downstairs bedroom. J Temporal Des Arch Environ 2:13–19

Kono S, Sone T, Nimura T (1982) Personal reaction to daily noise exposure. Noise Control Eng J 19:4–16

Korean Standard KS F 2810-2, Part 2 (2001) Field measurements of impact sound insulation of floors. Part 2: method using standard heavy impact sources

Kuwano S, Namba S, Hajime Miura H (1989) Advantages and disadvantages of a-weighted sound pressure level in relation to subjective impression of environmental noise. Noise Control Eng J 33:107–115

Lotz R (1977) Railroad and rail transit noise sources. J Sound Vib 51:319–336

May DN (1978) Basic subjective responses to noise. In: May DN (ed) Handbook of noise assessment. Van Nostrand Reinhold C, New York, pp 313–335

Meddis R, Hewitt M (1991) Virtual pitch and phase sensitivity of a computer model of the auditory periphery. I: pitch identification. J Acoust Soc Am 89:2866–2882

Mellet C, Letourneaux F, Poisson F, Talotte C (2006) High speed train noise emission: latest investigation of the aerodynamic/rolling noise contribution. J Sound Vib 293:535–554

Merthayasa IN, Hemmi H, Ando Y (1994) Loudness of a 1 kHz pure tone and sharply (1080 dB/Oct.) filtered noises centered on its frequency. Mem Grad School Sci Tech Kobe Univ 12A:147–156

Moritoh Y, Zenda Y, Nagakura K (1996) Noise control of high speed shinkansen. J Sound Vib 193:319–334

Nagakura K (2004) Problems of Shinkansen super-express railway noise. J Acoust Soc Jpn 60:284–289 (in Japanese)

Nagakura K (2006) Localization of aerodynamic noise sources of Shinkansen trains. J Sound Vib 293:547–556

Namba S, Kuwano S (1984) Psychological study on L_{eq} as a measure of loud-ness of various kinds of noises. J Acoust Soc Jpn (E) 5:135–148

Parmanen J (1995) Conclusions based on the Japanese loudness evaluations with artificial sounds. J Sound Vib 186:866–873

Remington PJ (1987) Wheel/rail squeal and impact noise: what do we know? What don't we know? Where do we go from here? J Sound Vib 116:339–353

Sato S, Ando Y (2002) Apparent source width (ASW) of complex noises in relation to the interaural cross-correlation function. J Temporal Des Arch. Environ 2:29–32

Sato S, Kitamura T, Ando Y (2002) Loudness of sharply (2068 dB/Octave) filtered noises in relation to the factors extracted from the autocorrelation function. J Sound Vib 250:47–52

Sato S, Kitamura T, Ando Y (2004) Annoyance of noise stimuli in relation to the spatial factors extracted from the interaural cross-correlation function. J Sound Vib 277:511–521

Schulte-Werning B, Beier M, Degen KG, Stiebel D (2006) Research on noise and vibration reduction at DB to improve the environmental friendliness of railway traffic. J Sound Vib 293:1058–1069

Shimokura R, Soeta Y (2011) Characteristics of train noise in ground and underground stations with side and island platforms. J Sound Vib 330:1621–1633

Soeta Y, Nakagawa S (2008) Effect of the repetitive components of a noise on loudness. J Temporal Des Arch Environ 8:1–7

Soeta Y, Shimokura R (2009) Comparison of noise characteristics in airplanes and high-speed trains. J Temporal Des Arch Environ 9:22–25

Soeta Y, Shimokura R (2011) The impact of external environments on noise inside a train car. Noise Control Eng J 59:581–590

Soeta Y, Shimokura R (2012) Change of acoustic characteristics caused by platform screen doors in train stations. Appl Acoust 73:535–542

Soeta Y, Shimokura R (2013a) Acoustic characteristics of noise in train sta-tions. In: Reinhardt C, Shroeder K (eds) Railways: types, design and safety is-sues. Nova Science Pub Inc, New York, pp 1–36

Soeta Y, Shimokura R (2013b) Survey of interior noise characteristics in various types of trains. Appl Acoust 74:1160–1166

Soeta Y, Maruo T, Ando Y (2004) Annoyance of bandpass filtered noises in relation to the factor extracted from autocorrelation function. J Acoust Soc Am 116:3275–3278

Stephens DG, Cazier FW (1996) NASA noise reduction program for advanced subsonic transports. Noise Control Eng J 44:135–140

Stüber C (1975) Air- and structure- borne noise of railway. J Sound Vib 43:281–289

Tachibana H, Hamada Y, Sato F (1988) Loudness evaluation of sounds transmitted through walls-basic experiment with artificial sounds. J Sound Vib 127:499–506

Talotte C, Gautier PE, Thompson DJ, Hanson C (2003) Identification, model-ling and reduction potential of railway noise sources: a critical survey. J Sound Vib 267:447–468

Tang SK, Wong MY (2004) On noise indices for domestic air conditioners. J Sound Vib 274:1–12

Thompson DJ, Jones CJC (2000) A review of the modelling of wheel/rail noise generation. J Sound Vib 231:519–536

Thompson DJ, Jones CJC (2006) Noise and vibration from railway vehicles. In: Iwnicki S (ed) Handbook of railway vehicle dynamics. Taylor & Francis, London, pp 279–326

Thron T, Hecht M (2010) The sonRAIL emission model for railway noise in Switzerland. Acta Acust United AC 96:873–883

Van Ruiten CJM (1988) Mechanism of squeal noise generated by trams. J Sound Vib 120:245–253

Versfeld NJ, Vos J (1997) Annoyance caused by sounds of wheeled and tracked vehicles. J Acoust Soc Am 101:2677–2685

Vincent N, Koch JR, Chollet H, Guerder JY (2006) Curve squeal of urban roll-ing stock-part 1: state of the art and field measurements. J Sound Vib 293:691–700

Yoshihisa N, Fukai M, Yamaguchi T (1968) Noise levels inside subway cars in Western Europe, U.S.A. and Japan. J Acoust Soc Jpn 24:69–75 (in Japanese)

Zwicker E, Fastl H (1999) Psychoacoustics: facts and models. Springer, Berlin

Chapter 7
Annoyance of Noise

Abstract Annoyance depends directory on the sound pressure level (SPL) when sounds are roughly equivalent in other attributes, such as timbre and duration. Therefore, a lot of efforts have been spent on noise reduction technologies in relation to reducing sound exposure level. For sound sources having widely different acoustical properties, however, this relationship may no longer hold. For example, a sound may exist that has a SPL below the exposure standards but that is perceived to be noisy or annoyed in a given situation (Kitamura et al. in J Temporal Des Arch Environ 2:13–19, 2002). This means that annoyance cannot be predicted by SPL alone. We need to consider the other factors influencing perceived annoyance. Basically, psychoacoustic annoyance depends on the loudness, timbre, and temporal structure of sounds. Loudness and annoyance have been distinguished and defined (Berglund et al. in J Acoust Soc Am 57:930–934, 1975; Hellman in J Acoust Soc Am 72:62–73, 1982). Loudness directly refers to SPL, and annoyance refers to how much people are bothered by the sound, whereas noisiness refers specifically to sound quality or sound clarity distinct from its annoyance or unacceptability. These authors relate annoyance to an individual's reaction to noise within the context of a given situation and loudness directly to sound intensity. To investigate subjective impression to noise, annoyance to noises in relation to factors extracted from autocorrelation function (ACF) and interaural cross-correlation function (IACF) is focused in this chapter.

Keywords Annoyance · Road traffic noise · Noise in a train station · Annoyance of noise in a train car · HVAC noise · Fluctuation

7.1 Annoyance of Noise with a Pure Tone Component

Pure tones are often present in sounds emitted by industrial equipment such as blowers and electrical generators. When pure tones are present in a noise spectrum, the A-weighted equivalent continuous sound pressure level (SPL), L_{Aeq} is not appropriate to predict human response because pure tones, especially at high

Y. Soeta and Y. Ando, *Neurally Based Measurement and Evaluation of Environmental Noise*, Mathematics for Industry 20,
DOI 10.1007/978-4-431-55432-5_7

frequencies, are much more annoying than a broadband noise of the same SPL (Hellman 1984).

The previous studies indicated that tonal components increase the perceived annoyance and noisiness more than the loudness of broadband noise. However, the magnitude of the effect varies widely from study to study (e.g., Kryter and Pearsons 1965; Little and Mabry 1969; Pearsons and Bennett 1971). Estimates of the contribution of pure tones to overall perceived loudness versus annoyance and noisiness range from none (Goulet and Northwood 1973), to nearly 15 dB (Kryter and Pearsons 1965). Hellman (1982) suggested that several factors may contribute to the lack of clear-cut results found in the previous study. First, some studies examined stimuli with tones below 80 dB SPL where tonal components may be subjectively less important than at SPLs greater than 80 dB. Second, even those studies which required evaluative judgments of sounds close to 90 dB SPL and above-stressed noisiness as opposed to annoyance. Third, many studies used ambiguous adjectives when giving instructions other than loudness so that the distinctions to listener among the judged attributes may be blurred.

Hellman (1982) assessed the effect of the overall SPL of the noise–tone complex and the effect of tone-to-noise ratio. Single tones centered within the noise spectrum were added to three different broadband spectra: flat, low pass, and high pass. Judgments of overall loudness, annoyance, and noisiness were obtained by absolute magnitude estimation.

Although judged, perceived magnitude (loudness, annoyance, and noisiness) does increase with overall SPL, the increase is clearly a nonmonotonic function of the overall SPL of the noise–tone complex as shown in Fig. 7.1. Further, the pattern of results is similar for the three judged attributes suggesting a common underlying basis, but not the extent of the excursions. The extent of the excursions depends on the judged attribute. It also seems to depend on the spectral characteristic of the noise.

Regardless of attribute judged, even when the overall SPLs are about the same, small tone-to-noise ratios (+5, +10, +15 dB) increase overall perceived magnitude more than large tone-to-noise ratios (+20 dB and greater) as shown in Fig. 7.2. More summation between tone and noise was found for relatively small tone-to-noise ratios (+5, +10, +15 dB), as measured in 1/3-octave bands, than for relatively large tone-to-noise ratios (+20 dB and greater). The relationship among the three perceived attributes determined by small tone-to-noise ratios depends on the overall SPL of the noise–tone complex, the frequency of the added tone, and the spectral shape of the noise. Data analysis suggests that the extent of the increments and decrements in perceived magnitude depends on the absolute loudness of the component stimuli, the interaction between a specific tone frequency and noise spectrum, and the attribute judged.

Hellman (1984) assessed the contribution of tonal components to overall perceived magnitude of noise–tone complex for tones located within the high- and low-frequency skirts of the noise spectra. The relation between overall perceived magnitude (loudness, annoyance, and noisiness) of noise–tone complex and the location of the tone within the spectrum was investigated by absolute magnitude

Fig. 7.1 Magnitude estimates of loudness, annoyance, and noisiness produced by **a** 1000-Hz and **b** 250-Hz tones added to broadband-flat noise as a function of the overall SPL of the noise–tone complex. Each point indicates the geometric mean of 20 judgments by the group of ten listeners (Hellman 1982)

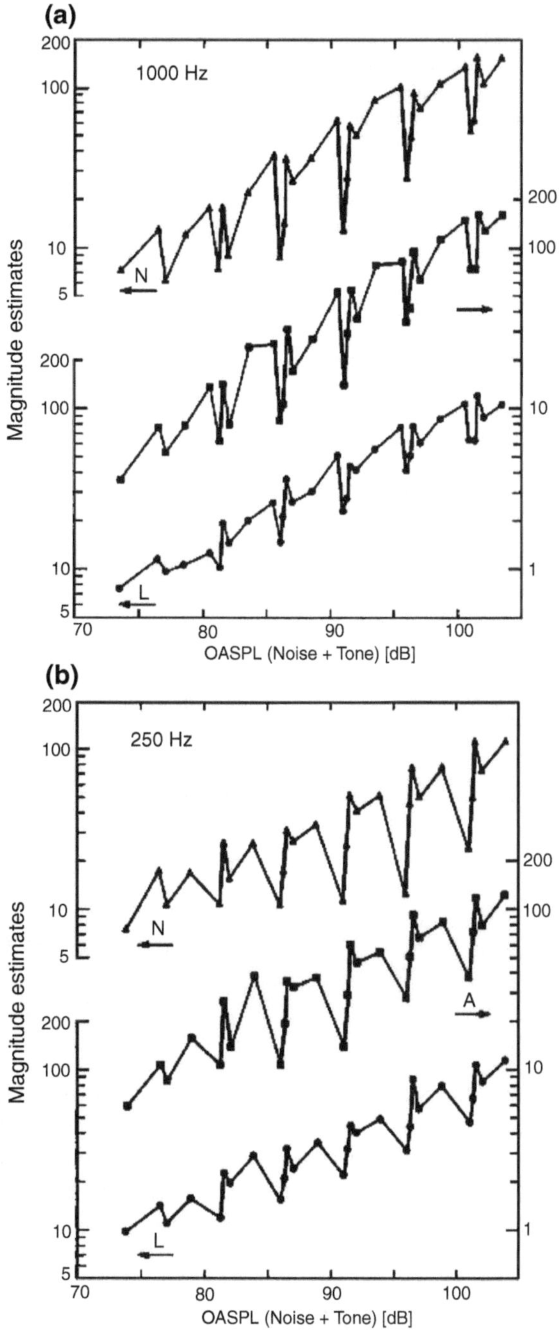

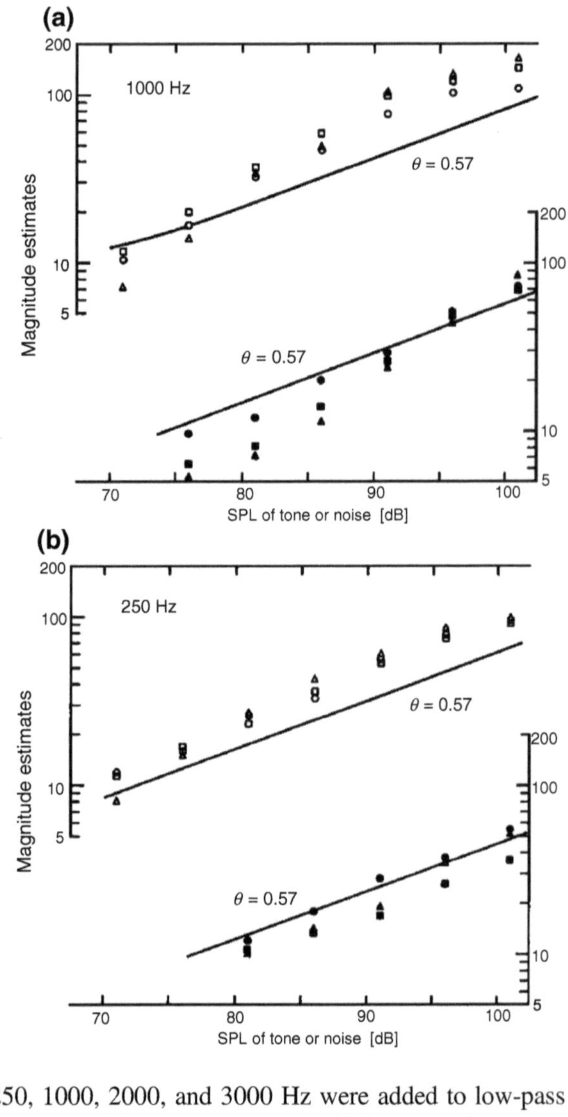

Fig. 7.2 Effect of small (+5, +10, +15 dB) and large (+20, +25, +30 dB) tone-to-noise ratios on loudness, annoyance, and noisiness judgments produced by **a** 1000-Hz and **b** 250-Hz tones added to broadband-flat noise as a function of the SPL of the tone or noise. *Unfilled* and *filled* symbols indicate results produced by small and large tone-to-noise ratios, respectively. *Circles* represent loudness judgments, *squares* represent noisiness judgments, and *triangles* represent annoyance judgments (Hellman 1982)

estimation. Single tones at 250, 1000, 2000, and 3000 Hz were added to low-pass noises and single tones at 250 and 3000 Hz were added to high-pass noises.

Figure 7.3a shows the results for a 250-Hz tone added to low-pass noise. Parallel curves were fitted to the data. They show that, for all three attributes, overall perceived magnitude peaks near a tone-to-noise ratio of +5 dB, and then decreases by one-half at a ratio of +25 dB. The decrease continues up to a tone-to-noise ratio of +30 dB. Figure 7.3b shows the results for a 250-Hz tone added to high-pass noise. At over all SPLs of 90 dB and above, noisiness decreases as tone-to-noise ratio increases beyond about +15 dB. Loudness and annoyance follow a different pattern and continue to increase up to a tone-to-noise ratio of at least 35 dB. For tone-to-noise ratios of +20 dB and greater, the loudness increase relative to

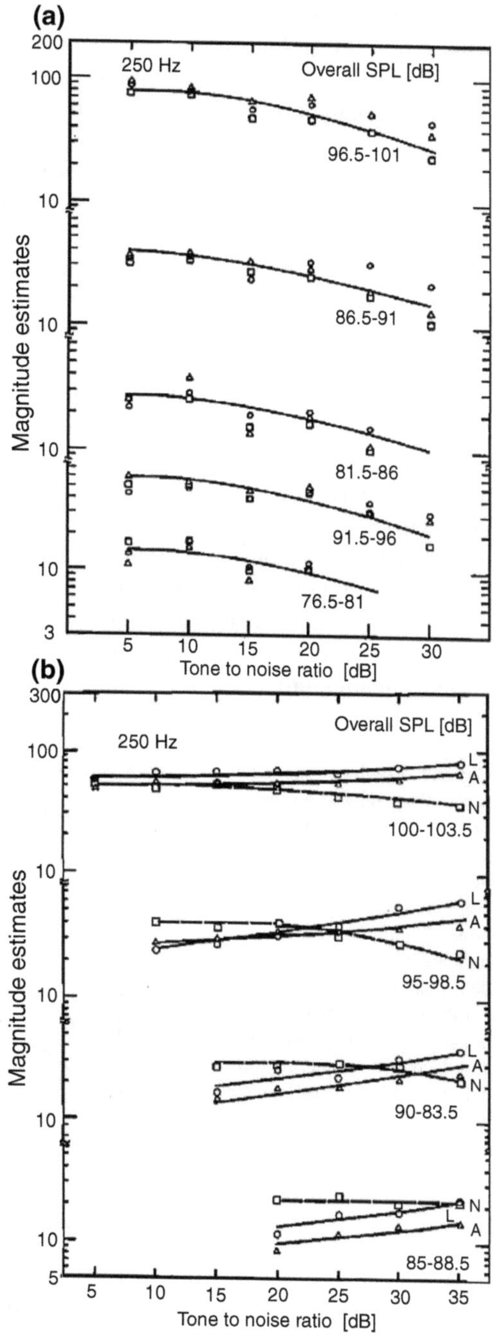

Fig. 7.3 Relationship between perceived magnitude and tone-to-noise ratio for a 250-Hz tone combined with **a** low-pass and **b** high-pass noises. Each point is based on the geometric mean of 20 judgments by the group of ten listeners. *Circles* indicate loudness judgments, *triangles* indicate annoyance judgments, and *squares* indicate noisiness judgments. The parameters are the overall SPL of the noise–tone complex. All points along each curve were obtained at about the same overall SPL (Hellman 1984)

annoyance is independent of the overall SPL of the noise–tone complex. Vertical cuts across SPL in Fig. 7.3 suggest that the slopes of the loudness and annoyance growth functions are dependent on tone-to-noise ratios when tones are added to the noise skirt, but not when they are centered within the spectrum. On the other hand, for those noisiness studied, the relation between noisiness growth rate and tone-to-noise ratio seems to be independent of the position of the tone in the noise.

Figure 7.4 summarizes loudness results observed across frequency. Three results are noted. First, when the tone is centered within the spectrum, the loudness exponent is invariant across tone-to-noise ratio. Second, as the tone's frequency is progressively shifted from 250 to 2000 Hz within the low-pass spectrum, the power-function exponents gradually increase as tone-to-noise ratio increases. The values of the exponents range from 0.6 to about 0.8. Third, the addition of a pure tone further along the tail of the noise skirt markedly alters the loudness slope. A least-squares fit to the data shows that a linear function describes the relationship between measured exponents and tone-to-noise ratio. The addition of a 3000-Hz tone to low-pass noise systematically increases the exponents as tone-to-noise ratio increases (r = 0.99), whereas the addition of a 250-Hz tone to high-pass noise decreases the exponents (r = −0.98). It also seems interesting to mention that the slopes of the regression functions are nearly equal, but opposite in sign.

To summarize, loudness and annoyance growth behavior depends on the relationship between frequency of the added tone and spectral shape of the noise. Tones centered in noise produce nonmonotonic loudness and annoyance growth functions;

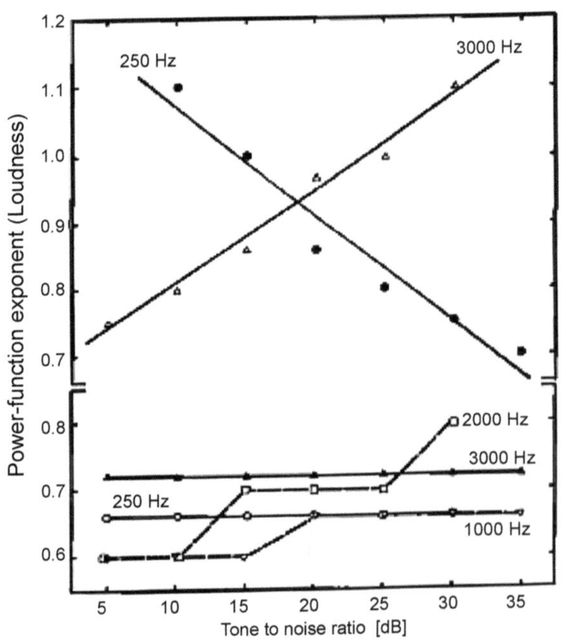

Fig. 7.4 Loudness exponents across frequency as a function of tone-to-noise ratio. *Filled symbols* indicate exponents determined by pure tones combined with high-pass noise. *Unfilled symbols* indicate exponents determined by pure tones combined with low-pass noise. The linearly increasing and decreasing functions were determined by a least-squares fit to the data (Hellman 1984)

those added to the skirt produce power functions. The measured exponents are invariant across tone-to-noise ratio when the tones are positioned within the spectrum, but not when they are added to the skirt. Moreover, for complexes at approximately the same overall SPL, the tone's position determines the functional relationship between loudness or annoyance and tone-to-noise ratio.

7.2 Annoyance of Band-Pass-Filtered Noise

Although changes in sound intensity are highly correlated with loudness changes, the relationship is not perfect. That is, changes in frequency, duration, and bandwidth all affect the perceived loudness of a stimulus even though the SPL of the stimulus remains fixed (Yost 2000). Previous studies have concluded that the loudness of a noise remains constant as the bandwidth of the noise increases until the bandwidth reaches the critical band. Loudness then increases with increasing bandwidth under the same SPL conditions (Zwicker et al. 1957; Greenwood 1961a, b; Scharf 1962). However, it was found that the loudness of a sharply (2068 dB/octave) filtered noise increases as the effective duration of the autocorrelation function (ACF), τ_e, increases, even when the bandwidth of the signal is within the critical band (Sato et al. 2002). τ_e represents repetitive features within the signal itself and increases as the filter bandwidth decreases. Not only loudness but also annoyance could be influenced by degree of repetitive features, which is characterized by τ_e. Thus, the annoyance of band-pass noises in terms of the factors extracted from the ACF was examined (Soeta et al. 2004).

Pure tone and band-pass noises with center frequencies of 1000 and 2000 Hz were used as source signals. The basic stimulus in this experiment was a maximum-length sequence (MLS) signal (order 21; sampling frequency 44,100 Hz) which was band-pass-filtered. The MLS signal used during this experiment is called pseudo-random white noise. To control the ACF factors of the band-pass noise, the filter bandwidth was varied at 0, 40, 80, 160, and 320 Hz by using a cutoff slope of 2068 dB/octave, obtained by a combination of two filters (Sato et al. 2002). In fact, the filter bandwidth of 0 Hz was the only slope component. The critical bandwidths of the center frequency of 1000 and 2000 Hz are approximately 160 and 300 Hz (Zwicker and Terhardt 1980). Figure 7.5 shows the power spectra of some of the sound used in this study. The source signals were characterized by ACF factors, τ_1, ϕ_1, and τ_e. τ_1 corresponds to the center frequency of the band-pass noise. ϕ_1 and τ_e increase as the filter bandwidth decreases as shown in Fig. 7.6, and there is a certain degree of coherence between ϕ_1 and τ_e.

The auditory stimuli were binaurally presented using headphones (Sennheiser HD-340). All stimuli were fixed at the same SPL (74 dBA). The SPL was calibrated by using a dummy head with 1/2-in condenser-type microphones at both ears. The IACC was kept constant at nearly unity because the signals fed to both ears were identical.

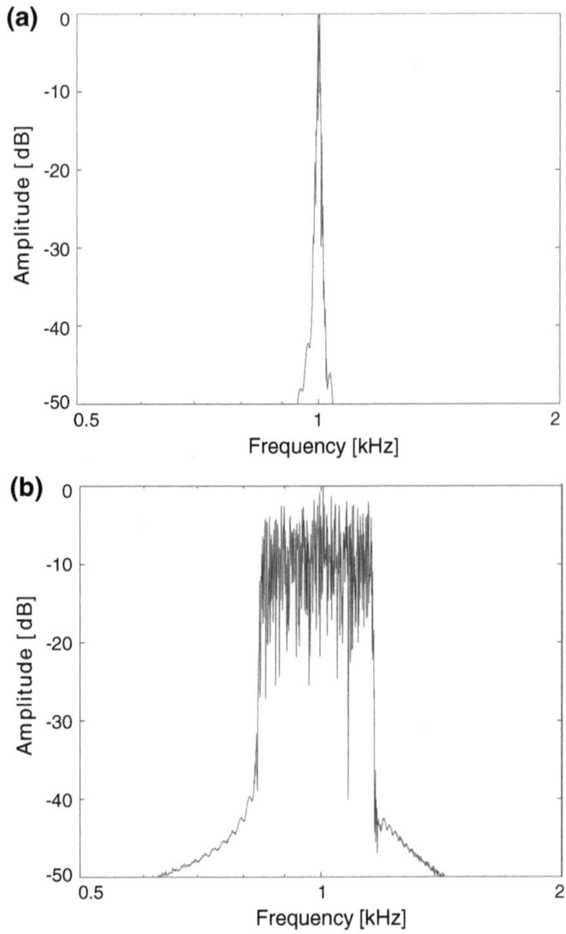

Fig. 7.5 Power spectra of band-pass noise with **a** 0 Hz bandwidth and **b** 320 Hz bandwidth of 1000 Hz center frequency (Soeta et al. 2004)

Eight 21- to 23-year-old listeners with no histories of hearing disorders took part in the experiment. They were seated in a dark, soundproof room with a comfortable thermal environment and heard the auditory stimuli. The paired-comparison tests were performed for all combinations of the pairs of pure tone and band-pass noise, i.e., 15 pairs ($N(N-1)/2$, $N = 6$) of stimuli interchanging the order in each pair per session, with the pairs presented in random order. A total of ten sessions were conducted for each listener. The duration of the stimuli was 2.0 s, the rise and fall time was 50 ms, the silent interval between stimuli was 1.0 s, and the interval between pairs was 3.0 s, which was the time allowed for the listeners to respond by pushing one of two buttons. They were asked to judge which of the two sound signals was more annoying. The scale values of the annoyance to each listener were calculated according to Case V of Thurstone's theory (Thurstone 1927) and the model of Case V for all data was reconfirmed by a goodness-of-fit test (Mosteller 1951).

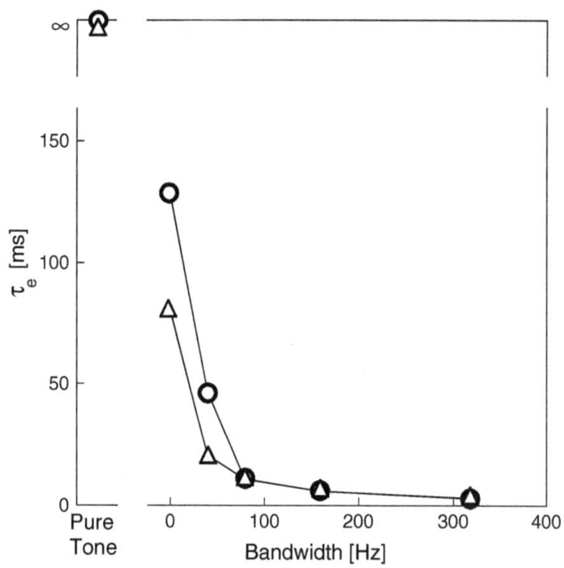

Fig. 7.6 Measured effective duration of ACF, τ_e, of the source signal as a function of the bandwidth. Different symbols indicate different frequencies: *open circle* 1000 Hz; *open triangle* 2000 Hz (Soeta et al. 2004)

The relationship between scale values of annoyance and filter bandwidths is shown in Fig. 7.7. Statistical significance was determined by Bonferroni test. For the center frequency of 1000 Hz tested, the averaged scale value of annoyance for pure tones was significantly greater than those of sharply filtered noises under the condition of equal SPLs ($p < 0.01$). For the center frequency of 2000 Hz tested, the scale value of annoyance for the pure tone was significantly greater than that of the band-pass noises with 40, 80, and 160 Hz bandwidths ($p < 0.05$). There was an agreement among listeners that the most annoying stimulus was a pure tone or band-pass noise with a 320 Hz bandwidth. Relatively large individual differences were found for band-pass noises with a 320 Hz bandwidth.

Averaged scale values of annoyance increased as the τ_e of sound source increased within the critical band for both center frequencies of 1000 and 2000 Hz as shown in Fig. 7.8. The τ_e represents the repetitive feature or tonal component of the source signals. The previous research suggests that tonal components increase the perceived annoyance and noisiness of broadband noise (e.g., Kryter and Pearsons 1965; Hargest and Pinker 1967; Fujii et al. 2002). These are consistent with our results. Two of the eight listeners had increasing annoyance from pure tone to band noise with widest bandwidth of a center frequency of 2000 Hz. This could indicate that the effects of τ_e on annoyance depend on the individual.

Loudness increases with increasing bandwidth under the same SPL conditions beyond the critical band (Zwicker et al. 1957; Greenwood 1961a, b; Scharf 1962). A significant increase of annoyance beyond the critical band was not found in this experiment. Relatively large inter-individual variations in annoyance were found for band-pass noises with a 320 Hz bandwidth. The loudness of band-pass noises with center frequencies of 1000 and 2000 Hz was also investigated. The same

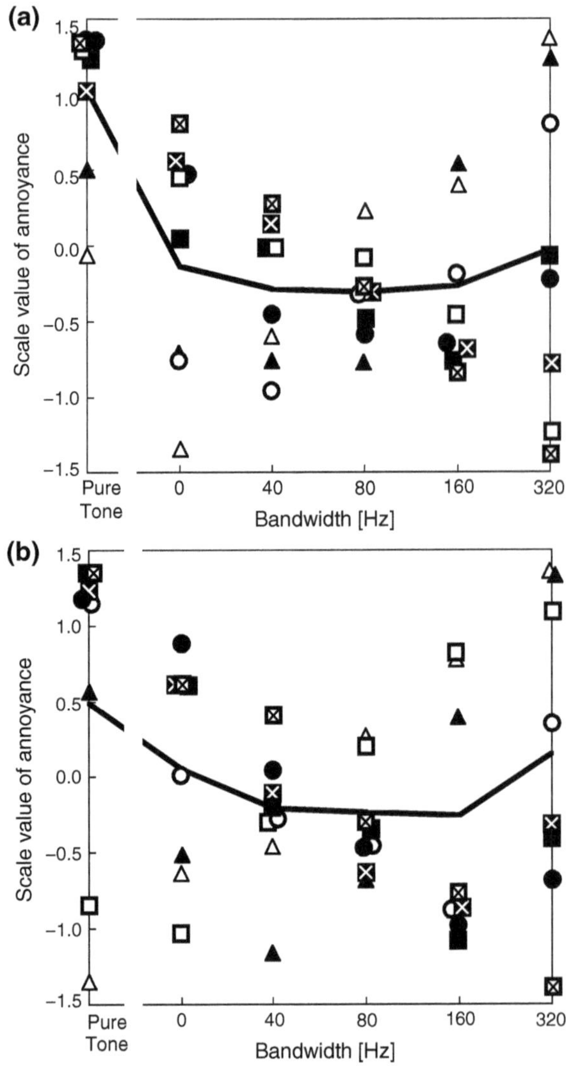

Fig. 7.7 Scale value of annoyance as a function of the bandwidth of band-pass noise with center frequency of **a** 1000 and **b** 2000 Hz. Each symbol represents one listener. The *line* represents the mean scale value of eight listeners (Soeta et al. 2004)

listeners who had joined in the annoyance tests also took part in the loudness tests. The auditory stimuli were presented in the same way as in the annoyance tests. The listeners were asked to judge which of two sound signals was the louder signal. The relationship between the scale values of loudness and the filter bandwidths is shown in Fig. 7.9. The averaged scale value of loudness for the band-pass noises with a 320 Hz bandwidth was significantly greater than those for the other band-pass noises for both center frequencies of 1000 and 2000 Hz, which is consistent with previous research by Zwicker et al. (1957), Greenwood (1961a, b), Scharf (1962), and Sato et al. (2002), one of which showed that the loudness for the pure tone was

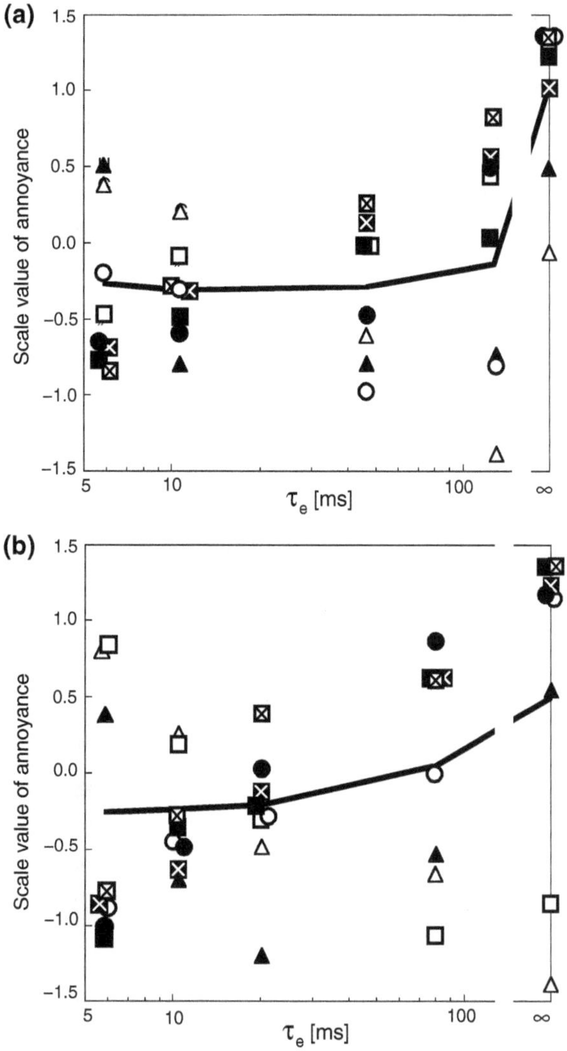

Fig. 7.8 Scale value of annoyance as a function of the τ_e of band-pass noise with center frequency of **a** 1000 and **b** 2000 Hz. Each symbol represents one listener. The *line* represents the mean scale value of eight listeners (Soeta et al. 2004)

significantly greater than that of the band-pass noises, and loudness increased with increasing τ_e, although one listener in the study had increasing loudness from pure tone to widest bandwidth (Sato et al. 2002). In this study, three of the eight listeners had increasing loudness with increasing bandwidth. The increase of loudness and annoyance by pure tone was not found in two listeners, and this is indicated with an open triangle and square in Figs. 7.7b and 7.8a, b. This indicates that there could have been a listener who was hardly affected by the τ_e on loudness or annoyance judgment. That is, the effect of τ_e for loudness or annoyance judgment seems to be partially dependent on the individual.

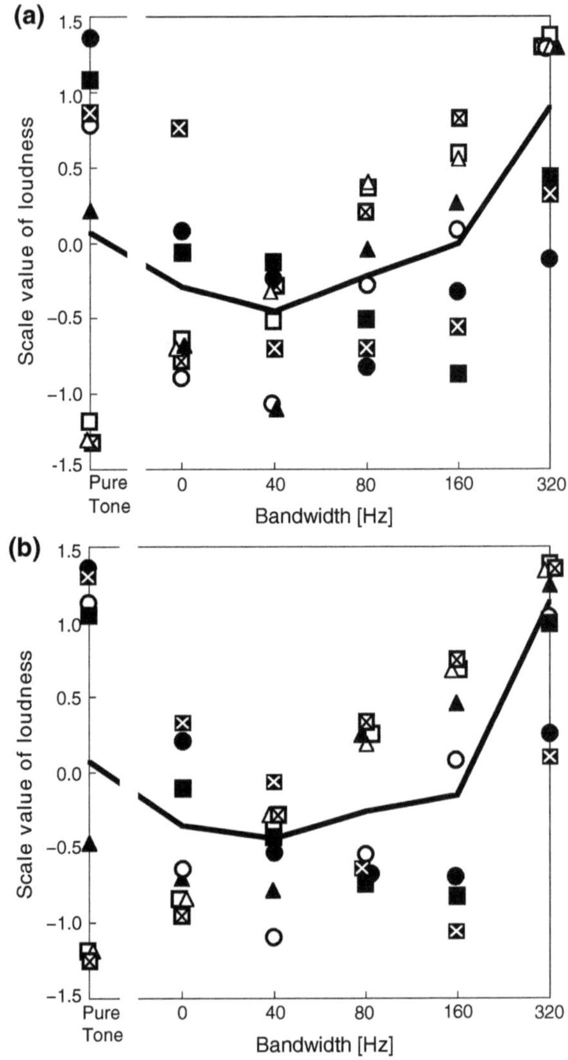

Fig. 7.9 Scale value of loudness as a function of the bandwidth of band-pass noise with center frequency of **a** 1000 and **b** 2000 Hz. Each symbol represents on listener. The *line* represents the mean scale value of eight listeners (Soeta et al. 2004)

Almost all the listeners showed an increase of loudness above the critical band. However, the increase of annoyance above the critical band showed individual variations. The increase in annoyance for pure tones was more evident than that of loudness. Loudness and annoyance produced different results. This is consistent with the findings obtained by Berglund et al. (1975), Hellman (1982, 1984), and Kuwano et al. (1988).

7.3 Annoyance of Noise in Relation to the Spatial Factors

In a three-floor apartment, located in a quiet area, one resident was very annoyed that their sleep was disturbed by the noise of a toilet flushing upstairs. The SPL for the flushing toilet noise was 5 dBA. However, the residents were annoyed by the noise. To investigate the reason why this flushing noise was so annoying, ACF and IACF factors were analyzed (Kitamura et al. 2002). The results showed that the ACF and IACF factors changed dramatically as a function of time. This may explain why the resident felt such an annoyance. To clarify the effects of dynamical change of spatial factors extracted from the IACF on annoyance of noise, annoyance to spatial moving sound images was investigated. The width and horizontal direction of the sound image were fluctuated (Sato et al. 2004).

7.3.1 Effect of IACC Fluctuation

Annoyance judgments were carried out by the paired-comparison method with changes in fluctuation of IACC and the SPL. The sound source was white noise. The moving sound images were simulated by the frontal direct sound (L_0) and two symmetrical lateral reflections (L_1 and L_2) in a soundproof chamber. The loudspeakers were at the level of the listener's ears. To produce incoherent conditions, the time delays between direct sound and two reflections were set at $\Delta t_1 = 20$ ms and $\Delta t_2 = 40$ ms. To simulate the fluctuation of IACC fixing the total SPL, the envelopes of the amplitude of the sounds of frontal and two reflections were modulated alternatively. The modulation periods, M_p, which is defined by the intervals between the times at maxima and minima of IACC, were set at 0.375, 0.75, 1.5 s, or ∞. The SPLs were set at 65, 70, and 75 dBA by measuring the square root of the ACF at the origin of the delay time, $\Phi(0)$. $\Phi(0)$ corresponds to SPL. The measured values of running IACC are shown in Fig. 7.10. To trace the fluctuation of the IACC, the IACF was analyzed for the integration intervals $2T = 0.1$, 0.25, 0.3, and 0.3 s for the stimuli of M_p of 0.375, 0.75, 1.5 s, and ∞, respectively. The running interval was 0.1 s. On the other hand, in order to trace the loudness, $2T$, for analyzing the $\Phi(0)$, was set at the constant at 0.1 s for all stimuli, because the $(\tau_e)_{min}$ of the stimuli was 3 ms (Mouri et al. 2001).

For more quantification of the fluctuation of IACC, Flu_IACC was obtained as follows: In general, the subjective scale value for IACC can be expressed by a nonlinear equation and can be formulated in terms of the 3/2 power of the IACC (Ando 1998)

$$S_{\text{IACC}} \approx -\alpha_1 \text{IACC}^{3/2}. \tag{7.1}$$

Let $S_{\text{IACC}}(t)$ be the subjective effects of the IACC as a function of the time, then the derivative of $S_{\text{IACC}}(t)$ is given by

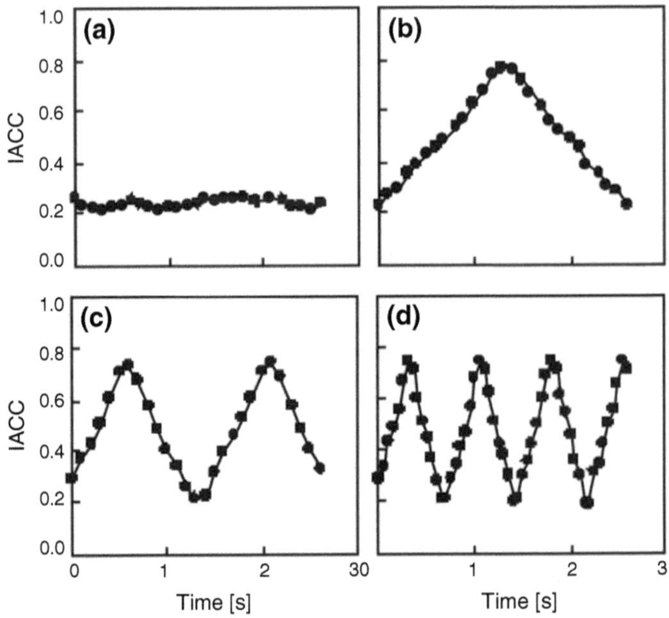

Fig. 7.10 Measured IACC for the experiment 1. **a** $E_p = \infty$ (integration interval for IACF, $2T = 0.3$ s; Flu_IACC = $0.17\alpha_1$); **b** $E_p = 1.50$ ($2T = 0.3$ s; Flu_IACC = $0.73\alpha_1$); **c** $E_p = 0.75$ ($2T = 0.25$ s; Flu_IACC = $1.33\alpha_1$); **d** $E_p = 0.375$ ($2T = 0.1$ s; Flu_IACC = $2.96\alpha_1$) (Sato et al. 2004)

$$\frac{\partial S_{\mathrm{IACC}}(t)}{\partial t} = \lim_{h \to 0} \frac{|S_{\mathrm{IACC}}(t) - S_{\mathrm{IACC}}(t-h)|}{h}. \tag{7.2}$$

The most subjectively effective piece of sound signal on annoyance, which is related to the IACC, may be obtained by $(\partial S_{\mathrm{IACC}}(t)/\partial t)_{\max}$. Thus, such a sampled data piece with constant time interval Δ is given by

$$\mathrm{Flu\_IACC} = \left(\frac{\partial S_{\mathrm{IACC}}(i)}{\partial i}\right)_{\max} = \frac{|S_{\mathrm{IACC}}(i\Delta) - S_{\mathrm{IACC}}((i-1)\Delta)|_{\max}}{\Delta}. \tag{7.3}$$

In the experimental conditions, $\Delta = 0.1$ s is small enough to obtain a convergent value of the Flu_IACC. The values of Flu_IACC for the stimuli of $M_p = 0.375$, 0.75, 1.5 s, and ∞ were 2.34, 1.21, 0.72, and 0.17, respectively. Fluctuations in other spatial factors, W_{IACC} was within -0.01 ± 0.01 ms, and τ_{IACC} was always within $\pm$ 0.02 ms. Hence, listeners always perceived the frontal sound image in the median plane.

The paired-comparison tests were conducted for 12 stimuli [Four levels of Flu_IACC and three levels of $\Phi(0)$]. Five listeners (four males and one female; 22–24 years old) with normal hearing ability joined. The listener was seated in a

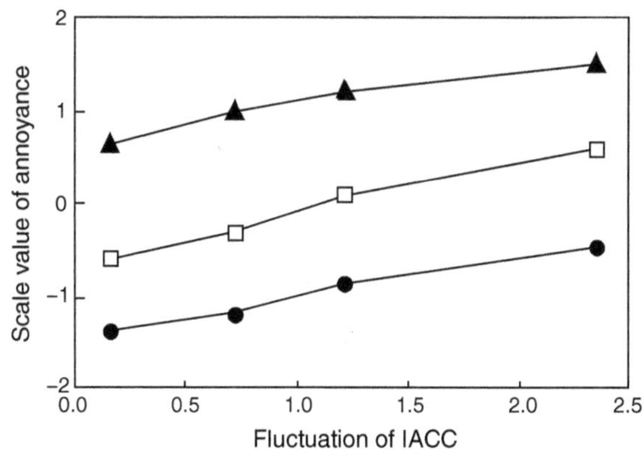

Fig. 7.11 Average scale values of annoyance as a function of Flu_IACC and as a parameter of SPL [Φ(0)]. *Filled circle* SPL = 65 dBA; *open square* SPL = 70 dBA; and *filled triangle* SPL = 75 dBA (Sato et al. 2004)

soundproof chamber and asked to judge which of the two stimuli was perceived as more annoying. The duration of the stimuli was 3 s, the rise and fall time was 50 ms, and the silent interval between stimuli was 1 s. Each pair of stimuli was presented in a random order separated by an interval of 3 s, which was the allotted time for the listener to respond. A single-test session consisted of 66 pairs ($N(N-1)/2$, $N = 12$) of stimuli. Ten sessions were performed for each listener.

Fifty responses (five listeners and 10 sessions) to each stimulus were obtained. As shown in Fig. 7.11, the scale values of annoyance increased with increasing Flu_IACC and Φ(0) as well. The moving sound images (Flu_IACC > 0) were always more annoying than the fixed sound image (Flu_IACC ≈ 0) when the Φ(0) was constant. The results of an analysis of variance (ANOVA) for scale values of annoyance indicate that the factors Flu_IACC and Φ(0) are significant ($p < 0.001$). The effects of the interaction between Flu_IACC and Φ(0) were not significant. Thus, Flu_IACC and Φ(0) contributed to the scale value of annoyance independently, so that the scale value of annoyance may be given by

$$SV_{\text{annoyance}} \approx f(\text{Flu\_IACC}) + f(\Phi(0)) \approx a(\text{Flu\_IACC}) + b(\Phi(0)). \quad (7.4)$$

The coefficients obtained by multiple regression analyses are: $a \approx 0.46$ and $b \approx 0.20$, which result in the best correlation between the measured and calculated scale values, 0.96 ($p < 0.01$).

7.3.2 *Effect of τ_{IACC} Fluctuation*

As similar procedure to previous experiment, annoyance judgments were carried out by the paired-comparison method while changing fluctuations of the τ_{IACC} and

$\Phi(0)$. The source signal was band-pass-filtered noise with the center frequency of 500 Hz (bandwidth = 160 Hz). The horizontally swaying sound images were simulated by two lateral symmetrical sounds (L_1 and L_2) in a soundproof chamber. To produce the fluctuation of τ_{IACC} fixing the total $\Phi(0)$, the envelopes of the amplitude of the sounds were modulated so that a sound image was altered between left and right. The M_p were set at 0.375, 0.75, 1.5 s, or ∞. The $\Phi(0)$ values were set at 65, 70, and 75 dBA. The measured values of τ_{IACC} are shown in Fig. 7.12. To allow the fluctuation of τ_{IACC}, the IACF was calculated for the integration intervals $2T$ = 0.1, 0.25, 0.3, and 0.3 s for the stimuli of M_p of 0.375, 0.75, 1.5 s, and ∞, respectively. The running interval was 0.1 s. To quantify the fluctuation of τ_{IACC}, Flu_ τ_{IACC} is defined as follows:

It is assumed that the scale value for τ_{IACC} is expressed by a linear equation.

$$S_{\tau IACC} \approx \alpha_2 \tau_{IACC}. \tag{7.5}$$

Similar to Eq. (7.2), we obtain

$$\frac{\partial S_{\tau_{IACC}(t)}}{\partial t} = \lim_{h \to 0} \frac{|S_{\tau_{IACC}C}(t) - S_{\tau_{IACC}}(t-h)|}{h}. \tag{7.6}$$

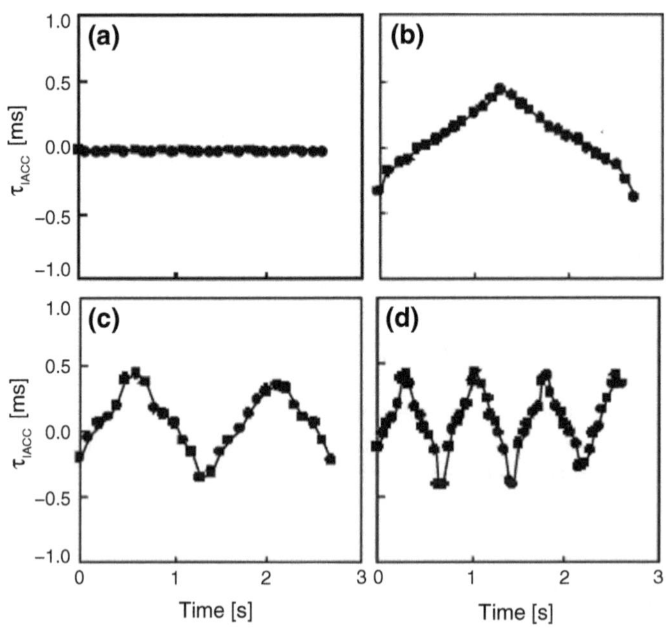

Fig. 7.12 Measured τ_{IACC} for the experiment 2. **a** $E_p = \infty$ ($2T$ = 0.3 s; Flu_τ_{IACC} = 0.07α_2); **b** E_p = 1.50 ($2T$ = 0.3 s; Flu_τ_{IACC} = 1.50α_2); **c** E_p = 0.75 ($2T$ = 0.25 s; Flu_τ_{IACC} = 2.07α_2); **d** E_p = 0.375 ($2T$ = 0.1 s; Flu_τ_{IACC} = 3.73α_2) (Sato et al. 2004)

Let Flu_τ_{IACC} can be defined as the maximum fluctuation of the subjective scale values related to the τ_{IACC}, $(\partial S_{IACC}(t)/\partial t)_{max}$, which may affect annoyance. Here, the fluctuation of the subjective scale value for τ_{IACC} is given by the sampled data strings with constant time interval Δ, thus,

$$\mathrm{Flu\_}\tau_{IACC} = \left(\frac{\partial S_{\tau_{IACC}}(i)}{\partial i}\right)_{max} = \frac{|S_{\tau_{IACC}}(i\Delta) - S_{\tau_{IACC}}((i-1)\Delta)|_{max}}{\Delta}. \quad (7.7)$$

Practically, $\Delta = 0.1$ s is small enough for the convergence of Flu_τ_{IACC}. The values of Flu_τ_{IACC} for the stimuli of $E_p = 0.375, 0.75, 1.5$ s, and ∞ were 2.34, 1.21, 0.72, and 0.17, respectively. Fluctuations in other spatial factors, W_{IACC} was within 0.91 ± 0.08 and 0.27 ± 0.01 ms, respectively. Hence, listeners always perceived the frontal sound image in the median plane. Similar to previous experiment, the paired-comparison tests were conducted for 12 stimuli [four levels of Flu_τ_{IACC} and three levels of $\Phi(0)$] with the same five listeners.

Fifty responses (five listeners and ten sessions) to each stimulus were also obtained. Figure 7.13 shows the scale values of annoyance increased as Flu_τ_{IACC} and $\Phi(0)$ increased. When the $\Phi(0)$ was constant, the moving sound images ($Flu\_\tau_{IACC} > 0$) were always more annoying than the fixed sound image ($Flu\_\tau_{IACC} \approx 0$). The results of an ANOVA for scale values of annoyance indicate that the factors Flu_τ_{IACC} and $\Phi(0)$ are significant ($p < 0.001$). The effects of the interaction between Flu_τ_{IACC} and $\Phi(0)$ are not significant. Accordingly, Flu_τ_{IACC} and $\Phi(0)$ contribute to the scale value of annoyance independently, so that

$$SV_{annoyance} \approx f(\mathrm{Flu\_}\tau_{IACC}) + f(\Phi(0)) \approx a(\mathrm{Flu\_}\tau_{IACC}) + b(\Phi(0)). \quad (7.8)$$

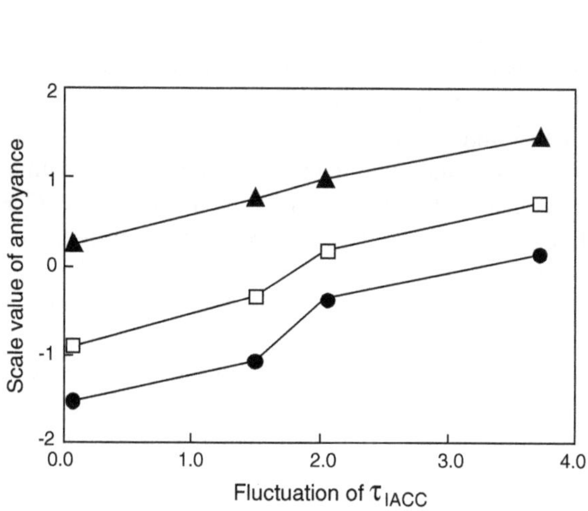

Fig. 7.13 Average scale values of annoyance as a function of Flu_τ_{IACC} and as a parameter of SPL [$\Phi(0)$]. *Filled circle* SPL = 65 dBA; *open square* SPL = 70 dBA; and *filled triangle* SPL = 75 dBA (Sato et al. 2004)

The coefficients obtained by multiple regression analyses are $a \approx 0.42$ and $b \approx 0.16$. The calculated scale values agree well with measured ones with a correlation coefficient of 0.98 ($p < 0.01$). In a comparison of the results from previous experiment, the effects of $\Phi(0)$ are relatively small in the present experimental conditions, so that effects of the Flu_τ_{IACC} cannot be ignored for the evaluation of annoyance.

The contributions of the Flu_τ_{IACC} and the Flu_IACC to the scale value of annoyance are significant with the reference to the $\Phi(0)$ difference of 10 dB. It is noteworthy that the variations in the ranges of $\Phi(0)$ were from 65 to 75 dBA in both experiments.

Since the condition, $\tau_{IACC} = 0$, is one of the preferred conditions for listening to sound (Sato et al. 1997), annoyance increased with the stimuli of Flu_$\tau_{IACC} > 0$. In these experiments, the horizontal angles of lateral sounds were ±54°, which is the most effective angle to obtain a low IACC for a sound source with the frequency range (Ando 1998). The value of IACC was alternated between 0.2 and 0.8 in previous experiment. However, the listeners found it hard to perceive a spatial fluctuation in respect of the IACC when it was lower than 0.4, because the just noticeable difference of IACC in the sound field with such a low IACC is larger than that with a higher IACC (Gabriel and Colburn 1981). In order to avoid these kinds of effects of nonlinearity, Eq. (7.1) was introduced. Under these experimental conditions, 3.73 of the Flu_IACC is equivalent to an increase of 4.9 dB in the $\Phi(0)$, and 2.96 of the Flu_τ_{IACC} is equivalent to an increase of 9.7 dB in the $\Phi(0)$.

On the other hand, in the preset experiment, the values of τ_{IACC} were changed from –0.4 to 0.4 ms and M_p was 0.375 s at minimum. The threshold of the interaural time delay (for the 1000-Hz tone) is 10 ms (Klumpp and Eady 1956). The M_p of 0.375 s is long enough to perceive the movement of the sound source (Grantham 1986; Chandler and Grantham 1992). In addition, all stimuli had clear sound images because the values of IACC for all stimuli in the present experiment were greater than 0.82. Therefore, listeners equally perceived the fluctuation of τ_{IACC} during the whole period of the stimuli.

The results of the study lead to the following conclusions: (1) Moving sound images were always more annoying than fixed sound images with a constant $\Phi(0)$. (2) The annoyance increased with increasing fluctuation of IACC as well as $\Phi(0)$. (3) The fluctuations of IACC and $\Phi(0)$ independently contribute to the scale value of annoyance. (4) The annoyance increased with increasing fluctuation of τ_{IACC} as well as $\Phi(0)$. (5) The fluctuations of τ_{IACC} and $\Phi(0)$ independently contribute to the scale value of annoyance.

For subjective evaluations of moving sound sources, it is recommended that binaural measurements are to be conducted to obtain the spatial factors extracted from the IACF, in addition to the temporal factors extracted from the ACF.

7.4 Annoyance of Road Traffic Noise

In the previous literature, there is a sizable study describing acoustic parameters relating to annoyance. They include, for example, the frequency distribution, tonality, temporal fluctuation, and impulsivity. As for tonal color, or timbre of sound, previous studies were intended to quantify sound qualities in terms of their spectral shapes. For example, Versfeld and Vos (1997) and Cermak and Cornillon (1976) proposed several measures for describing the shapes of the frequency spectra, such as the location and the level difference between high-frequency (500–2000 Hz) and low-frequency (50–100 Hz) peaks in the spectrum. Also, as for the frequency distribution of noise, it is known that both of the high- and the low-frequency parts of sound have effect on annoyance. The high-frequency components of a sound make it sound sharp, and sharpness of sound increases annoyance (Zwicker and Fastl 1999). In contrast, some studies have demonstrated the much effects of low-frequency noise on annoyance (Berglund et al. 1996).

As for the tonality of sound, a number of tone corrections are proposed for the evaluation of the perceived noise level. This is based on the finding that the tone-to-noise ratio in the spectrum increases annoyance of noise. Generally, a corrected value for extracted tonal component is added to the "Perceived Noise Level" (PNL) to give the "Tone Corrected Perceived Noise Level" (PNLT). However, the calculation for this correction is lengthy, and their accuracy is not well established yet (May 1978).

As for fluctuation of SPL, annoyance to temporally fluctuating white noise was investigated (Hiramatsu et al. 1978, 1983). It was found that standard deviation, equivalent SPL (L_{eq}), and L_{10} were good measure for annoyance fluctuating sound. Also, it was found recently that the fluctuation in the sound location and the diffuseness also much affects subjective annoyance (Jeon 2001; Kitamura et al. 2002).

A laboratory experiment was designed to examine the possible procedures for describing noise annoyance (Fujii et al. 2002). Traffic noise was used as the example of noise that we are exposed in our daily life. It is considered that ACF and IACF factors may contribute to annoyance in a complex manner. To simplify the problem, only the effect of ACF factors and their temporal fluctuation were investigated. For particular purpose of the experiment, overall SPL was adjusted to be roughly equal to emphasize the other physical properties. This manipulation could reveal the potential importance of further properties or show that no additional parameters besides SPL are significant. To investigate only the effect of the ACF factors of sounds, the spatial properties were kept constant.

Nine recordings of noise sounds were used in the experiment (Fujii et al. 2002). They are passenger car, two buses, truck, three scooters, motor bike, and another motorbike. Each stimulus was edited on computer software to have a 4-s duration and contains single vehicle's passage. The maximum SPL was adjusted to be equal (73 ± 2 dBA) and to occur near the middle of the sound. By this manipulation, the average SPL was also equated (69 ± 1 dBA). To make the envelope of sounds equal, a 0.5-s rise and fall time was added to all stimuli.

The traffic sounds were reproduced in an anechoic room through a laptop computer, a DA converter, a power amplifier, and a loudspeaker. A single loudspeaker was used to keep the spatial properties of the sound field constant. Based on the running short-time SPL (dBA) measured for the experimental condition, eight standard measures were calculated for nine stimuli: (1) mean SPL, (2) variance σ^2 of the SPL, (3) maximum SPL, (4) minimum SPL, (5)–(7) the SPL values exceeded 10 % of the time (L_{10}), 50 % of the time (L_{50}), and 90 % of the time (L_{90}), and (8) equivalent sound level, L_{eq}. Most of these standard measures were highly intercorrelated. Clearly, all of these factors contain information about the overall SPL and its variability. Therefore, only the median (L_{50}) and variance (σ^2) of the SPL were considered in the subsequent analysis.

The cumulative frequency of measured SPL and four ACF factors is shown in Fig. 7.14. $\Phi(0)$ corresponds to SPL. These factors were calculated at intervals of 0.1 s for each sound. It can be seen that the range of $\Phi(0)$ is reasonably controlled. Other parameters except for $\Phi(0)$ were not controlled systematically, but it seems that these parameters covered a wide range suitable for the purpose of the experiment. To characterize the acoustical properties of a stimulus, the median and variance of each factor within stimulus durations were used.

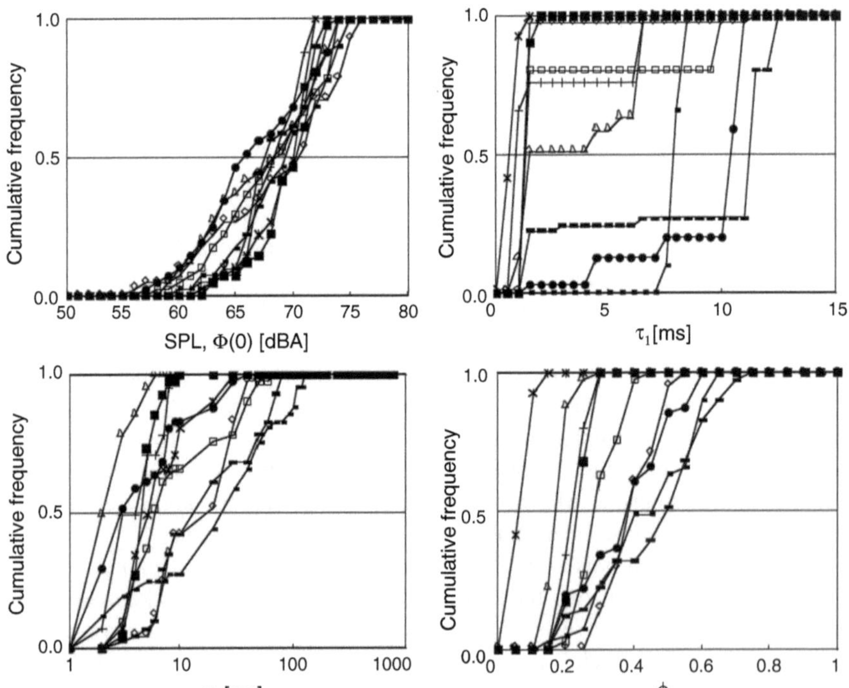

Fig. 7.14 Cumulative frequencies of SPL ($\Phi(0)$) and ACF factors measured for stimuli used in the experiment (Fujii et al. 2002)

Subjective annoyance was measured by a paired-comparison method. The listeners sat 1.0 m in front of the loudspeaker. Ten listeners (nine males and one female) took part in the experiment. They were between the ages of 23 and 27, in good health, with normal auditory acuity. All possible pairs from the nine sounds (36 pairs) were presented to the listeners in a random order in one session. After the presentation of paired stimuli, the listeners were asked to judge which of the two sound signals was more annoying. All listeners had four series of sessions, giving a total of 144 comparisons. Collected data were processed by applying the law of comparative judgment (case V; Thurstone 1927).

Considering the consistency between listeners' response, the scale values (SV) of annoyance were averaged across listeners so that there is a single SV for each sound. Then, the correlation coefficients, r, were calculated between SV and the physical measures. As expected, the $\Phi(0)$ and annoyance are not related. It is considered that the range of $\Phi(0)$ among the stimuli was too small (4 dBA) to affect annoyance. Instead, the variance of $\Phi(0)$ had much effect on annoyance ($r = 0.64$, $p < 0.05$). Although Cermak and Cornillon (1976) did not find a significant contribution of measures other than L_{eq}, the results may suggest the importance of other acoustical factors when the $\Phi(0)$ is relatively constant.

The values of τ_e and ϕ_1 were significantly correlated with annoyance ($r = 0.56$ and 0.57, $p < 0.05$). This result shows that a sound having a strong tonal component was perceived to be more annoying than the un-tonal noise. The listeners' comment also indicated that they judged a sound having a clear pitch to be more annoying. In the evaluation of the PNL for a tonal sound as used in the experiment, a number of tone corrections are proposed, such as PNL and PNLT. However, the calculation for this correction is lengthy, and their accuracy is not well established (May 1978). Instead, by using the value of τ_e and ϕ_1, the effect of the tonal component on perceived annoyance is clearly extracted.

Before the experiment, it was assumed that sound with a small value of τ_1 (i.e., it represents high-pitch sound) might be more annoying. It means that annoyance and τ_1 should have negative correlation. But the result was contrary ($r = 0.30$). It is probably because the high-pitch sounds tested in the experiment had weak pitch sensation (it is represented by small values of ϕ_1). The high correlation between τ_1 and ϕ_1 ($r = 0.82$) implies that low-pitch sounds are tonal and high-pitch sounds are un-tonal sounds. This tendency was generally observed for measured traffic noise in this study. As shown in Fig. 6.7b, the un-tonal sounds tend to be estimated to have high pitch in the current analysis, because the structure of the ACF does not have dominant periodical peaks. Consequently, it is not distinguishable from the high-pitched tonal sound like an aircraft noise (Fujii et al. 2001) by only seeing the pitch itself. The result suggests that the effects of pitch on annoyance should be considered together with its pitch strength.

Considering the results above, it is considered that the ACF factors and the variance of $\Phi(0)$ affect perceived annoyance with a complex manner. To calculate the effect of each factor on perceived annoyance more precisely, multiple regression analysis using a linear combination of eight variables was carried out. To obtain an optimal equation, all possible combinations were examined. The correlation coefficients and significance levels were used to determine the goodness of fit. The best combination of variables was found as the variance of Φ(0), the median of τ_e, and the variance of τ_1, so that

$$SV_{\text{annoyance}} \approx a_1 * \text{Var\_}\Phi(0) + a_2 * \tau_e + a_3 * \text{Var\_}\tau_1 + c. \quad (7.9)$$

Using these tentative values and constant $c = -1.62$ in Eq. (7.9), the total correlation coefficient 0.91 was obtained with the significance level ($p < 0.05$). Multi-colinearity is avoided in this prediction because these three parameters are not correlated with each other. This result shows that the temporal fluctuation of Φ(0) and tonal component had a major effect on annoyance. For sounds with pitch variation, listeners also perceived more annoyance. Standardized partial regression coefficients of each variable a_1, a_2, and a_3 in Eq. (7.9) were 0.64, 0.50, and 0.36, respectively and these coefficients were statistically significant ($p < 0.05$ for a_1 and a_2, $p < 0.1$ for a_3).

The purpose of this study was to describe the perceived annoyance of traffic noise by ACF factors and the temporal fluctuations. It was concluded that perceived annoyance is greatly affected by the variation of the Φ(0), τ_e, which corresponds to tonal components included in the noise, and the variation of τ_1, which corresponds to pitch fluctuation of the noise, when the difference of overall SPL is small.

7.5 Annoyance of Noise in a Train Station

Train noise in a station (TNIS) can annoy passengers and reduce the speech intelligibility of the public address system (Bhattachaya 1994; Endoh et al. 2007; Isumi and Fujii 2007). Thus, clarifying the acoustic characteristics of TNIS that causes annoyance is important for comfort of passengers, transit workers, and operators. Recent surveys showed that TNIS does not seem to cause hearing loss from the L_{Aeq} in subway stations (Gershon et al. 2005; Shimokura and Soeta 2011). However, it is necessary to take into account not only the noise levels (e.g., the L_{Aeq}) but also the sound quality described by factors extracted from the ACF and IACF for the evaluation of environmental noise. To find the factors that affect annoyance for TNIS, subjective annoyance was determined by paired-comparison methods and predicted by the ACF and IACF factors (Soeta and Shimokura 2013a).

Eight stimuli were selected from noise in above-ground and underground stations for each type of platform and platform screen doors (PSDs). PSDs can be principally classified in two types, i.e., mobile full-height (MFH) and mobile half-height (MHH). The stimuli were presented binaurally through headphones (HD650, Sennheiser). Eight listeners aged 20 to 47 years with no history of hearing disorders participated in the experiment. They sat in a comfortable thermal environment in a soundproof room and heard the auditory stimuli. Paired-comparison tests were performed for all combinations of pairs (i.e., 28 pairs ($N(N - 1)/2$, $N = 8$)) of stimuli, interchanging the order that the stimuli in each pair were presented in each session and presenting the pairs in random order. Six sessions were conducted for

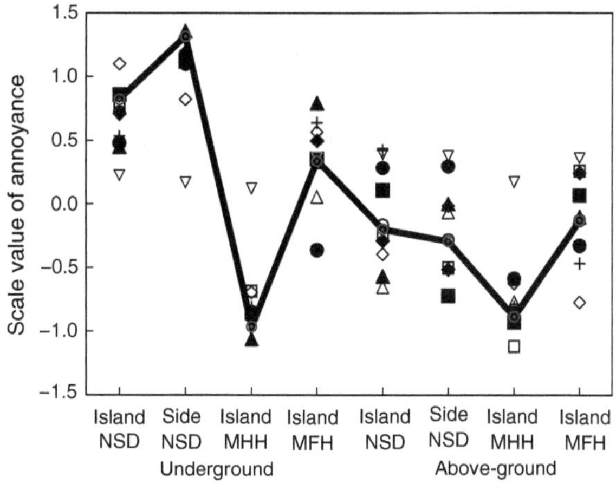

Fig. 7.15 The scale values of the annoyance for the eight noises in above-ground and underground stations with each platform and with MFH, MHH, or without PSDs (NSD). The *thick line* indicates averaged value for all listeners. Each symbol indicates each listener

each listener. The duration of the stimuli was 5.0 s, the rise and fall time was 100 ms, and the silent interval between stimuli was 1.0 s. After the presentation of the two stimuli, listeners were asked to judge which of the two sound signals was more annoying. The scale values of the annoyance were calculated according to Case V of Thurstone's theory (case V; Thurstone 1927).

The averaged scale values of the annoyance for the eight noises in above-ground and underground stations for each type of platform and PSDs are presented in Fig. 7.15. The relationships between scale value of annoyance and each ACF/IACF factor are presented in Fig. 7.16. Annoyance was found to increase with increasing $\Phi(0)$. Since τ_1 values were not normally distributed, they were ruled out of correlation analysis.

To calculate the effects of each objective factor on averaged scale values of the annoyance for all listeners, multiple regression analyses were conducted using a linear combination of ACF/IACF factors and their variations by stepwise procedures. The final model indicated $\Phi(0)$ was the significant factor:

$$SV_{\text{annoyance}} \approx a * \Phi(0) + c. \tag{7.10}$$

The relationships between the measured and calculated annoyance are shown in Fig. 7.17. The correlation coefficient was 0.95 ($p < 0.01$).

In a similar way, scale values of annoyance for each listener were predicted by multiple regression analyses using a linear combination of ACF/IACF factors and their variations by stepwise procedure. The significant factors, the standardized regression coefficients, and the correlation coefficients between measure and calculated scale value of annoyance are listed in Table 7.1. $\Phi(0)$ had the significant factor on subjective annoyance for six out of eight listeners. τ_e, IACC, variation of ϕ_1, τ_e, and $W_{\phi(0)}$ were the significant factors for some of the listeners. Negative correlation between annoyance and IACC suggests that reflections from many directions may cause greater annoyance. This is consistent with previous findings on the annoyance of heavyweight floor impact sounds (Jeon et al. 2009).

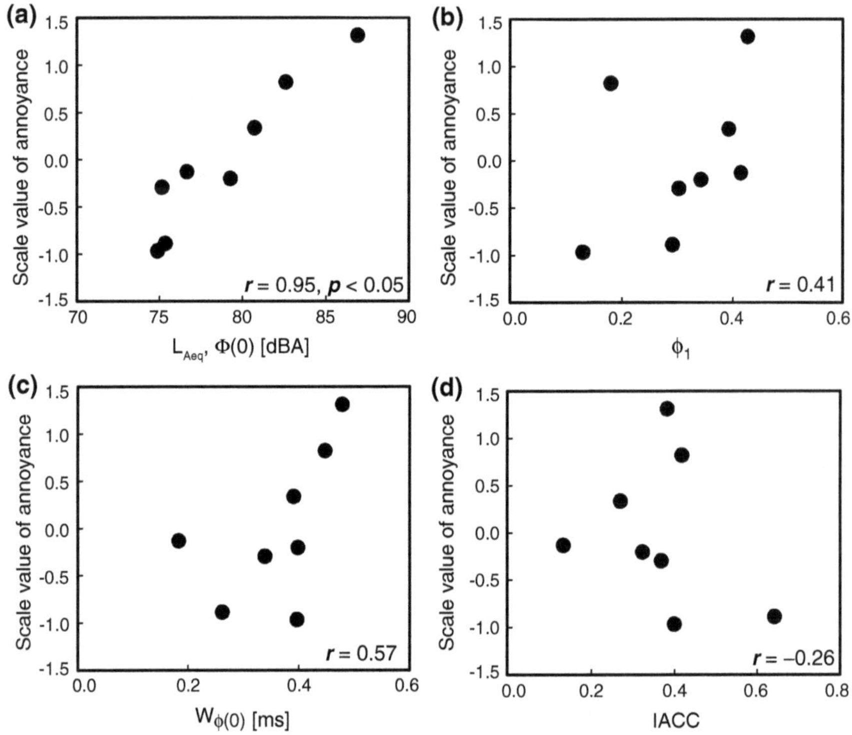

Fig. 7.16 The relationships between the scale value of annoyance and each ACF/IACF factor (Soeta and Shimokura 2013a)

Fig. 7.17 The relationships between the measured and calculated annoyance for ACF/IACF factors (Soeta and Shimokura 2013a)

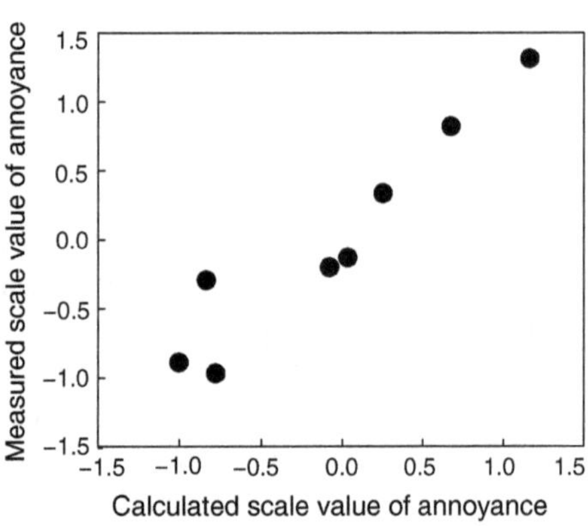

Table 7.1 Individual and total results for significant factors, the standardized regression coefficients, and the correlation coefficients between measure and calculated scale value of annoyance for noise in train stations

Listener	Factor 1	Factor 2	Factor 3	r
A	$\Phi(0)$ 0.91	IACC −0.24	VAR_τ_e 0.18	0.99
B	τ_e 1.01	VAR_τ_e −0.47		0.94
C	$\Phi(0)$ 0.86	VAR_ϕ_1 −0.41		0.94
D	$\Phi(0)$ 0.91			0.91
E	$\Phi(0)$ 1.22	τ_e −0.35		0.99
F	τ_e 0.99	VAR_τ_e 0.34		0.98
G	$\Phi(0)$ 0.85	IACC −0.36		0.96
H	$\Phi(0)$ 0.75	VAR_τ_1 0.50	VAR_$W_{\Phi(0)}$ 0.18	0.99
Total	$\Phi(0)$ 0.95			0.95

7.6 Annoyance of Noise in a Train Car

Trains are used as a commuting tool by a lot of people for a long time in big cities of Japan. Although some studies have investigated the noise characteristics inside running train cars (Yoshihisa et al. 1968; Chang and Hermann 1974; Eade and Hardy 1977; Kono et al. 1982; Hardy 2000; Soeta and Shimokura 2009, 2011), the factors influencing subjective annoyance to noises in train cars have not been evaluated. To find the factors that affect annoyance for noises in train cars, subjective annoyance was determined by paired-comparison methods and predicted by the ACF and IACF factors.

Eight stimuli were selected from noise in various types of trains (Soeta and Shimokura 2013b). The stimuli were presented binaurally through headphones (HD650, Sennheiser). Eight listeners aged 20 to 47 years with no history of hearing disorders participated in the experiment. They sat in a comfortable thermal environment in a soundproof room and heard the auditory stimuli. Paired-comparison tests were performed for all combinations of pairs [i.e., 28 pairs ($N(N-1)/2$, $N = 8$)] of stimuli, interchanging the order that the stimuli in each pair were presented in each session and presenting the pairs in random order. Six sessions were conducted for each listener. The duration of the stimuli was 5.0 s, the rise and fall time was 100 ms, and the silent interval between stimuli was 1.0 s. After the presentation of the two stimuli, listeners were asked to judge which of the two

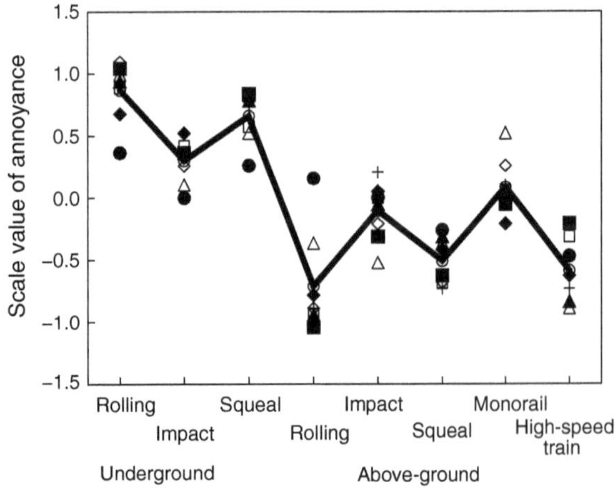

Fig. 7.18 The averaged scale values of the annoyance for the eight types of noises in trains. The *thick line* indicates averaged value for all listeners. Each symbol indicates each listener

sound signals was more annoying. The scale values of annoyance were calculated according to Case V of Thurstone's theory (case V; Thurstone 1927).

The averaged scale values of the annoyance for the eight noises in train cars are presented in Fig. 7.18. The relationships between the scale value of annoyance and each ACF/IACF factor are presented in Fig. 7.19. Annoyance was found to increase

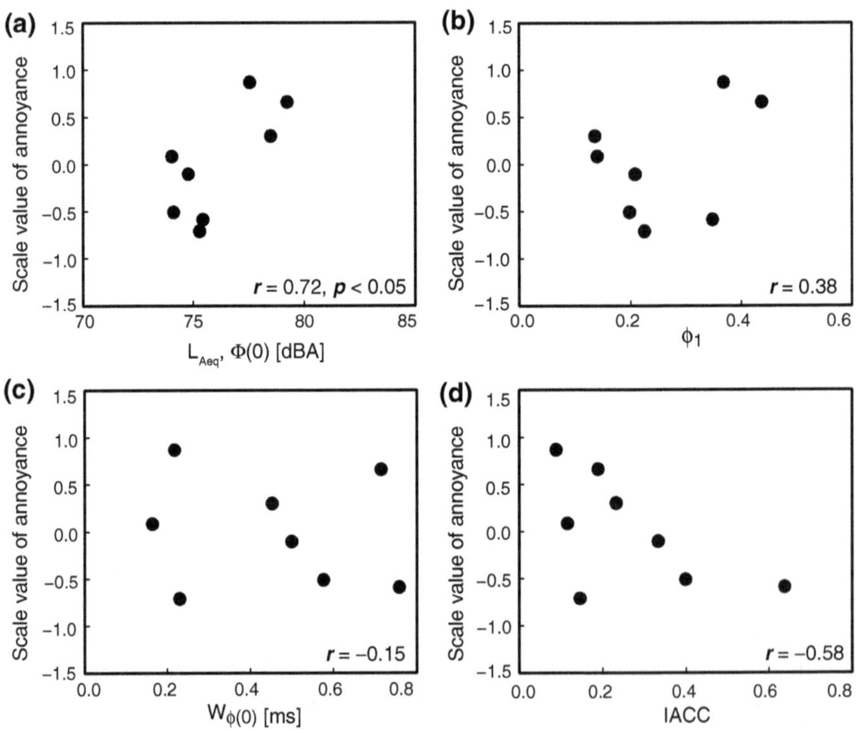

Fig. 7.19 The relationships between the scale value of annoyance and each ACF/IACF factor

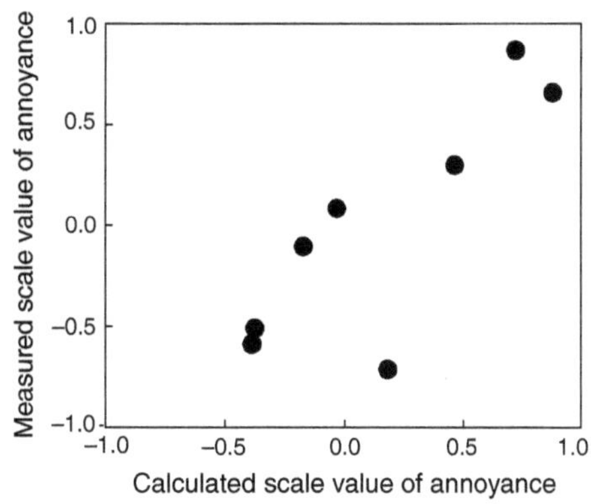

Fig. 7.20 Relationship between measured scale values of annoyance and scale values calculated by Eq. (7.11) for ACF and IACF factors

with increasing Φ(0). Since τ_1 values were not normally distributed, they were ruled out of the correlation analysis.

To calculate the effects of each objective factor on averaged scale values of annoyance for all listeners, multiple regression analyses were conducted using a linear combination of ACF/IACF factors and their variations by stepwise procedures. The final model indicated Φ(0) was the significant factor:

$$\mathrm{SV}_{\mathrm{annoyance}} \approx a_1 * \Phi(0) + c. \tag{7.11}$$

The relationships between the measured and calculated annoyance for different ACF/IACF factors are shown in Fig. 7.20. The correlation coefficient was 0.72 ($p < 0.01$). Φ(0) has the greatest effect on averaged scale value of subjective annoyance for all listeners. Perceived annoyance increased with increasing Φ(0). This is consistent with previous findings on the annoyance of noise in train stations (Soeta and Shimokura 2013a).

In a similar way, scale values of annoyancee for each listener were predicted by multiple regression analyses using a linear combination of ACF/IACF factors and their variations by stepwise procedure. The significant factors, the standardized regression coefficients, and the correlation coefficients between measure and calculated scale value of annoyance are listed in Table 7.2. Φ(0) had the significant factor on subjective annoyance for three out of eight listeners. IACC and variation of τ_e had the significant factor on subjective annoyance for two out of eight listeners. Negative correlation between annoyance and IACC suggests that reflections from many directions may cause greater annoyance. This is consistent with previous findings on the annoyance of heavyweight floor impact sounds (Jeon et al. 2009).

Table 7.2 Individual and total results for significant factors, the standardized regression coefficients, and the correlation coefficients between measure and calculated scale value of annoyance for noise in train cars

Listener	Factor 1	Factor 2	Factor 3	r
A	$\Phi(0)$ 0.77			0.77
B	IACC −0.91	ϕ_1 0.38		0.95
C				
D	$\Phi(0)$ 1.23	VAR_τ_e −0.64	τ_1 −0.35	0.99
E				
F	IACC −0.81			0.81
G	$\Phi(0)$ 1.09	VAR_τ_e −0.66		0.89
H				
Total	$\Phi(0)$ 0.72			0.72

7.7 Annoyance of Floor Impact Sound

Loudness and annoyance from floor impact sounds for different sound insulation treatments applied in rooms of apartment units have been investigated in terms of the impact SPL (Jeon et al. 2004). Their study showed that sound insulation in both the floors and walls in a box frame-type reinforced concrete structure reduces loudness and annoyance. However, for sounds with widely different acoustical properties, annoyance cannot be reduced by the SPL reduction alone (Kitamura et al. 2002). To clarify the effects of ACF factors on the annoyance, subjective annoyance of heavyweight floor impact sounds in the rooms of apartments with different types of sound insulation treatments was evaluated (Jeon and Sato 2008).

Eight floor impact sounds generated by the bang machine and the impact ball were chosen to cover a wide range of ACF factors. The source signals were presented through headphones (Sennheiser HD600). The average differences of the $\Phi(0)$, τ_e, τ_1, and ϕ_1 for the signals recorded in rooms and reproduced by headphones were 0.17 dB, 5.7 ms, 0.13 ms, and 0.007, respectively. The head-related transfer functions of the dummy head and each listener's head were not exactly the same. However, it was shown that the differences between the acoustical parameters measured using the dummy head and human heads were within the difference limens except at the high-frequency ranges (Sakai et al. 2004). Therefore, the sounds presented by the headphones in this study are regarded as actual floor impact sounds.

A test session consisted of 28 pairs of stimuli for the two heavyweight impact sources (bang machine and impact ball) at two SPLs (45 and 55 dB) realized as minimum and maximum jumping noise levels. The duration and silence interval between stimuli, which consisted of two repeated noises, were approximately

4.3 and 0.5 s, respectively. Each pair of stimuli was presented in random order separated by an interval of 3 s, which was the allotted time for the listener to respond. Test sessions lasted approximately 5.6 min, and a total of 40 listeners (20 university students and 20 housewives) who had normal hearing participated in the test.

The SPL for this experiment was fixed at a constant $L_{i,\mathrm{Fmax,Aw}}$. Listeners were seated in a sound-proof chamber and asked to judge which one of two stimuli they perceived to be more annoying as floor impact sounds. A scale value of annoyance was developed by applying the law of comparative judgment (case V; Thurstone 1927). To calculate the effects of each objective factor on annoyance, multiple regression analyses were conducted using a linear combination of ACF factors.

There was a positive correlation between the ACF factors Φ(0) and VAR_Φ(0), and scale values, and a negative correlation between VAR_ϕ_1 and the scale values. Thus, these three variables were selected for the multiple regression analysis

$$SV_{\mathrm{annoyance}} \approx a_1 * \Phi(0) + a_2 * \mathrm{VAR\_}\Phi(0) + a_3 * \mathrm{VAR\_}\phi_1. \qquad (7.12)$$

If the resonance frequency and its harmonics indicate sharp amplitude peaks, ϕ_1 becomes large. In this case, a listener perceives a clearer pitch. It was confirmed that the regression model using these three parameters gave the highest regression coefficient. The relationship between the scale value obtained by annoyance judgments and the calculated scale value for ACF factors are shown in Fig. 7.21. The scale values of annoyance were calculated from Eq. (7.12) ($r = 0.80$, $p < 0.01$). The standardized partial regression coefficients for variables a_1, a_2, and a_3 in Eq. (7.12) were 0.61, 0.15, and 0.46, respectively. These coefficients were statistically significant ($p < 0.05$ for a_1, a_2, and a_3).

The results of the multiple regression analysis indicated that the effect of Φ(0) was the largest among the ACF factors. Even though the value of $L_{i,\mathrm{Fmax,Aw}}$ was fixed in the subjective tests, Φ(0) changed and affected the subjective evaluations. Different sound insulation treatments affected the sound reduction at different frequency ranges. It was found that Φ(0), which consider all ranges of frequencies,

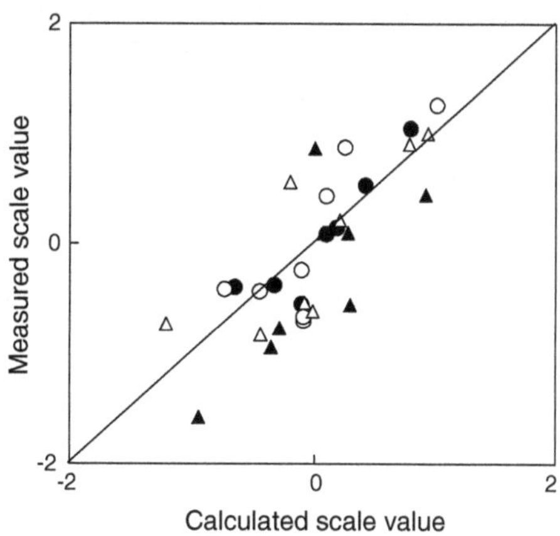

Fig. 7.21 Relationship between scale values obtained by annoyance judgments and scale values calculated by Eq. (7.12) for ACF factors. (*Filled circle* ball, 45 dB; *open circle* ball, 55 dB; *filled triangle* bang, 45 dB; and *open triangle* bang, 55 dB) (Jeon and Sato 2008)

was more suitable than $L_{i,\mathrm{Fmax,Aw}}$ for evaluating these effects. In addition, larger $\Phi(0)$ fluctuation, which corresponds to the fluctuation of loudness, resulted in higher annoyance.

Another finding obtained from the multiple regression analysis was how VAR_ϕ_1 affects annoyance. Pitch strength is represented by ϕ_1 (i.e., tonality). Variation in this value indicates the ratio of the initial and the tail parts of the impact sound. The coefficient a_3 for VAR_ϕ_1 in Eq. (7.12) is a negative value. Thus, a smaller variance resulted in higher annoyance. When the impact sound is not effectively reduced at the initial stage by insulation treatment, the difference between the initial and the tail parts of the impact sound becomes smaller. This means that annoyance cannot be decreased if sound properties in the initial part of the impact sound remain longer than in the tail part.

To investigate the effect of ACF factors on the annoyance of heavyweight floor impact sounds, subjective evaluations were conducted. Annoyance due to heavyweight floor impact sound was highly correlated with $\Phi(0)$, even though the SPL was fixed at a constant $L_{i,\mathrm{Fmax,Aw}}$; in addition to the noise level, ACF parameters are measures for calculating annoyance. The factors important for evaluating annoyance were $\Phi(0)$ and fluctuations in $\Phi(0)$ and ϕ_1.

7.8 Annoyance of Heating, Ventilation, and Air Conditioning (HVAC) Noise

Subjective annoyance of HVAC noises was evaluated to clarify the effects of ACF factors on the annoyance. Eight stimuli were selected from noise in various types of HVAC systems, that is, two cassette type (CT1 and CT2), five split type air-conditioners (ST1, ST2, ST3, ST4 and ST5), and one central air conditioning system (CA). The stimuli were presented binaurally through headphones (HD650, Sennheiser). Ten listeners aged 20 to 39 years with no history of hearing disorders participated in the experiment. They sat in a comfortable thermal environment in a soundproof room and heard the auditory stimuli. Paired-comparison tests were performed for all combinations of pairs [i.e., 28 pairs ($N(N-1)/2, N=8$)] of stimuli, interchanging the order that the stimuli in each pair were presented in each session and presenting the pairs in random order. Six sessions were conducted for each listener. All stimuli were presented at the same SPL (58 dB). The duration of the stimuli was 5.0 s, the rise and fall time was 100 ms, and the silent interval between stimuli was 1.0 s. After the presentation of the two stimuli, listeners were asked to judge which of the two sound signals was more annoying. The scale values of the annoyance were calculated according to Case V of Thurstone's theory (case V; Thurstone 1927).

The averaged scale values of the annoyance for the eight noises in train cars are presented in Fig. 7.22. The relationships between the scale value of annoyance and each ACF factor are presented in Fig. 7.23. Annoyance was found to increase with increasing $\Phi(0)$.

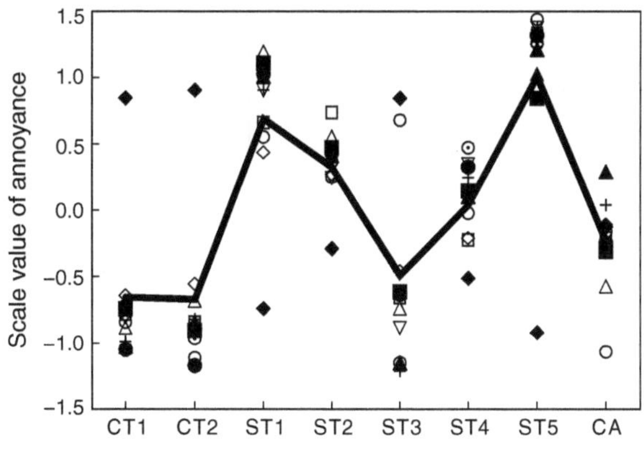

Fig. 7.22 The scale values of the annoyance for the eight HVAC noises. The *thick line* indicates averaged value for all listeners. Each symbol indicates each listener

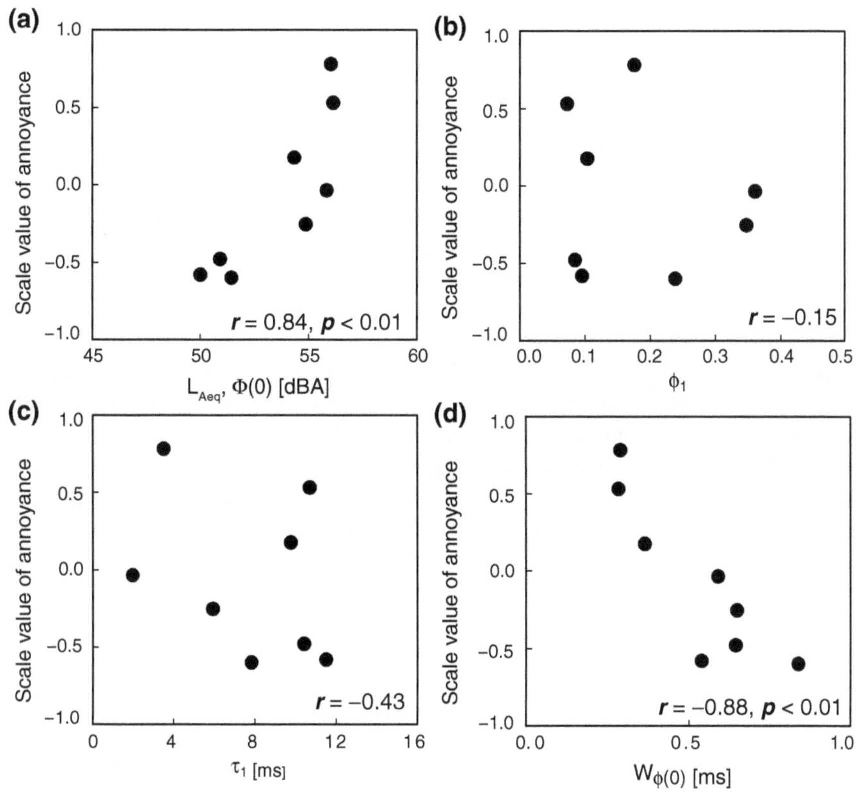

Fig. 7.23 The relationships between the scale value of annoyance and each ACF/IACF factor

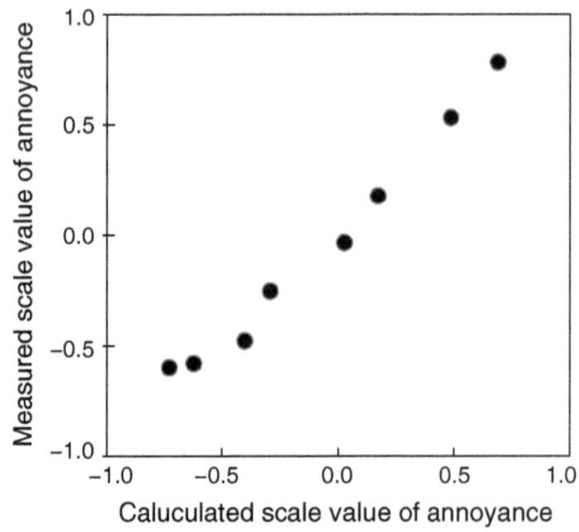

Fig. 7.24 The relationships between the measured and calculated annoyance for ACF/IACF factors

To calculate the effects of each objective factor on annoyance, multiple regression analyses were conducted using a linear combination of ACF factors and their variations by stepwise procedures. The final model indicated $\Phi(0)$ and $W_{\phi(0)}$ were significant factors:

$$\mathrm{SV}_{\text{annoyance}} \approx a_1 * \Phi(0) + a_2 * W_{\phi(0)} + c. \tag{7.13}$$

The relationships between the measured and calculated annoyance for different ACF factors are shown in Fig. 7.24. The correlation coefficient was 0.96 ($p < 0.01$). Standardized partial regression coefficients of each variable a_1 and a_2 in Eq. (7.13) were 0.50 and −0.58, respectively. $W_{\phi(0)}$ has the greatest effect on subjective annoyance. Perceived annoyance increased with decreasing $W_{\phi(0)}$ and increasing $\Phi(0)$.

In a similar way, scale values of annoyance for each listener were predicted by multiple regression analyses using a linear combination of ACF factors and their variations by stepwise procedure. The significant factors, the standardized regression coefficients, and the correlation coefficients between measure and calculated scale value of preference are listed in Table 7.3. $W_{\phi(0)}$ had the significant factor on subjective annoyance except for a listener. Negative correlation between annoyance and $W_{\phi(0)}$ suggests that lower frequency components of HVAC noises cause greater annoyance.

7.9 A General Equation for Annoyance

So far, how the central auditory signal processing model can be applied for some specific applications describing for annoyance. We are now these results put into a single equation based on the temporal and spatial factors for describing annoyance for a general case.

Table 7.3 Individual and total results for significant factors, the standardized regression coefficients, and the correlation coefficients between measure and calculated scale value of annoyance for HVAC noises

Listener	Factor 1	Factor 2	Factor 3	r
A	$W_{\Phi(0)}$ −0.42	$\Phi(0)$ 0.62	VAR_ϕ_1 −0.23.	0.99
B	$\Phi(0)$ 0.57	$W_{\Phi(0)}$ −0.51		0.97
C	$\Phi(0)$ 0.98	VAR_τ_e −0.32		0.98
D	$\Phi(0)$ −0.56	$W_{\Phi(0)}$ 0.44	τ_1 0.23	0.99
E	$\Phi(0)$ 0.63	$W_{\Phi(0)}$ −0.44		0.96
F	$W_{\Phi(0)}$ −0.86			0.86
G	$W_{\Phi(0)}$ −0.96	VAR_$\Phi(0)$ 0.38		0.95
H	$W_{\Phi(0)}$ −0.85	τ_1 −0.45		0.94
I	$W_{\Phi(0)}$ −0.74			0.74
J	$\Phi(0)$ 0.56	$W_{\Phi(0)}$ −0.52		0.96
K	$\Phi(0)$ 0.66	$W_{\Phi(0)}$ −0.42		0.97
Total	$W_{\Phi(0)}$ −0.58	$\Phi(0)$ 0.50		0.96

According to Figs. 2.22 and 7.25, a general equation describing for annoyance scale value ($SV_{annoyance}$) is given by the left hemisphere evaluations SV_{left} and the right hemisphere evaluations SV_{right}

$$SV_{annoyance} \approx SV_{left} + SV_{right} \tag{7.14}$$

where

$$\begin{aligned} SV_{left} = {} & aW_{\phi(0)} + b\tau_1 + c\phi_1 + d\tau_e + e\Delta(W_{\phi(0)})/\Delta t \\ & + f\Delta(\tau_1)/\Delta t + g\Delta(\phi_1)/\Delta t + h\Delta(\tau_e)/\Delta t \end{aligned} \tag{7.15}$$

$$\begin{aligned} SV_{right} = {} & i\mathrm{LL} + j\mathrm{IACC} + kW_{IACC} + l\tau_{IACC} + m\Delta(\mathrm{LL})/\Delta t \\ & + n\Delta(\mathrm{IACC})/\Delta t + o\Delta(W_{IACC})/\Delta t + p\Delta(\tau_{IACC})/\Delta t. \end{aligned} \tag{7.16}$$

LL is the geometrical mean of the sound energies arriving at the two ears, $\Phi_{ll}(0)$ and $\Phi_{rr}(0)$, corresponds to the binaural listening level, and Δ signifies the running

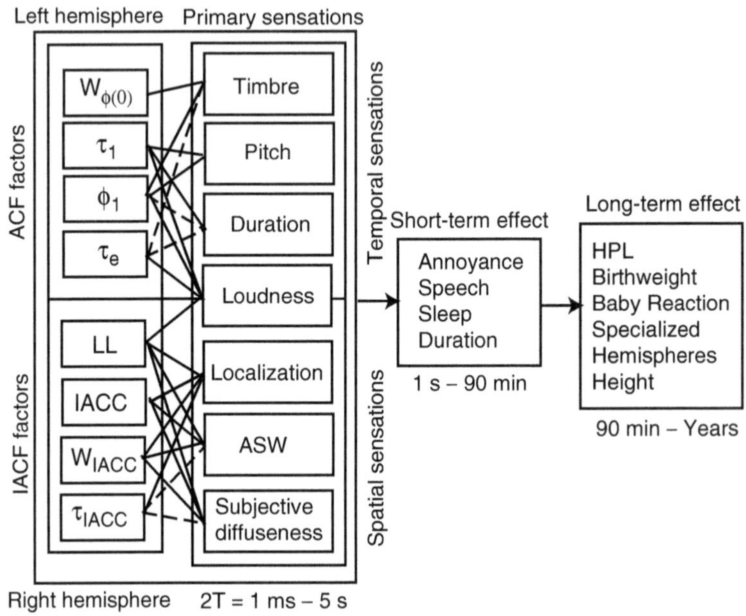

Fig. 7.25 General relationships between the ACF/IACF factors and primary sensations and their short-term and long-term effects on human

interval. Δ(ACF/IACF factor)/Δt means the rate of change. Therefore, a general single graph indicating the calculated values and measured or evaluated the scale values of annoyance may be given.

The previous sections indicate that the significant factors for annoyances are dependent on the noise source and the listener. By accumulating the data, such as just noticeable difference of ACF/IACF factors and its variations, the equation can be simplified based on the characteristics of noise sources and listeners.

References

Ando Y (1998) Architectural acoustics: blending sound sources, sound fields, and listeners. AIP Press Springer, New York

Berglund B, Berglund U, Lindvall T (1975) Scaling loudness, noisiness, and annoyance of aircraft noise. J Acoust Soc Am 57:930–934

Berglund B, Hassmen P, Job RFS (1996) Sources and effects of low-frequency noise. J Acoust Soc Am 99:2985–3002

Bhattachaya SK (1994) Assessment of noise environment in a major railway station in India. Ind Health 32:187–192

Cermak GW, Cornillon PC (1976) Multidimensional analysis of judgments about traffic noise. J Acoust Soc Am 59:1412–1420

Chandler DW, Grantham DW (1992) Minimum audible movement angle in the horizontal plane as a function of stimulus frequency and bandwidth, source azimuth, and velocity. J Acoust Soc Am 91:1624–1636

Chang HC, Hermann ER (1974) Acoustical study of a rapid train system. Am Ind Hyg Assoc J 35:640–653

Eade PW, Hardy AEJ (1977) Railway vehicle internal noise. J Sound Vib 57:403–415

Endoh H, Suzuki H, Izumi Y, Hio Y (2007) Appropriate sound level of broadcasting on station platforms under noisy environment. RTRI REPORT 21:41–46 (In Japanese)

Fujii K, Atagi J, Ando Y (2002) Temporal and spatial factors of traffic noise and its annoyance. J Temporal Des Arch Environ 2:33–41

Fujii K, Soeta Y, Ando Y (2001) Acoustical properties of aircraft noise measured by temporal and spatial factors. J Sound Vib 241:69–78

Gabriel KJ, Colburn HS (1981) Interaural correlation discrimination: I. Bandwidth and level dependence. J Acoust Soc Am 69:1394–1401

Gershon RRM, Qureshi KA, Barrera MA, Erwin MJ, Goldsmith F (2005) Health and safety hazards associated with subways: a review. J Urban Health 82:10–20

Goulet P, Northwood TD (1973) Subjective rating of broadband noises containing pure tones. J Acoust Soc Am 53:365

Grantham DW (1986) Detection and discrimination of simulated motion of auditory targets in the horizontal plane. J Acoust Soc Am 79:1939–1949

Greenwood DD (1961a) Auditory masking and the critical band. J Acoust Soc Am 33:484–502

Greenwood DD (1961b) Critical bandwidth and the frequency of the basilar membrane. J Acoust Soc Am 33:1344–1356

Hardy AE (2000) Measurement and assessment of noise within passenger trains. J Sound Vib 231:819–829

Hargest TJ, Pinker RA (1967) The influence of added narrow band noises and tones on the subjective response to shaped white noise. J Acoust Soc Am 71:428–430

Hellman RP (1982) Loudness, annoyance and noisiness produced by single-tone-noise complexes. J Acoust Soc Am 72:62–73

Hellman RP (1984) Growth rate of loudness, annoyance, and noisiness as a function of tone location within the noise spectrum. J Acoust Soc Am 75:209–218

Hiramatsu K, Yamanaka K, Takagi K, Yamamoto T (1978) Annoyance of fluctuating noise (Effects of standard deviation and fluctuation frequency of sound level). J Acoust Soc Jpn 34:376–386 (in Japanese)

Hiramatsu K, Takagi K, Yamamoto T (1983) Experimental investigation on the effect of some temporal factors of nonsteady noise on annoyance. J Acoust Soc Am 74:1782–1793

Isumi Y, Fujii K (2007) An investigation into the actual condition and subjective evaluation test of acoustical environment in railway station. RTRI REPORT 21:4147–5246 (In Japanese)

Jeon JY (2001) Subjective evaluation of floor impact noise based on the model of ACF/IACF. J Sound Vib 241:147–155

Jeon JY, Sato S (2008) Annoyance caused by heavy-weight floor impact sounds in relation to the autocorrelation function and sound quality metrics. J Sound Vib 311:767–785

Jeon JY, Jeong JH, Vorländer M, Thaden R (2004) Evaluation of floor impact sound insulation in reinforced concrete buildings. Acta Acust United AC 90:313–318

Jeon JY, Lee PJ, Kim JH, Yoo SY (2009) Subjective evaluation of heavy-weight floor impact sounds in relation to spatial characteristics. J Acoust Soc Am 125:2987–2994

Kitamura T, Sato S, Shimokura R, Ando Y (2002) Measurement of temporal and spatial factors of a flushing toilet noise in a downstairs bedroom. J Temporal Des Arch Environ 2:13–19

Klumpp RG, Eady HR (1956) Some measurements of interaural time-difference thresholds. J Acoust Soc Am 28:859–860

Kono S, Sone T, Nimura T (1982) Personal reaction to daily noise exposure. Noise Control Eng J 19:4–16

Kryter KD, Pearsons KS (1965) Judged noisiness of a band of random noise containing an audible pure tone. J Acoust Soc Am 38:106–112

Kuwano S, Namba S, Fastl H (1988) On the judgment of loudness, noisiness and annoyance with actual and artificial noise. J Sound Vib 127:457–465
Little JW, Mabry JE (1969) Empirical comparisons of calculation procedures for estimating annoyance of jet aircraft flyovers. J Sound Vib 10:59–68
May DN (1978) Basic subjective responses to noise. In: May DN (ed) Handbook of noise assessment. Van Nostrand Reinhold C, New York, pp 313–335
Mosteller F (1951) Remarks on the method of paired comparisons III. Psychometrika 16:207–218
Mouri K, Akiyama K, Ando Y (2001) Preliminary study on recommended time duration of source signals to be analyzed, in relation to its effective duration of the auto-correlation function. J Sound Vib 241:87–95
Pearsons KS, Bennett RL (1971) Effects of temporal and spectral combinations on the judged noisiness of aircraft sounds. J Acoust Soc Am 49:1076–1082
Sakai H, Sato S, Prodi N (2004) Orthogonal factors for the stage and pit inside a historical opera house. Acta Acust United AC 90:319–334
Sato S, Mori Y, Ando Y (1997) The subjective evaluations of source locations on the stage by listeners. In: Ando Y, Noson D (eds) Music and concert hall acoustics. Academic Press, London
Sato S, Kitamura T, Ando Y (2002) Loudness of sharply (2068 dB/Octave) filtered noises in relation to the factors extracted from the autocorrelation function. J Sound Vib 250:47–52
Sato S, Kitamura T, Ando Y (2004) Annoyance of noise stimuli in relation to the spatial factors extracted from the interaural cross-correlation function. J Sound Vib 277:511–521
Scharf B (1962) Loudness summation and spectrum shape. J Acoust Soc Am 34:228–233
Shimokura R, Soeta Y (2011) Characteristics of train noise in ground and underground stations with side and island platforms. J Sound Vib 330:1621–1633
Soeta Y, Shimokura R (2009) Comparison of noise characteristics in airplanes and high-speed trains. J Temporal Des Arch Environ 9:22–25
Soeta Y, Shimokura R (2011) The impact of external environments on noise inside a train car. Noise Control Eng J 59:581–590
Soeta Y, Shimokura R (2013a) Acoustic characteristics of noise in train stations. In: Reinhardt C, Shroeder K (eds) Railways: types, design and safety issues. Nova Science Pub Inc, New York, pp 1–36
Soeta Y, Shimokura R (2013b) Survey of interior noise characteristics in various types of trains. Appl Acoust 74:1160–1166
Soeta Y, Maruo T, Ando Y (2004) Annoyance of bandpass filtered noises in relation to the factor extracted from autocorrelation function. J Acoust Soc Am 116:3275–3278
Thurstone LL (1927) A law of comparative judgement. Psychol Rev 34:273–289
Versfeld NJ, Vos J (1997) Annoyance caused by sounds of wheeled and tracked vehicles. J Acoust Soc Am 101:2677–2685
Yoshihisa N, Fukai M, Yamaguchi T (1968) Noise levels inside subway cars in Western Europe, USA and Japan. J Acoust Soc Jpn 24:69–75 (in Japanese)
Yost WA (2000) Fundamentals of hearing: an introduction. Academic Press, San Diego
Zwicker E, Fastl H (1999) Psychoacoustics: facts and models. Springer, Berlin
Zwicker E, Terhardt E (1980) Analytical expressions for critical-band rate and critical bandwidth as a function of frequency. J Acoust Soc Am 68:1523–1525
Zwicker E, Flottorp G, Stevens SS (1957) Critical bandwidth in loudness summation. J Acoust Soc Am 29:548–557

Chapter 8
Short-Term Effects of Noise

Abstract In Chap. 7, it has been discussed annoyance of noise in relation to the temporal and spatial factors extracted from the running autocorrelation function (ACF) and interaural cross-correlation function (IACF) of the sound field, respectively. In this chapter, instantaneous effects of environmental noise on speech disturbance by different directional noise, sleep disturbance due to spatial factors, and model duration experience are discussed. First, we shall show that speech disturbance from different directional noise is well described by the temporal and spatial factors, respectively. Second, sleep disturbance of weak flushing toilet noise from upstairs changing spatial factors is discussed. Third, after proposing a neural model, duration experience of aircraft noise for people from a quiet living area and a noisy living area is explained.

Keywords Speech disturbance · Sleep disturbance · Duration experience

8.1 Speech Disturbance by Different Directional Noise

In the hearingexperiment, fourteen single syllables, /pa/ /pu/ /te/ /zo/ /bo/ /yo/ /mi/ /ne/ /kya/ /kyo/ /pya/ /gya/ /nya/ /zya/, with 4 s intervals between syllables were presented to each participant by the frontal loudspeaker ($\xi = 0°$, the distance to the center of the listener's head, $d = 70 \pm 1$ cm) in an anechoic chamber. The white noise used as a disturbance was continuously produced by one of loudspeakers located at different horizontal angles: $\xi = 30°$, 60°, 90°, 120°, or 180° ($d = 70$ cm). The sound pressure level (SPL) of both speech signals and the continuous white noise were fixed 65.0 dBA at the peak level. Ten listeners participated in the experiment, which was asked what syllable was heard.

According to the model shown in Fig. 2.22, three temporal factors associated with the left hemisphere together with the SPL were extracted from the autocorrelation function (ACF) of the sound signal arriving at one of the ear entrances. In addition, three spatial factors associated with the right hemisphere were extracted

Y. Soeta and Y. Ando, *Neurally Based Measurement and Evaluation of Environmental Noise*, Mathematics for Industry 20,
DOI 10.1007/978-4-431-55432-5_8

from the interaural cross-correlation function (IACF) of sound signals arriving at the two ear entrances. The running ACF and the running IACF with the integration interval $2T = 30$ ms with the running step of 10 ms were analyzed.

For the identification of the speech signals itself, the psychological distance between characteristics of single syllables due to each of the four factors extracted from the ACF, which are associated with the left cerebral hemisphere excluding the sound energy, is calculated by

$$D_{\tau_e}(X,K) = \left[\sum_{i=1}^{I} \left|\log \tau_e^i\cdot\left(\begin{smallmatrix}\mathrm{SF}\\X\end{smallmatrix}\right) - \log \tau_e^i\cdot\left(\begin{smallmatrix}T\\K\end{smallmatrix}\right)\right|\right] \Big/ I$$

$$D_{\phi_1}(X,K) = \left[\sum_{i=1}^{I} \left|\log \phi_1^i\left(\begin{smallmatrix}\mathrm{SF}\\X\end{smallmatrix}\right). - \log \phi_1^i\left(\begin{smallmatrix}T\\K\end{smallmatrix}\right).\right|\right] \Big/ I$$

$$D_{\tau_1}(X,K) = \left[\sum_{i=1}^{I} \left|\log \tau_1^i\left(\begin{smallmatrix}\mathrm{SF}\\X\end{smallmatrix}\right). - \log \tau_1^i\left(\begin{smallmatrix}T\\K\end{smallmatrix}\right).\right|\right] \Big/ I$$

$$D_{\Phi(0)}(X,K) = \left[\sum_{i=1}^{I} \left|\log \Phi(0)^i.\left(\begin{smallmatrix}\mathrm{SF}\\X\end{smallmatrix}\right)/\Phi(0)^{\max}\left(\begin{smallmatrix}\mathrm{SF}\\X\end{smallmatrix}\right) - \log \Phi(0)^i.\left(\begin{smallmatrix}T\\K\end{smallmatrix}\right)/\Phi(0)^{\max}\left(\begin{smallmatrix}T\\K\end{smallmatrix}\right)\right|\right] \Big/ I. \tag{8.1}$$

where K and X represent the syllable number of template (T) and the syllable in a sound field (SF), respectively. Also, i is the frame number of the running ACF and I is the total frame number.

In addition, finding effects of the different directional noise, three spatial factors are extracted from the IACF, which are associated with the right cerebral hemisphere. The distances due to the spatial factors, D_{IACC}, $D\tau_{\mathrm{IACC}}$, and $D_{W_{\mathrm{IACC}}}$, respectively, are given by

$$\begin{aligned} D_{\mathrm{IACC}}(X,K) &= \left[\sum_{i=1}^{I} \left|\mathrm{IACC}^i\left(\begin{smallmatrix}\mathrm{SF}\\X\end{smallmatrix}\right) - \mathrm{IACC}^i\left(\begin{smallmatrix}T\\K\end{smallmatrix}\right)\right|\right] \Big/ I \\ D_{\tau_{\mathrm{IACC}}}(X,K) &= \left[\sum_{i=1}^{I} \left|\tau_{\mathrm{IACC}}^i\left(\begin{smallmatrix}\mathrm{SF}\\X\end{smallmatrix}\right) - \tau_{\mathrm{IACC}}^i\left(\begin{smallmatrix}T\\K\end{smallmatrix}\right)\right|\right] \Big/ I \\ D_{W_{\mathrm{IACC}}}(X,K) &= \left[\sum_{i=1}^{I} \left|W_{\mathrm{IACC}}^i\left(\begin{smallmatrix}\mathrm{SF}\\X\end{smallmatrix}\right) - W_{\mathrm{IACC}}^i\left(\begin{smallmatrix}T\\K\end{smallmatrix}\right)\right|\right] \Big/ I. \end{aligned} \tag{8.2}$$

In general, the shorter distance between the template syllable and the syllables with noise disturbance signifies the higher intelligibility. According to the multiple regression analysis, the nonidentification (NI) rate of the syllable, which was not matched with the template, has been directly calculated, so that

$$\mathrm{NI}(S_0, S_X) = S_L + S_R = [aD\tau_e + bD\tau_1 + cD\phi_1]_L + [dD\Phi(0) + eD_{IACC} + fD\tau_{IACC} + gD_{W_{IACC}}]_R \quad (8.3)$$

where $S_L = [aD\tau_e + bD\tau_1 + cD\phi_1]_L$, $S_R = [dD\Phi(0) + eD_{IACC} + fD\tau_{IACC} + g\,D_{W_{IACC}}]_R$, and $\Phi(0)$ is measured in (dBA). Note that the listening level or $\Phi(0)$ is associated with the right hemisphere (Table 8.1). Weighting coefficients a through g in Eq. (8.3) were determined by maximizing NI with experimental data.

For example, values of τ_e extracted from the running ACF for the signal /mi/ with and without the noise ($\xi = 90°$) as a function of time are shown in Fig. 8.1. The important initial half parts of the speech signal indicating $\Phi(0) > 0.5$ as shown in Fig. 8.2 of both template and test syllables with the noise were applied in computation by Eqs. (8.1) to (8.3).

Results of the NI rate for some single syllables as a function of the horizontal angle ξ of the noise disturbance are shown in Fig. 8.3. Almost similar tendencies, NI of these syllables was found. When the noise arrived from 30°, the NI indicated the maxima in the horizontal angle range tested, and when the noise was presented from 120°, it was the minima. The same was true for the averaged NI rate as shown in Fig. 8.4.

Since the direct speech sound arrived from the frontal direction to the listener, the value of τ_{IACC} is always close to zero being an invariable. Thus, this factor was eliminated from the analysis by Eq. (8.3) (Table 8.1). The minima of the psychological distance were always found for the noise disturbance from 120°, so that the

Table 8.1 The distance calculated due to each of six factors

Horizontal angle of noise	$D_{\Phi(0)}$	$D\tau_e$	$D\tau_1$	$D\phi_1$	D_{IACC}	DW_{IACC}
30°	0.064	0.420	0.164	0.442	0.248	0.052
60°	0.056	0.351	0.247	0.355	0.266	0.049
90°	0.063	0.348	0.162	0.401	0.292	0.049
120°	0.058	0.279	0.157	0.376	0.270	0.043
180°	0.074	0.383	0.171	0.494	0.247	0.071

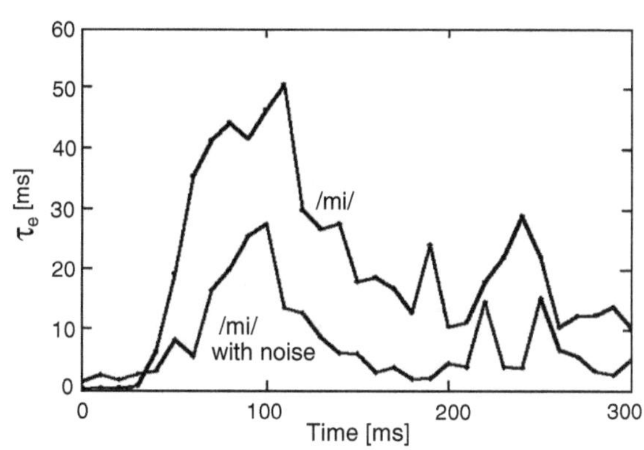

Fig. 8.1 Values of effective duration τ_e extracted from the running ACF for the frontal signal /mi/ only, and the /mi/ with the white noise from $\xi = 90°$

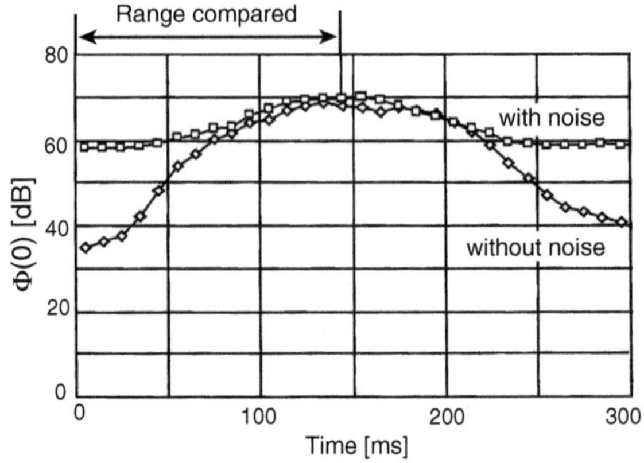

Fig. 8.2 For making comparison, initial pieces analyzed of a frontal single syllable with and without the white noise from $\xi = 90°$

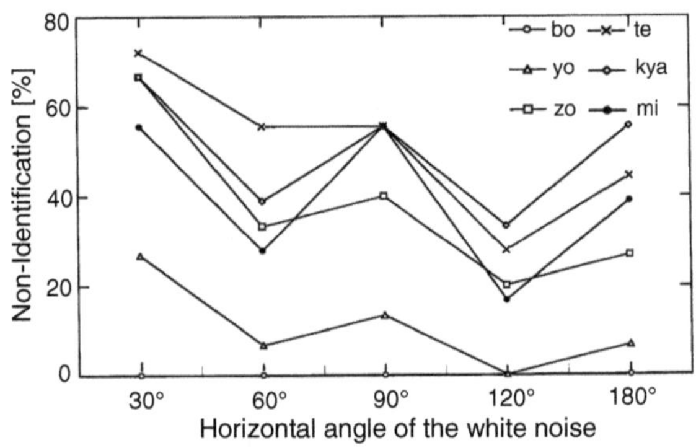

Fig. 8.3 Examples the percentage of nonidentification for single syllables as a function of the horizontal angle of the white noise from different horizontal angles ξ. At the horizontal angles $\xi = 120°$, the percentage of nonidentification was minimum for the single syllables

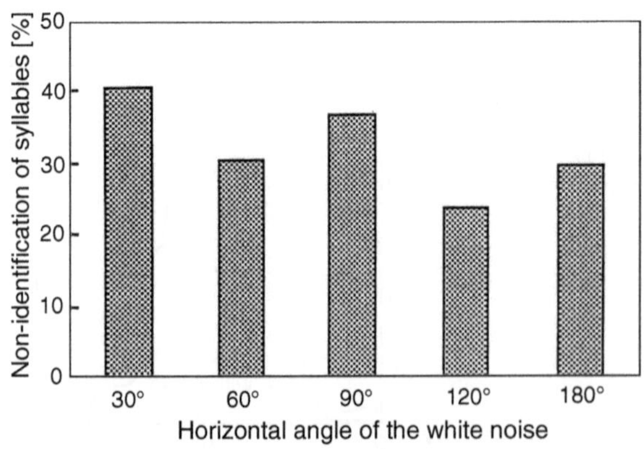

Fig. 8.4 Averaged percentile of nonidentified with all single syllables tested obtained by the listening test for different angles ξ of the white noise incidence as a disturbance

Table 8.2 Weighting coefficients determined

	$\Phi(0)$	τ_e	τ_1	ϕ_1	IACC	W_{IACC}
Coefficient	0.053	0.335	0.028	0.136	0.086	0.384

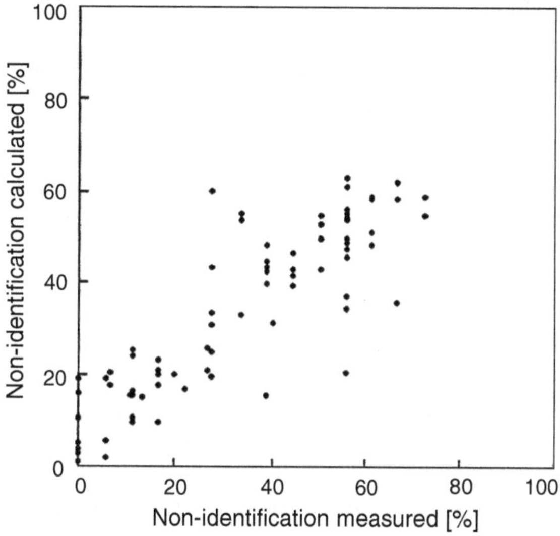

Fig. 8.5 Relationship between the calculated percentile of nonidentified syllables by Eq. (8.3) and that obtained by listening tests ($r = 0.86$, $p < 0.01$)

NIs were minima. On the other hand, when the noise disturbance arrived from 30°, the distance due to τ_e for all of the syllables commonly indicated the maxima in six factors.

The weighting coefficients in Eq. (8.3) for the six factors are listed in Table 8.2. According to the weighting coefficients obtained here, the factors of τ_e and W_{IACC} contributed significantly to the NI. For each single syllable, the relationship between the calculated values by Eq. (8.3) and the measured values is shown in Fig. 8.5. Obviously, the linear relationship was achieved ($r = 0.86$, $p < 0.01$). If a new possible factor $\Delta\phi_1/\Delta t$ (Ando 2015) is taken into account to Eq. (8.3), then much more reliable results could be obtained, because these factors are significant for speech recognition.

In this study, the most effective and significant temporal factor was the τ_e value in the temporal factors. In order to obtain effects of the different direction of the noise disturbance, the spatial factors may be taken into consideration. Conclusions are as follows:

1. The syllable identification (SI = 1 – NI) in the sound field may be calculated by both temporal factors extracted from the ACF and spatial factors extracted from the IACF.

2. Particularly in the condition of this experiment, the value of τ_e as the temporal factor is the most significant as is similar to previous results (Ando et al. 1999a), in addition the W_{IACC} in the spatial factor contributes significantly to the speech identification.

8.2 Sleep Disturbance by Upstairs Toilet Noise Changing Spatial Factors

The purpose of this study is to identify factors for sleep disturbance by the flushing noise of an upstairs toilet, which despite having a low SPL <36 dBA at the peak. We analyzed the temporal and spatial factors of the flushing noise from an upstairs toilet at the head position on bed, because the resident was very annoyed and disturbed while sleeping. The noise signals were picked up by 1/2-in. condenser microphones placed at two ear entrances of the spherical dummy head placed on the bed.

The measurement time the noise from upstairs flushing toilet noise was 5 s. The values of all factors were obtained with the integration interval of 0.5 s and the running interval of 100 ms. As shown in Fig. 8.6, the SPL was 29–35 dBA measured by $\Phi_{ll}(0)$ extracted from running ACF, and the background noise was less than about 23 dBA (Kitamura et al. 2002). Thus, the maximum signal-to-noise ratio was 12 dBA. If the difference between the background noise level and the noise signal level is greater than 10 dB, the background noise does not affect significantly the noise signal measurement (Beranek 1971). As shown in Fig. 8.6, the τ_e value for the flushing toilet noise exceeded 100 ms with $\phi_1 > 0.5$, while $\tau_e < 0.1$ ms and $\phi_1 < 0.01$ of the background noise throughout the measurement time. Thus, the flushing toilet noise had much more repetitive features than the background noise. It has been reported that loudness increases in proportion to the value τ_e (Merthayasa and Ando 1996). The value of τ_e for the flushing toilet noise was the largest near the peak of $\Phi_{ll}(0)$, as shown in Fig. 8.6. The value of τ_1 for the flushing toilet noise had a discrete value at 3.6 ms, which means that the perceived pitch was 275 Hz. The background noise did not have any clear pitch and tonal components, similar to the white noise.

In addition, the measured special factors, extracted from the running IACF, are shown in Fig. 8.7a–c. The IACC value for the flushing toilet noise was much higher than that for the background noise (below 0.05) in the measurement time. Thus, no specific directions may be perceived for the background noise. On the contrary, all spatial factors, the IACC, τ_{IACC}, and W_{IACC}, of the flushing noise changed dramatically as a function of time. These results signify that subjective diffuseness, localization of sound source, and the apparent source width (ASW) of the flushing toilet noise, greatly changed as a function of time. Particularly, values of τ_{IACC} were varied from −0.6 to +0.6 that signify the noise perceived as rotating around the

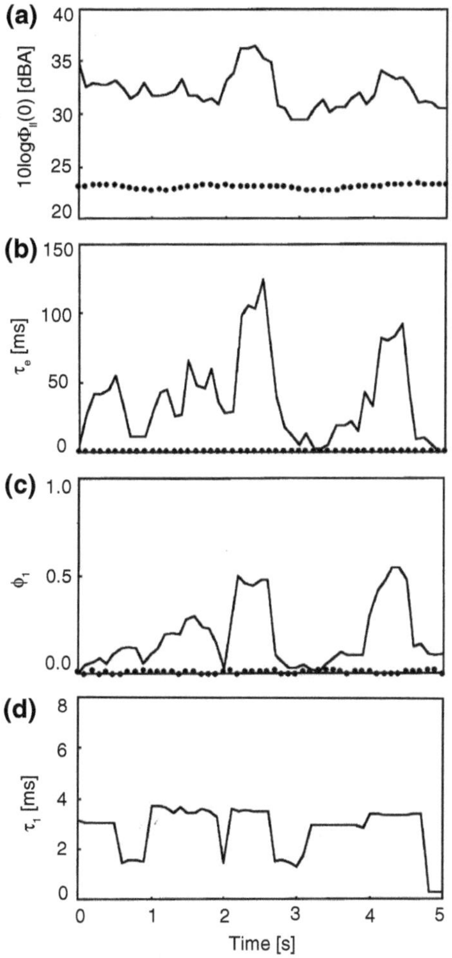

Fig. 8.6 Measured factors extracted from the running ACF. *Solid line* indicates values for the flushing toilet noise and *dotted line* indicates the background noise. These values of factors were obtained every 100 ms with an integration interval of 0.5 s. **a** $\Phi_{ll}(0)$, **b** τ_e, **c** ϕ_1, and **d** τ_1. The maximum level of the flushing toilet noise was about 36 dBA

head of sleeper. This phenomenon could be caused by the hexagon of the sleeping room.

So far, it was discussed that the temporal and spatial factors extracted from the ACF and IACF of the flushing toilet noise had specific characteristics. These facts imply that both temporal and spatial sensations of the flushing toilet noise changed dramatically. According to the model of auditory-brain system, the temporal information is mainly processed in the left hemisphere and the spatial information is mainly processed in the right hemisphere (Ando 1998). Thus, flushing noise from an upstairs toilet may stimulate both the left and right hemispheres of this resident at the same time. This may explain why the resident felt that the flushing noise of an upstairs toilet was very annoying and disturbing while sleeping despite its low SPL.

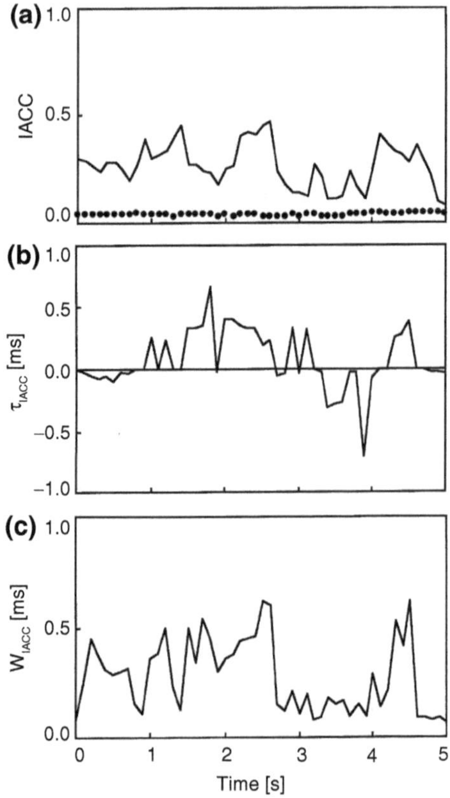

Fig. 8.7 Measured factors extracted from the running IACF. *Solid line* indicates values for the flushing toilet noise and *dotted line* indicates the background noise. The values of all factors were obtained every 100 ms with an integration interval of 0.5 s. **a** IACC, **b** τ_{IACC}, and **c** W_{IACC}. Remarkably, τ_{IACC} dynamically changed from +0.6 to −0.7, and the noise was perceived as round and round around the head

8.3 Model of Duration Experience Due to Noise

The present study is concerned with duration experience regarding time in mind. We previously investigated unconscious duration experience of university students covering three-year periods of their school lives from when they were 3–18 years old (Ando et al. 1999b). Results showed that in regard to duration experience for elementary school periods (below 12 years old), students reported time passed slowly (the mean value was about 1.2 times), in reference to that for the three years (13–15 years of age) of their junior high school ($p < 0.01$). In addition, study shows that students of senior high school periods (16–18 years old) felt the time was shorter (the mean value was about 0.8 times), than that spent by their junior high school periods ($p < 0.01$). These results imply that time seems to pass faster as the age of students increases.

To understand such an "unconscious" duration experience conducted without notifying the purpose of investigation to participants, a model of duration experience is proposed here as shown in Fig. 8.8. Endogenous clock oscillators with characteristic periods can be synchronized with, for example, brain waves, pulsation, and breathing rate inside the body in addition to daily, monthly, and annual

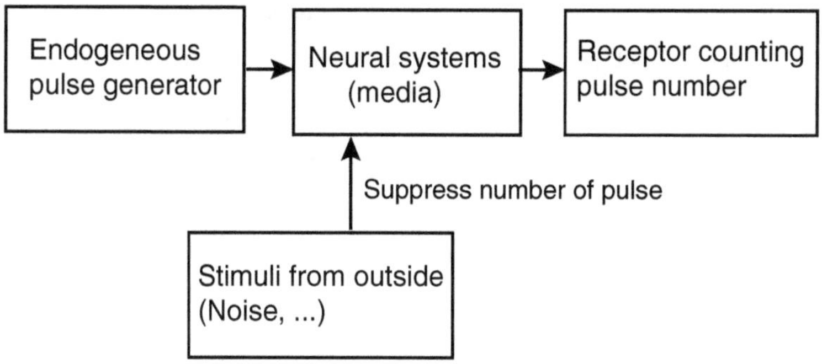

Fig. 8.8 A model of duration experience consisting of endogenous oscillators, media, and a receptor in the brain. According to the model of an auditory-brain system and specialization of cerebral hemispheres (Ando 1998; Ando and Cariani 2009), it is acceptable that the receptor exists in the left hemisphere (Polzella et al. 1977). External stimuli may act as a suppressor of a number of clock pulses from the oscillators, so that the receptor may get less clock pulses and person may feel a short duration (Luce 1971; Ando 1977)

cyclical changes in the external environment (Luce 1971). That is to say morning light may reset a circadian oscillator, and a long-term temperature change may reset a seasonal oscillator. We assume that clock signals from oscillators propagate through a system (media) to a receptor in the brain. Activities of the receptor in the brain depend on a person being awake or being sleep. In the awake state, there are two levels, i.e., "conscious" or "unconscious." In the conscious stage, a person always notices time passing; in the unconscious state, they do not notice time passing. When a person is in a deep sleep, for example, their receptor gets less endogenous clock signals and they may feel time passes faster unconsciously than it does, while they are awake. Much has been conducted for conscious duration experience for a long time; for example, asking participants counting for 60 at the speed they thought were about one per second in total 60 s.

A question arises whether or not environmental noise influences an unconscious duration experience. There has only been one study on the effects of sound on an unconscious duration experience (Ando 1977). Most of the listeners showed that time can be accelerated by noise around the level of 75 dBA, feeling that time passes faster than in silence. When the level increased to 90 dBA, then many listeners had an increase in the duration experience. The experiments on unconscious impressions of duration during jet plane noise presentations to junior high school pupils (13–14 years of age) from quiet and noisy residential areas are reviewed here in detail.

To ensure that listeners provided information on the unconscious duration experience in regard to noise, a masker investigation (in which their urine was collected) to determine effects of noise on internal secretion was performed in parallel. In the first hour of the experiment, picture slides of a world tour were shown in a classroom to make the listeners relax. During the following hour, cycles composed of 60 s of jet plane noise followed by 60 s of silence were repeated 30 times.

Table 8.3 Percentage of listeners who rated the duration of noise and silence

Group	Living area	SPL at the peak (dBA)	Number of listeners	R_{50}	Null hypothesis R_{50}	P
A	Noisy	~75	20	0.4	=1.0	<0.05
B	Quiet	~75	28	0.5	=1.0	<0.05
C	Noisy	~90	20	0.7	=1.0	<0.05
D	Quiet	~90	25	0.9	=1.0	–

As listed in Table 8.3, the peak SPL in all the classrooms of the junior high school in the study was 75 ± 2 dBA (group A: 20 participants; group B: 28 participants) and 90 ± 2 dBA (group C: 20 listeners; group D: 25 listeners). Group A and group C listeners lived in a noisy residential area, and group B and group D listeners lived in a quiet residential area. During the hour of noise–silence cycles, the listeners were permitted to look at picture books. For the last hour, the slide show was presented again to make the listeners relax again. Afterward, the listeners were given a questionnaire asking for the ratio of the subjective duration of the period of noise to that of period silence. That is, they were asked "how much longer was the noise period than silence period?" The question was carefully explained so that all listeners understood the question and could answer it meaningfully, choosing a ratio from 0.1, 0.2, 0.3, … 5.0. Note that the objective ratio of time is unity, i.e., 60/60 s.

A notable result is that about 79 % of group A and group B (75 ± 2 dBA) rated the subjective duration of the noise period shorter than that of the silent period ($p < 0.01$; the null hypothesis: 50 %). The rated value at 50 % of listeners (R50) of group A and group B, respectively, is 0.4 and 0.5 ($p < 0.05$). About 80 % of listeners of group C (90 ± 2 dBA) (from a noisy residential area) rated the subjective duration of the noise period shorter than the silent period ($p < 0.05$). However, only about 52 % of listeners of group D (from a quiet residential area) rated it shorter than the silent period. This result may be explained as "conscious" duration due to the fact that listeners from the quiet residential area perceived as much more louder at about 90 dBA than did listeners from the noisy area. In the noisy living area, listeners' receptor is subject to get fewer number of pulses due to the long-term noise, and duration is shorter than those in quiet area.

If such a chronic effect is true for people living in noisy area of 70 dBA or its below, the long-term duration could be about half of the time being, for example, the span of life 80 years will be only 40 years.

References

Ando Y (1977) Effects of noise on duration experience. J Sound Vib 55:600–603

Ando Y (1998) Architectural acoustics: blending sound sources, sound fields, and listeners. AIP Press, Springer, New York

Ando Y (2015) Autocorrelation-based features for speech representation. Acta Acust United AC 101:145–154

Ando Y, Cariani P (eds) (2009) Auditory and visual sensations. Springer, New York

Ando Y, Sato S, Sakai H (1999a) Fundamental subjective attributes of sound fields based on the model of auditory brain system. In: Sendra JJ (ed) Computational acoustics in architecture. WIT Press, Southampton, pp 63–99

Ando Y, Tsuruta H, Motokawa A, Matsushima T, Saifuddin K (1999b) Subjective duration of every three-year period for 3 to 18 years of age, estimated by students. J Hum Ergol 28:33–37

Beranek LL (1971) Noise and vibration control. McGraw-Hill Inc., New York

Kitamura T, Sato S, Shimokura R, Ando Y (2002) Measurement of temporal and spatial factors of a flushing toilet noise in a downstairs bedroom. J Temporal Des Arch Environ 2:13–19

Luce GG (1971) Biological rhythms in human and animal physiology. Dover Publications, New York

Merthayasa IN, Ando Y (1996) Variation in the autocorrelation function of narrow band noises; their effect on loudness judgment. In: Proceedings of the 3rd Japanese-Swedish noise symposium on medical effects

Polzella DJ, DaPolito F, Hinsman MC (1977). Cerebral asymmetry in time perception. Percept Psychophys 21:187–192

Chapter 9
Long-Term Effects of Noise

Abstract The brain vividly responses to instantaneous changes of the environment with the period less than several seconds, because the auditory temporal window is usually less than 5 s. Therefore, annoyance and loudness are responded in such a short time and much have been investigated. However, a long-term temporal sensation for a stimulus changing slowly in several minutes is weak enough and usually does not come to aware of it. Thus, the long-term effects of noise more than one year may not be perceived. Integrated effects on the body development of unborn babies, born children, and specialization of hemispheres of growing children are not known, and we shall make an attempt to discuss for the future of body and mind of human being in this chapter. Since 1968, effects of aircraft noise on unborn babies and children were investigated around the Osaka International Airport, Japan, which is located in Itami City. We have found that: (1) Effects on the mind: Postnatal effects of aircraft noise on sleep of babies were depend on the period when their mother came into the noise area in reference to the pregnancy period. It has been appeared as an adaptation to aircraft noise of newborn babies (Ando and Hattori in J Acoust Soc Am 47:1128–1130, 1970; Ando and Hattori in J Sound Vib 27:101–110, 1973; Ando and Hattori in J Acoust Soc Am 62:199–204, 1977a). (2) Effects on the body: Time-integrated effects of the aircraft noise on developments of the body of unborn babies have been found in terms of human placental lactogen (HPL), in turn appeared decreasing the birthweight (Ando and Hattori in Br J Obstet Gynaecol 84:115–118, 1977b). (3) After the birth, an effect of development of the height has been investigated (Schell and Ando in J Sound Vib 151:371–382, 1991). (4) Time-integrated effects on development of the specialization of cerebral hemispheres of children have been found, which were clearly described by the results of testing two different mental works, which are called the left and the right hemisphere tasks. Typical effects of the daily noise appeared as a temporal abandonment in the adding task (Ando et al. in J Sound Vib 43:683–691, 1975; Ando in J Sound Vib 127:411–417, 1988; Ando in J Sound Vib 241:129–140, 2001). This may considered that integrated effects of noise are appeared as decrease of the spiritual power. (5) Results obtained here have allowed us to design the human healthy environment of urban, so that many airports constructed after 1977 in the world have been moved out from residential areas.

Y. Soeta and Y. Ando, *Neurally Based Measurement and Evaluation of Environmental Noise*, Mathematics for Industry 20,
DOI 10.1007/978-4-431-55432-5_9

Keywords Human placental lactogen · Birthweight · Reaction of baby · Development of height

9.1 Human Placental Lactogen (HPL) in Maternal Plasma

In the Itami district around the Osaka International Airport (noise area), blood samples were taken from 343 randomly selected expectant mothers, between 22- and 41-week gestation, who were attending the Itami City Hospital (Ando and Hattori 1977b). As shown in Fig. 9.1, area living mothers in Itami, the aircraft noise level were 75–95 WECPNL. In the Kobe district, a district with little aircraft noise (reference area), blood samples were obtained from 112 randomly selected mothers of equivalent gestation who were attending either the Kobe University Hospital or the Kanabo Hospital. Other environmental conditions such as atmosphere pollution, density of population, standard living, weather, and traffic conditions were almost same for the two areas. The noise area was residential area and commercial, while the reference area was commercial and industrial.

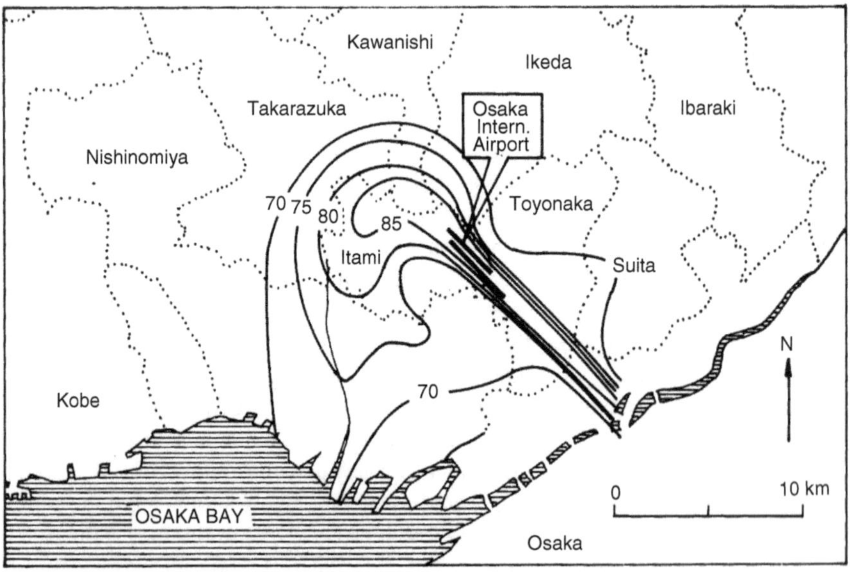

Fig. 9.1 Aircraft noise level in weighted equivalent continuous perceived noise level (WECPNL) around the Osaka International Airport. All living area in Itami City was WECPNL > 70. WECPNL = dBA + 10log $N - 27$, where dBA is the average value of peak noise level and N is the number of flight in a day $N = N_1 + 3N_2 + 10N_3$. N_1, N_2, and N_3 are the number of flights between 07:00 and 18:59 h, 19:00 and 21:59 h, and 22:00 and 06:59 h, respectively

One 5-ml sample of venous blood was taken from each participant and refrigerated overnight. On the following day, about 2 ml of serum was separated from each sample and stored in a deep freeze. The human placental lactogen (HPL) levels were measured at Kobe University Hospital (Ando and Hattori 1977b). In addition, the birthweights of the babies were obtained from hospital records.

Figure 9.2 shows the HPL levels in mothers in the noise area at varying stages of gestation. The graph also shows the mean, +2 Standard deviation (SD), and −2SD levels (Lindberg and Nilsson 1973), and the fetal danger zone (FD Zone) is outlined. In the reference area, it is apparent that in most of the HPL levels, the mothers remained within the normal range. The incidence of HPL levels more than 1SD below the mean at various stage of pregnancy is shown in Fig. 9.3. Remarkably, after the 32nd week of pregnancy, the HPL levels of more than 40 % of participants in the noise area had fallen to more than 1SD below the mean and that this percentage was higher for participants at later stages of pregnancy. Significant differences between the percentages of the noise area and the reference area have been achieved. Also, the percentage of mothers with HPL levels of 4 g/ml or less, a measure of fetal danger (Lindberg and Nilsson 1973), tended to be higher in the noise area than in the reference area.

We also reconfirmed the facts that the birthweight of babies born to mothers with low HPL levels tended to be low (Letchworth et al. 1971; Spellacy et al. 1971). The decrease in the HPL level indicated lower birthweights for babies whose mothers lived in the noise area as is discussed in the following section (Ando and Hattori 1973, 1974; Ando 1988).

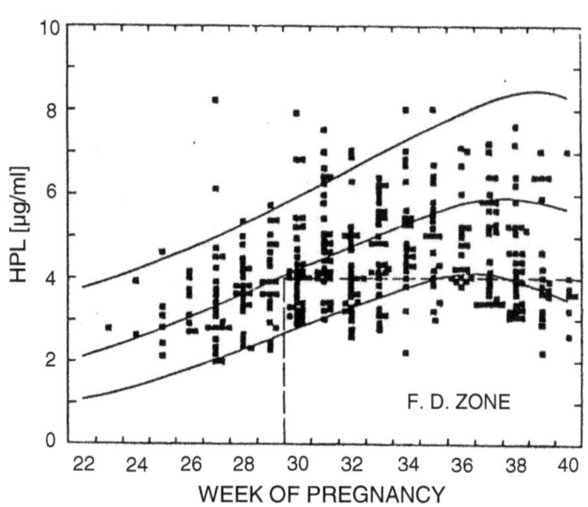

Fig. 9.2 HPL levels of 343 mothers in the noise area (Itami City) by the stage of pregnancy (Ando and Hattori 1977b). The *three lines* show the 2SD of mean values for normal HPL levels from Lindberg and Nilsson (1973)

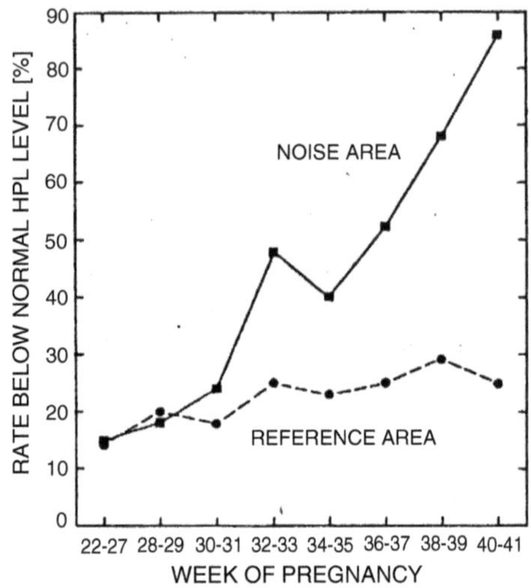

Fig. 9.3 The percentage of participants with HPL levels more than 1SD below the mean in the stage of pregnancy (Ando and Hattori 1977a). The noise area was corrected at the Itami City Hospital in Itami City, and the reference area was collected at the Kanebo Hospital and the Kobe University Hospital in Kobe City

9.2 Birthweight

A total of 713 participants who lived in the noise area were randomly chosen. To study the birthweight of babies, a factor analysis including 12 items was conducted (Ando and Hattori 1973). It was seen that the noise environment is a significant factor, and no differences in birthweight were found between the subcategories in items including classification of the family by total annual income index, birth order, working mothers, age of mothers, and others.

Relative birthweight under 3000 g in reference to that of a neighboring quiet area was −3 % for the 3 years (1961–1963), during which no jet planes flew, as shown in Fig. 9.4. As soon as jets began to fly regularly in 1964, however, the relative birthweight under 3000 g in the noisy areas increased clearly and after 1965, was over 5 % in average for both males and females in the Itami district, WECPNL > 70 (Ando 1988). These results are significant; the number of participants was between 1900 (1961) and 3400 (1969) in the noise area; and between 11,000 (1961) and 21,000 (1969) in the relatively quiet area. It is worth noticing that the noise level of the jet plane was roughly 20–30 dB greater than that of the propeller plane. After 1965, the level could be expressed as

$$\text{WECPNL} \approx \text{dB}(A) + 10 \log N - 27 \tag{9.1}$$

where N is the number of jets weighted according to each hour of the day; the noise number index is NNI ≈ WECPNL − 30. Next, relative birthweight under 2500 g in

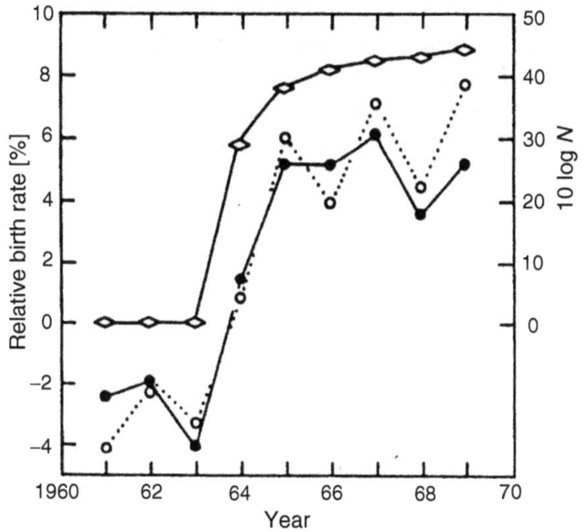

Fig. 9.4 Relative birthrate under 3000 g in the noisy area (Itami City) and in the reference to that in the neighboring low noise area, which relate to the number of jet planes using the Osaka International Airport. *Filled circle* Male. *Open circle* female. *Open diamond* Number of jet planes parameter 10log *N*, *N* being the number of jet planes using the airport

reference to the noise level was investigated with 12,468 participants who were born in Itami City in 1969 (Ando and Hattori 1973). Results are shown in Fig. 9.5 against equivalent continuous perceived noise level (ECPNL). In this area at that time, smoking mothers were few in Japan.

Later in the USA, airport noise exposure has been associated with shorter gestational ages for males and females, and smaller birthweights for females (Schell 1981). It has been found that indications of reduced prenatal growth in response to

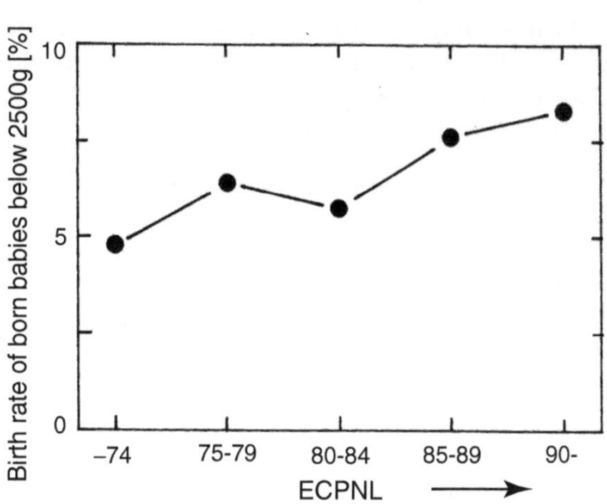

Fig. 9.5 Birthrate of born babies below 2500 g in relation to the area of noise level measured by ECPNL (Ando and Hattori 1973)

noise (Knipschild et al. 1981). And comparing mothers of similar socioeconomic status, they observed that those exposed to extremely loud noise from Schiphol Airport near Amsterdam had a significantly higher frequency of newborn babies weighting less than 3000 g. Also, a report near de Gaulle Airport outside of Paris suggested that lower birthweight was more common where there was greater noise (Coblentz and Martel 1986).

9.3 Reactions in Sleeping Babies

Reactions in sleeping babies (2–4 months old) to aircraft noise were studied by means of electroplethysmography (PLG) and electroencephalography (EEG) in relation to the periods when those mothers had moved in the noisy area in reference to birth. Sleeping babies whose mothers had begun living in a noisy area before conception or during the first five months of pregnancy did not react to the daily aircraft noise "prenatal adaptability", but did react to music. The reactions were evidenced in the PLG and EEG records of 78 sleeping babies. It has been explained that these differential responses by the difference between the temporal factors extracted from the running autocorrelation function (ACF) of the noise and music signals (Ando 2001).

9.3.1 Reaction of Babies on PLG and EEG

Previously, the manner in which babies reacted to aircraft noise during their sleep has been investigated using questionnaires to mothers. Results showed that reactions differed according to the length of stay of the mothers in a noise area in reference to birth of the babies (Ando and Hattori 1970, 1973, 1974).

In this study, 71 participants were chosen among the babies of two to four months of age in the area of the Osaka International Airport. The participants were classified into five groups according to the length of stay of the mothers in reference to birth of the babies (Ando and Hattori 1977a).

Group I: Babies whose mothers had come to the noisy area before conception (33 participants).
Group II: Babies whose mothers came to the noisy area during the first five months of pregnancy (17 participants).
Group III: Babies whose mothers came to the noisy area during the last four months of pregnancy (10 participants).
Group IV: Babies whose mothers had come to the noisy area after giving birth to these babies (3 participants).
Group V: Babies whose mothers lived in a quiet area where there were no aircraft flying over either during the pregnancy or after birth (8 participants).

PLG and EEG records were obtained during the baby's natural sleep (without the aid of a soporific) in the baby's regular sleeping rooms at their houses with a testing car as shown in Fig. 9.6. The aircraft noise reproduced in the sleeping rooms was that of a typical jet airplane. The order of exposure was 70, 80, and 90 dBA at a peak level, and the reactions of babies were recorded. If babies did not react to these noise levels, then music piece (segment immediately preceding the chorus of Beethoven's Symphony No. 9) was presented. The spectra at the peak levels measured were almost the same (Ando 2001).

The aircraft noise reproduced by loudspeaker was located at the distance 1.5 m from the babies head. Percentage reactions at the peak level of 70 and 80 dBA for Groups I through V from the data listed in Table 9.1 are shown in Fig. 9.7. The 1 %

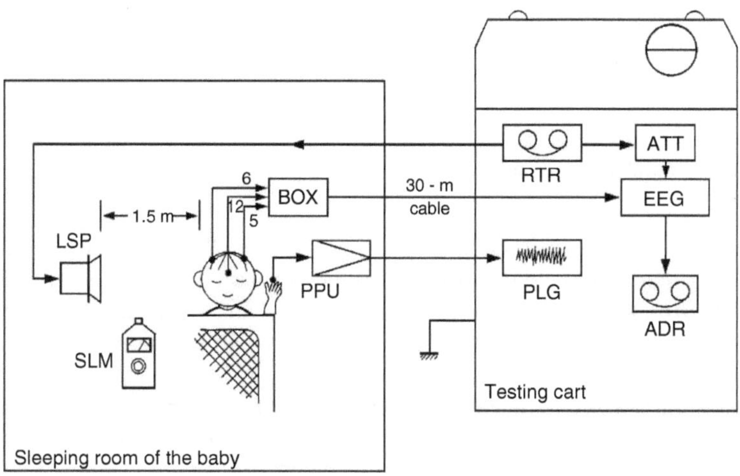

Fig. 9.6 Equipment for recording is the PLG and EEG at a house where the participant regularly sleeps (Ando and Hattori 1977a). *LSP* Loudspeaker reproducing sound stimuli. *SLM* Sound level meter. *PPU* Optical pulse pick-up. *EEG* Electroencephalograph. *PLG* Plethysmograph (X-Y recorder). *RTR* Reproduction tape recorder. *ATT* Attenuator. *ADR* Analog data recorder (4ch)

Table 9.1 PLG reactions in response to aircraft noise

Group	Number of participants reacting			No reaction 70–90 N	Percentage reaction 70 and 80 N	Total
	70 N	80 N	90 N			
I	1	4	1	10(2)[a]	28	18
II	1	2	1	5(1)	30	10
III	4	4	0	0	100	8
IV	1	1	0	0	100	2
V	7	1	0	0	100	8

[a]Number in parentheses is the number of participants who did not react to 70 and 80 N, because the test could not be done at 90 N as the participants were awakened by other factors

The total number of participants was 46 (varied data)

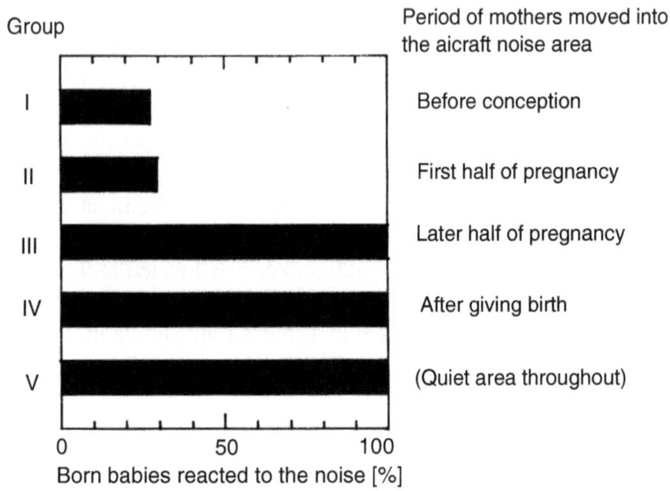

Fig. 9.7 Percentages of PLG reactions of each group (I–V) in response during sleep to the jet plane noise at peak levels 70 and 80 dBA

significance level between Groups (I–II) and Groups (III–V) was achieved. Since majority of participants in Groups I and II did not react aircraft noise at peak level of 70–90 dBA and continued to sleep, then music with similar sound pressure level (SPL) as a function of time was reproduced. For babies, this music was a new sound.

Results of percentage of cumulative frequency of reactions on PLG for Groups I and II and Groups III–V to each stimulus sound level are shown in Fig. 9.8. Also, percentages of cumulative frequency of reactions on EEG for Groups I and II to each stimulus SPL are shown by dotted lines. The differential responses of the Groups I and II babies to the aircraft noise and the music can be explained as a "selective adaptation" to the aircraft noise during pregnancy of the mother. However, we should know a negative effect of the noise on growth of unborn babies as appeared as decrease in HPL level (Sect. 9.1) associating with a lower birthweight for babies whose mothers lived in the noise area (Sect. 9.2; Ando and Hattori 1973).

It is clear that regarding reactions to noise and music, vasoconstriction responses found using PLG was more sensitive than the brain responses found using EEG.

9.3.2 Remarks of ACF Factors of the Noise and Music

So far, a remarkable finding is that most of participants in Groups I and II showed both EEG and PLG reactions to the music but not to the noise. To explain this selective ability of the babies, the temporal factors were extracted from the running

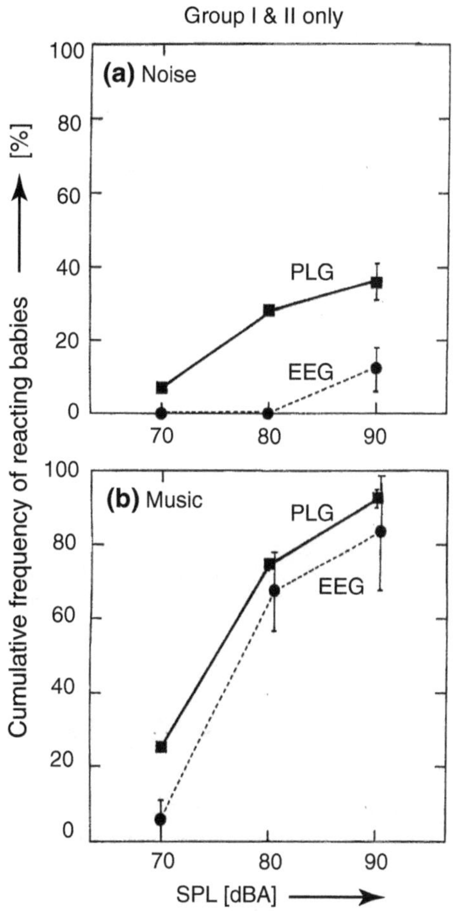

Fig. 9.8 Percentage of cumulative frequency of reactions to each stimulus SPL. Range: possible reactions. **a** Results for jet plane noise stimuli. *Filled square* Reaction of babies in PLG in Groups I and II (28 participants). *Filled triangle* Reaction of babies on PLG in Groups III–V (18 participants). *Filled circle* Reactions of babies on EEG in Groups I and II (17 participants). **b** Results to music stimuli. *Filled square* Reactions of babies on PLG in Groups I and II (20 participants). *Filled triangle* Reaction of babies on PLG in Groups III–V (10 participants). *Filled circle* Reactions of babies on EEG in Groups I and II (9 participants)

ACF by the model (Fig. 2.22). Results for the sound signals during first 25 s are shown in Fig. 9.9a–d. The SPL change in time is not differed; however, remarkable differences are found in the delay time of the first peak of ACF, τ_1, corresponding to pitch (Inoue et al. 2001), and in the effective duration, τ_e, corresponding to similar repetitive features. The pitch of music changes with time, over a range of 750–3300 Hz, and that of noise changes over a much lower range of 300–700 Hz. The value of τ_e for noise varies smoothly between 4 and 10 ms, while those for music fluctuates sharply between 5 and 60 ms. These suggest that music may associate with the left hemisphere of babies than have not been adapted to the daily noise associated with the right hemisphere without any meaning. Therefore, such distinctive reactions of babies to noise and music could be observed.

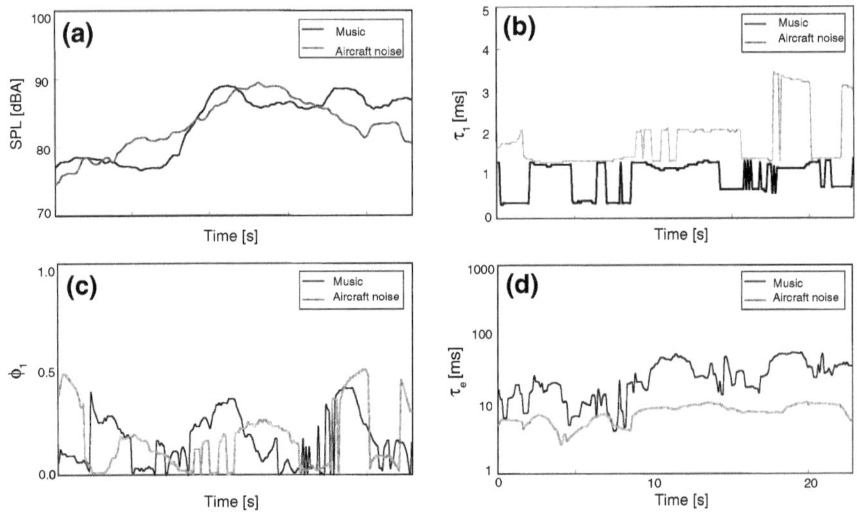

Fig. 9.9 Differences of the aircraft noise and music analyzed in temporal factors extracted from the running ACF. **a** SPL [dBA], no significant differences was observed. **b** τ_1, pitch of music was much higher than that of the noise. **c** ϕ_1, no significant differences was observed. **d** τ_e, much more activities of music were observed than that of the noise

9.4 Development of Specialization of Cerebral Hemispheres

In order to evaluate differential effects of the jet plane noise and music reproduced on two kinds of mental tasks, namely a left hemispheric task (addition) and a right hemispheric task (pattern search), we investigated children from noisy and quiet living areas. The tasks carried out in their classrooms by children under conditions of no stimulus; the noise stimulus and music stimulus are reproduced by two loudspeakers. Comprehensive evidences obtained by children from noisy and quiet living areas may be described by development of the specialization of hemispheres.

Tests were carried out in classrooms (the reverberation time 0.5–0.9 s in the 500 Hz octave band) of two schools in noisy living area near the international airport and two schools in a quiet living area (Ando et al. 1975). The total number of participants joined in the experiments was 1285 (Table 9.2). The no-stimulus group was tested in a normal classroom without any reproduced sound. The noise group was tested while being exposed to the jet plane noise of 95 ± 5 dBA peak. The music group was tested while listening to an excerpt of music from the fourth movement of Beethoven's Ninth Symphony (85 ± 5 dBA peak). The time pattern (spectrum also) of the music was similar to those of jet noise (Ando 2001). The sound stimulus was reproduced from two loudspeakers set at the front of the classroom, during every alternative period during the tasks given by,

Table 9.2 Number of participants from both quiet and noisy living areas monitored while performing metal tasks (Total: 1286)

Task	Age (years)	Living area	No-stimulus group	Noise group	Music group	Total
Addition (left hemisphere)	9–10	Noisy	146	151	34	331
		Quiet	120	123	36	279
Pattern search (right hemisphere)	7–8	Noisy	183	179	34	396
		Quiet	123	119	38	280

$$i = 2n \tag{9.2}$$

where $n = 1, 2, \ldots, 7$ for the adding task and $n = 1, 2, \ldots, 5$ for the search task. Examples of task of one period are shown in upper part of Fig. 9.10 (60 s/period) and Fig. 9.11 (30 s/period). The individual work produced in each period, called the "working curve," was drawn for all test results. The mean work performance is not discussed here, because there were no significant differences between the different conditions. Of particular importance in evaluating the tests results is the rate of number of "V-type relaxation." Such relaxation is thought to be caused by an abandonment of effort when mental functions are disturbed.

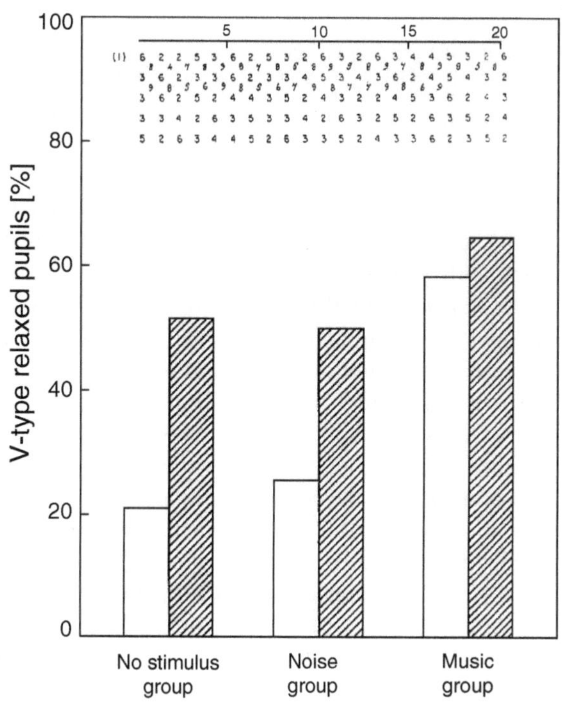

Fig. 9.10 Proportion of V-type relaxed children during adding task. *Upper part* indicates the task of one period in N (=15). *Unshaded bars* results for children from quiet living areas. *Shaded bars* show results for children from noisy living area

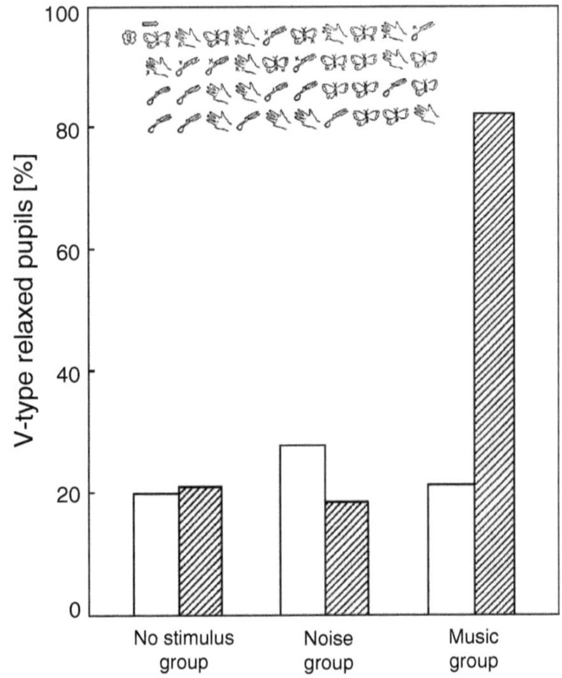

Fig. 9.11 Proportion of V-type relaxed children during the search task. *Upper part* indicates the task of one period in N (=10). *Unshaded bars* show results for children from quiet living areas. *Shaded bars* show results for children from noisy ling area

This score is classified into two categories according to the occurrence of a sudden large fall in the working curve during each task. This is assessed by $M_i < M - (3/2)W,\ i = 1, 2, \ldots, N$, where M is the mean work performance and W is the average variation of the curve excluding an initial effect at the first period, $i = 1$.

As shown in Fig. 9.10, in a quiet area, the percentage of V-type relaxed children given the additional task ($N = 15$) was much greater in the music group than in either the no-stimulus group or the noise group ($p < 0.01$). In a noisy area, however, the percentage of relaxed children was always greater than that in the quiet area, particularly in the no-stimulus group ($p < 0.01$).

As shown in Fig. 9.11 for pattern search task ($N = 10$), the percentage of relaxed children in a quiet area was similar under all test conditions, except for a slight increase in the noise group. The only significant change between participants from different living areas was found in the music group ($p < 0.01$). The results of the mental tasks were not dependent on the sex, birth order, or birthweight of child, on whether or not the mother was working, during or suffered from toxicosis during her pregnancy, on the child's gestational age when the mother moved into the noisy area, or on the child's feelings about aircraft noise (Ando et al. 1975).

Significant differences in the factors τ_1 and τ_e extracted from the running ACF of the noise and the music as a function of time may be found in measured results as shown in Fig. 9.9 (Ando 2001). Due to the auditory-brain model (Fig. 2.22), these temporal factors may stimulate the left hemisphere activated by the fluctuation of these temporal factors.

In the quiet living area, interference effects of music and noise stimuli reproduced on mental tasks were closely related to the content of the task being performed and specialization of cerebral hemispheres. In the case of the addition task, there were no significant differences between the noise group and the no-stimulus group in the percentage of V-type relaxed children. This may support the theory that noise and calculation tasks are independently processed in the right and left hemispheres, respectively (Ando 1988). Thus, as illustrated in Fig. 9.12 for children from quiet living area, no interference effects of the noise were evident in the adding task. However, the percentage of relaxed children in the music stimulus group differs significantly from that in the noise group and the no-stimulus group. This may be explained as an interference effect in the left hemisphere: Music perception and calculation being processed sequentially in this hemisphere. On the other hand, music perception and search task may be independently processed in the left and right hemispheres, respectively. In the search task, therefore, although no significant differences in the number of V-type relaxed children could be observed under the no-stimulus and music conditions. But, a difference was observed ($p < 0.1$), so that interference of the noise and the search task in the right hemisphere seems to be discernible.

For children living in the noise area as shown in Fig. 9.13, however, effects of the temporary noise and music reproduced on mental tasks are quite different. In the case of the addition task, the proportion of V-type relaxed children was always increased, even under no-stimulus conditions; this may thus be considered as a chronic effect on mental performance of the daily noise. As mentioned above, such a relaxation is caused by an abandonment of effort when mental functions are disordered. In the search task, the percentage of relaxed children increased greatly in the music group, but not in the noise group.

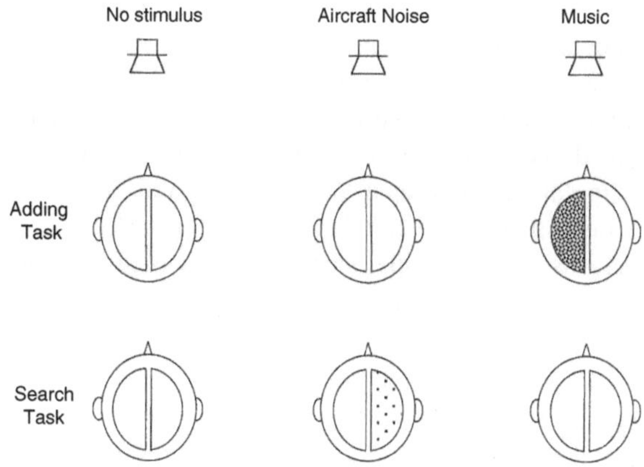

Fig. 9.12 Explanations of interference between mental tasks and sound stimuli for children from **a quiet living area**. Interference effects shown by shaded areas differ remarkably from noisy and quiet living areas. According to the auditory-brain system model, noise and music are, respectively, associated with the right hemisphere and the left hemisphere. And the adding task and search task, respectively, are associated with the left hemisphere and the right hemisphere

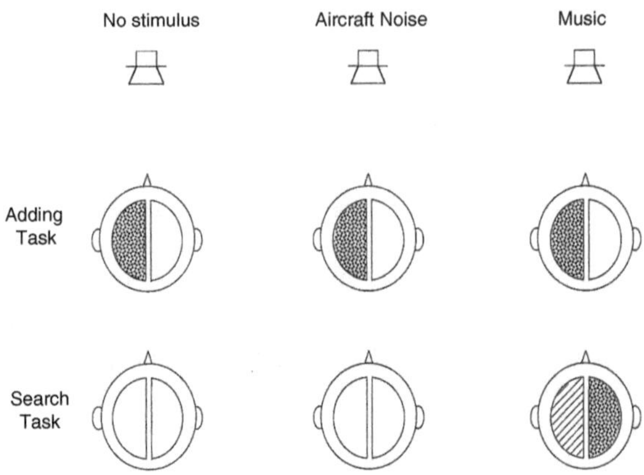

Fig. 9.13 Explanations of interference between mental tasks and sound stimuli for children from **a noisy living area**. According to the auditory-brain system model, noise and music are, respectively, associated with the right hemisphere and the left hemisphere. And the adding task and search task, respectively, are associated with the left hemisphere and the right hemisphere

Differences in interference effects as shown in Fig. 9.12 for children from the quiet living areas and Fig. 9.13 for children from the noisy living area may conclude that the daily noise affects the development of the left and right hemisphere specialization in children.

9.5 Development of Height

Participants of 6686 three-year-old children were examined by noise exposure according to contours of WECPNL (Schell and Ando 1991). Information on child growth was taken from records of health examinations at three years of age conducted at the public health centers in the five cities around the airport (Table 9.3). We have calculated the percentages of children who are very short (less than 87.0 cm) and very tall (greater than 99.0 cm). Results are shown in Fig. 9.14 according to the rank of noise level. Also, mean heights according to the rank are shown in Fig. 9.15. Significant differences were observed that the height of both

Table 9.3 Sample size of three-year-old children of 6686 in total by exposure group, broken down by sex and year of health examination

Rank	WECPNL	Male	Female
4	85–	545	539
3	80–84.9	730	733
2	75–79.9	799	680
1	70–74.9	472	482
0	–69.9	827	716
Total		3373	3313

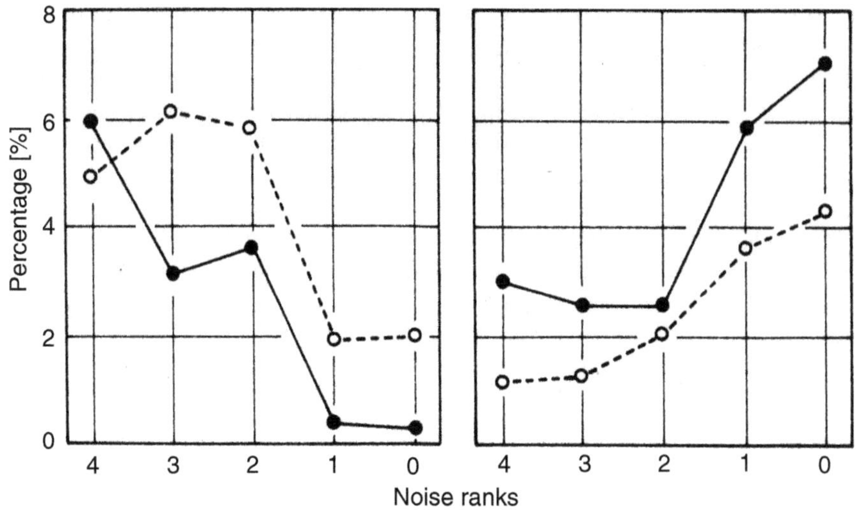

Fig. 9.14 Percentages of boys (*filled circle*) and girls (*open circle*) who are very short less than 87 cm tall *left panel*, and percentage of boys and girls who are very tall greater than 99 cm *right panel*, in each noise living area

Fig. 9.15 Mean height of boys (*filled circle*) and girls (*open circle*) in each noise living area relative to the mean for the same sexed children (Schell and Ando 1991). Smaller circles denote means of Kawanishi children, while enclosed circles denote means of Toyanaka children

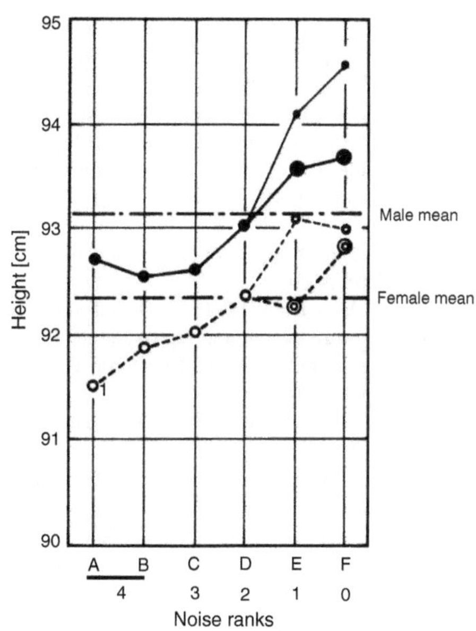

girls and boys shifted to be lower as the noise level was increased ($p < 0.02$). It is worth noticing that socioeconomic status varied little among the noise exposure groups being compared.

It is considered that noise stress could be integrated for three years to decrease the height or the skeletal structure (Fox et al. 1970; Macmahon and Pugh 1970; Tanner 1989). In a study of 5–13-year-old children living near an international airport in USA (Schell and Norelli 1983; Schell and Hodges 1985), ten anthropometric dimensions were compared after adjustment for social, economic, and demographic characteristics of each child. They found children in the noise-exposed community were smaller overall.

From laboratory studies of animals exposed to noise stress, and from studies of endocrine influences on physical growth in human, they supported for the causal relationship with the noise exposure (Sackler et al. 1959; Zondek and Tamari 1967; Autrum and Von Holst 1968; Iturrian and Fink 1968; Welch and Welch 1970; Fell et al. 1976).

References

Ando Y (1988) Effects of daily noise on fetuses and cerebral hemisphere specialization in children. J Sound Vib 127:411–417

Ando Y (2001) Differential effects of noise and music signals on the behavior of children. J Sound Vib 241:129–140

Ando Y, Hattori H (1970) Effects of intense noise during fetal life upon postnatal adaptability (Statistical study of the reactions of babies to aircraft noise). J Acoust Soc Am 47:1128–1130

Ando Y, Hattori H (1973) Statistical studies on the effects of intense noise during human fetal life. J Sound Vib 27:101–110

Ando Y, Hattori H (1974) Reaction of infants to aircraft noise and effect of the noise on human fetal life. Practica Oto-Rhino-Laryngologica 67:129–136 (in Japanese)

Ando Y, Hattori H (1977a) Effects of noise on sleep of babies. J Acoust Soc Am 62:199–204

Ando Y, Hattori H (1977b) Effects of noise on human placental lactogen (HPL) levels in maternal plasma. Br J Obstet Gynaecol 84:115–118

Ando Y, Nakane Y, Egawa J (1975) Effects of aircraft noise on the mental work of pupils. J Sound Vib 43:683–691

Autrum H, Von Holst D (1968) Social stress in the tree shrew (Tupaia glis) and its effect on growth, body weight and reproduction. J Comp Physiol 58:347–355

Coblentz A, Martel A (1986) Effects of fetal exposition to noise on the birthweight. Am J Phys Anthropol 69:188

Fell RD, Ellis CJ, Griffith DR (1976) Thyroid responses to acoustic stimulation. Environ Res 12:208–213

Fox JP, Hall CE, Elveback LR (1970) Epidemiology: man and disease. Macmillan, London

Inoue M, Ando Y, Taguti T (2001) The frequency range applicable to pitch identification based upon the auto-correlation function mode. J Sound Vib 241:105–116

Iturrian WB, Fink GB (1968) Effects of noise in animal house on seizure susceptibility and growth mice. Lab Anim Care 18:557–560

Knipschild P, Meijer H, Salle H (1981) Aircraft noise and birth weight. Int Arch Occup Environ Health 48:131–136

Letchworth AT, Boardman RJ, Bristow C, Landon J, Chard T (1971) A rapid semi-automated method for the measurement of human chorionic somatomammotrophin. The normal range in the third trimester and its relation to fetal weight. J Obstet Gynaecol Br Commonw 78:542–548

Lindberg BS, Nilsson BA (1973) Variations in maternal plasma levels of human placental lactogen (HPL) in normal pregnancy and labour. J Obstet Gynaecol Br Commonw 80:619–626

Macmahon B, Pugh TF (1970) Epidemiology: principles and methods. Little Brown Book, Boston
Sackler AM, Weltman M, Bradshaw M, Justshuk P (1959) Endocrine change due to auditory stress. Acta Endocrinologia 31:405–418
Schell LM (1981) Environmental noise and human prenatal growth. Am J Phys Anthropol 56:63–70
Schell LM, Ando Y (1991) Postnatal growth of children in relation to noise from Osaka International Airport. J Sound Vib 151:371–382
Schell LM, Hodges D (1985) Longitudinal study of growth status and airport noise exposure. Am J Phys Anthropol 66:383–389
Schell LM, Norelli RJ (1983) Airport noise exposure and the postnatal growth of children. Am J Phys Anthropol 61:473–482
Spellacy WN, Buhi WC, Schram JD, Birk SA, McCreary SA (1971) Control of human chorionic somatomammotropin levels during pregnancy. Obstet Gynecol 37:567–573
Tanner JM (1989) Fetus into man: physical growth from conception to maturity. Harvard University Press, Massachusetts
Welch BL, Welch AS (1970) Physiological effects of noise. Plenum Press, New York
Zondek B, Tamari I (1967) Effects of auditory stimuli on reproduction. In: Wolstenholme GEW, O'Connor M (eds) Effects of external stimuli on reproduction. Little Brown, Boston, p 4–16

Chapter 10
Application to Sound Design

Abstract In previous chapters, noise characteristics and annoyance to noise are evaluated by the auditory system model that includes factors extracted from autocorrelation function (ACF) and interaural cross-correlation function (IACF). In this chapter, other applications to sound design are discussed based on the ACF and IACF factors. They are subjective diffuseness of music signals, listening level of music through headphones in noise environments, subject preference for birdsongs, and urban soundscape design.

Keywords Sound design · Subjective diffuseness · Listening level · Subjective preference · Soundscape

10.1 Subjective Diffuseness of Music Signals Convolved with Binaural Impulse Responses

Spatial impression is an important subjective attribute of a sound field, which can be separated into two subjective dimensions: apparent source width (ASW) and listener envelopment (LEV), and has been evaluated from sound field factors calculated from binaural impulse responses (BIRs) measured in hall (Bradley and Soulodre 1995). The ASW is defined as the width of a sound image fused from direct sound and early reflections. The LEV is defined as the sense of being surrounded by late arriving sound from a diffused array of distant sound images. The ASW has a high correlation with the maximum peak amplitude (IACC) and the width of the maximum peak (W_{IACC}) of the interaural cross-correlation function (IACF) (Ando et al. 1999; Sato and Ando 2002). The LEV has a high correlation with the late-arriving relative lateral sound level (LG or G_{LL}) and the late lateral energy fraction (LF_L) (Bradley 1994; Bradley et al. 2000). The common experimental method observed in these studies for the spatial impression is a fixed source signal with producing different sound fields. For example, for simulated sound fields, the source signal used in terms of ASW was Mozart's "Jupiter"

Y. Soeta and Y. Ando, *Neurally Based Measurement and Evaluation of Environmental Noise*, Mathematics for Industry 20,
DOI 10.1007/978-4-431-55432-5_10

(Barron 1971; Barron and Marshall 1981), and the source signal used in terms of LEV was Handel's "Water Music" (Bradley 1994: Bradley et al. 2000).

On the other hand, the subjective diffuseness is a spatial impression considering the effect of the source signals. Originally, subjective diffuseness was evaluated by the IACC of white noise which arrived from a different horizontal angle using multichannel loudspeaker reproduction (Damaske and Ando 1972). A large IACC makes the listener to perceive the well-defined direction of the incoming sound. A small IACC corresponds to subjectively diffused sound, and the listener has no impression of the clear direction of the sound. Researchers have investigated both the effects of the source signals and source locations on subjective diffuseness (Ando and Kurihara 1986; Singh et al. 1994; Okano et al. 1998). Octave band-pass noise with different center frequencies were used as source signal, while the sound field was altered by changing the locations of the loudspeakers (Ando and Kurihara 1986; Singh et al. 1994). In the results, the IACC changed according to the center frequency even if the location of the loudspeakers was the same, and the subjective diffuseness showed a strong correlation with the IACC.

The IACF is defined by the correlation between the signals at the left and right ears as a function of delay time (Ando 1998, 2002). When the binaural signals are a BIR, the calculated IACC is a sound field factor used to evaluate the spatial impression of a hall. Since the BIRs indicate the transfer function of sound fields, an anechoic source convolved with the BIRs represents the signal that listeners hear in that sound field. Therefore, when the binaural signals are music convolved with a BIR, the IACC is a combined factor of both sound field and source signal used to evaluate the subjective diffuseness. This study defines the former as $IACC_{IR}$ (i.e., the IACC of impulse response) and the latter as $IACC_{SR}$ (i.e., the IACC of other signal response through a hall). For example, the subjective rank ordering of concert halls has been determined by $IACC_{IR}$ (Beranek 1996), while the subjective preference of concert halls, as an overall impression of sound fields, has been evaluated by $IACC_{SR}$ (Ando 1998).

During a musical performance, people listen to the music not to the impulse signal. Therefore, in the actual situation, sound arriving at listeners is not determined only by the BIR of a hall but also by the music signal. The $IACC_{SR}$ changes according to the music even when the source signal is presented by the same loudspeaker location (Ando 1977). However, it has not been clarified yet which factors extracted from the music signal affect the $IACC_{SR}$ and the subjective diffuseness. The aims of this study are to estimate the $IACC_{SR}$ by the $IACC_{IR}$ and factors obtained from music signal and to evaluate the subjective diffuseness by these factors. First, the relationship between $IACC_{IR}$ and $IACC_{SR}$ is investigated using eight BIRs and five music signals. Second, the subjective diffuseness was evaluated in a simulated sound field using four BIRs and three music signals. It is hypothesized that the subjective diffuseness has a correlation with the calculated $IACC_{SR}$ by the $IACC_{IR}$ and the factors from the music signal (Shimokura et al. 2011).

10.1.1 Relationship Between $IACC_{IR}$ and $IACC_{SR}$

The cross-correlation of two real functions $f(t)$ and $g(t)$ of a real variable t, denoted Φ_{fg}, is defined by

$$\Phi_{fg} = f(-t) * g(t), \tag{10.1}$$

where * denotes convolution. Correlation and convolution are the same except for the flip of t in one of the two functions. When a music signal, $m(t)$, is convolved with a BIR at the left, $h_{\mathrm{l}}(t)$, and right, $h_{\mathrm{r}}(t)$, ears, the echoic music arriving at left and right ears can be expressed as

$$mh_{\mathrm{l}}(t) = m(t) * h_{\mathrm{l}}(t), \tag{10.2}$$

$$mh_r(t) = m(t)^* h_r(t). \tag{10.3}$$

The IACF of the music signal response through a hall ($\mathrm{IACF_{SR}}$) is calculated by Eqs. (10.1)–(10.3) as follows:

$$\begin{aligned}\mathrm{IACF_{SR}} &= \Phi_{mh_{\mathrm{l}}mh_{\mathrm{r}}} = mh_{\mathrm{l}}(-t) * mh_{\mathrm{r}}(t) = m(-t) * m(t) * h_{\mathrm{l}}(-t) * h_{\mathrm{r}}(t) \\ &= \Phi_{\mathrm{mm}} * \Phi_{\mathrm{lr}},\end{aligned} \tag{10.4}$$

where Φ_{mm} is the autocorrelation function (ACF) of the music signal, and Φ_{lr} is the IACF of the BIR ($\mathrm{IACF_{IR}}$). The $\mathrm{IACC_{IR}}$ and $\mathrm{IACC_{SR}}$ are calculated by

$$\mathrm{IACC_{IR}} = \frac{\Phi_{\mathrm{lr}}(\tau_{\mathrm{IACC}})}{\sqrt{\Phi_{\mathrm{ll}}(0)\Phi_{\mathrm{rr}}(0)}}, \tag{10.5}$$

$$\mathrm{IACC_{SR}} = \frac{\int_{-T}^{T} \Phi_{\mathrm{mm}}(t)\Phi_{\mathrm{lr}}(t+\tau_{\mathrm{IACC}})\mathrm{d}t}{\sqrt{\int_{-T}^{T} \Phi_{\mathrm{mm}}(t)\Phi_{\mathrm{ll}}(t)\mathrm{d}t \int_{-T}^{T} \Phi_{\mathrm{mm}}(t)\Phi_{\mathrm{rr}}(t)\mathrm{d}t}}, \tag{10.6}$$

where τ_{IACC} is the interaural delay time of the maximum peak of the $\mathrm{IACF_{IR}}$. $\Phi_{\mathrm{ll(rr)}}$ is the ACF of the BIR at the left (right), and $2T$ is the integral time of the ACF and IACF. When $m(t)$ is a white noise, the ACF becomes a Dirac delta function; thus, the $\mathrm{IACF_{SR}}$ (or $\mathrm{IACC_{SR}}$) is equal to the $\mathrm{IACF_{IR}}$ (or $\mathrm{IACC_{IR}}$). When $m(t)$ is a pure tone, the ACF becomes a cosine function, so the $\mathrm{IACF_{SR}}$ is also a cosine function, and $\mathrm{IACC_{SR}}$ is always 1 regardless of the $\mathrm{IACC_{IR}}$. Therefore, the periodicity included in the signal affects the $\mathrm{IACC_{SR}}$.

The periodicity of music lies between those of white noise (no periodicity) and pure tone (perfect periodicity). The degree of periodicity can be expressed by the effective duration of the ACF, τ_{e} (Ando 1998). The τ_{e} is defined by the delay time at which the envelope along the early decay of the normalized ACF falls by −10 dB. The τ_{e} can be controlled by using noise with different bandwidths (Ando and Alrutz 1982). When the bandwidth of the noise becomes narrower, the τ_{e} becomes longer.

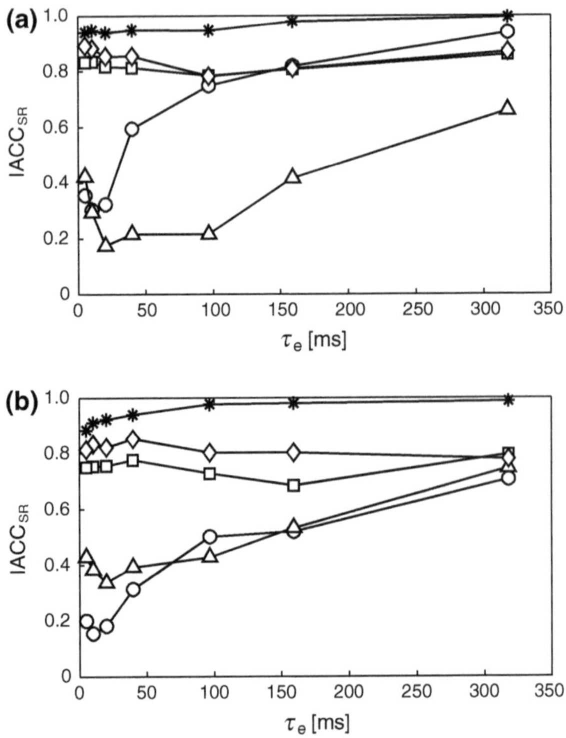

Fig. 10.1 $IACC_{SR}$ from band-pass noise convolved with BIRs as a function of τ_e from the band-pass noise. The center frequencies of the band-pass filters were **a** 500 Hz and **b** 1 kHz, and the bandwidths were 640, 320, 160, 80, 40, 20, and 10 Hz. The different symbols indicate BIR1 (*open circle*), BIR2 (*open triangle*), BIR3 (*open square*), BIR4 (*open diamond*), and BIR5 (*division times*) (Shimokura et al. 2011)

Figure 10.1 shows the relationship between $IACC_{SR}$ and τ_e of band-pass noise. The $IACC_{SR}$ of band-pass noise becomes higher as the τ_e becomes longer. Thus, the $IACC_{SR}$ of music may also become higher as the τ_e calculated from the music becomes longer.

Eight music signals generated using musical instrument digital interface were used as source signals. Two melodies (A and B) were played by four musical instruments. The duration of music was 10 s. Music signals are characterized by ACF factors, $\Phi(0)$, τ_1, ϕ_1, τ_e, and $W_{\phi(0)}$. The $\Phi(0)$, τ_1, and ϕ_1 are related to perceived loudness, pitch, and pitch strength of a music signal, respectively (Ando 1998; Ando et al. 1999; Ando and Cariani 2009). Because ACF factors of music vary as a function of time, the ACF of the music was calculated in the integral interval, $2T$, that slides along the duration of music. The running ACF calculated using a $2T$ of 0.5 s with 0.1 s sliding steps is shown in Fig. 10.2.

Five BIRs measured in Kirishima International Concert Hall (BIR 1 and 2), Tsuyama Concert Hall (BIR 3 and 4) in Japan, and Delphi Ancient Theater (BIR 5) in Greece were used as sound fields. In the measurements, an omnidirectional loudspeaker was located on the center of the stage, and all BIRs were measured with a dummy head in a central seat of the hall. To examine only the effect of the IACC, BIRs that have different $IACC_{IR}$ but similar IACF factors, $(W_{IACC})_{IR}$ and $(\tau_{IACC})_{IR}$, were selected.

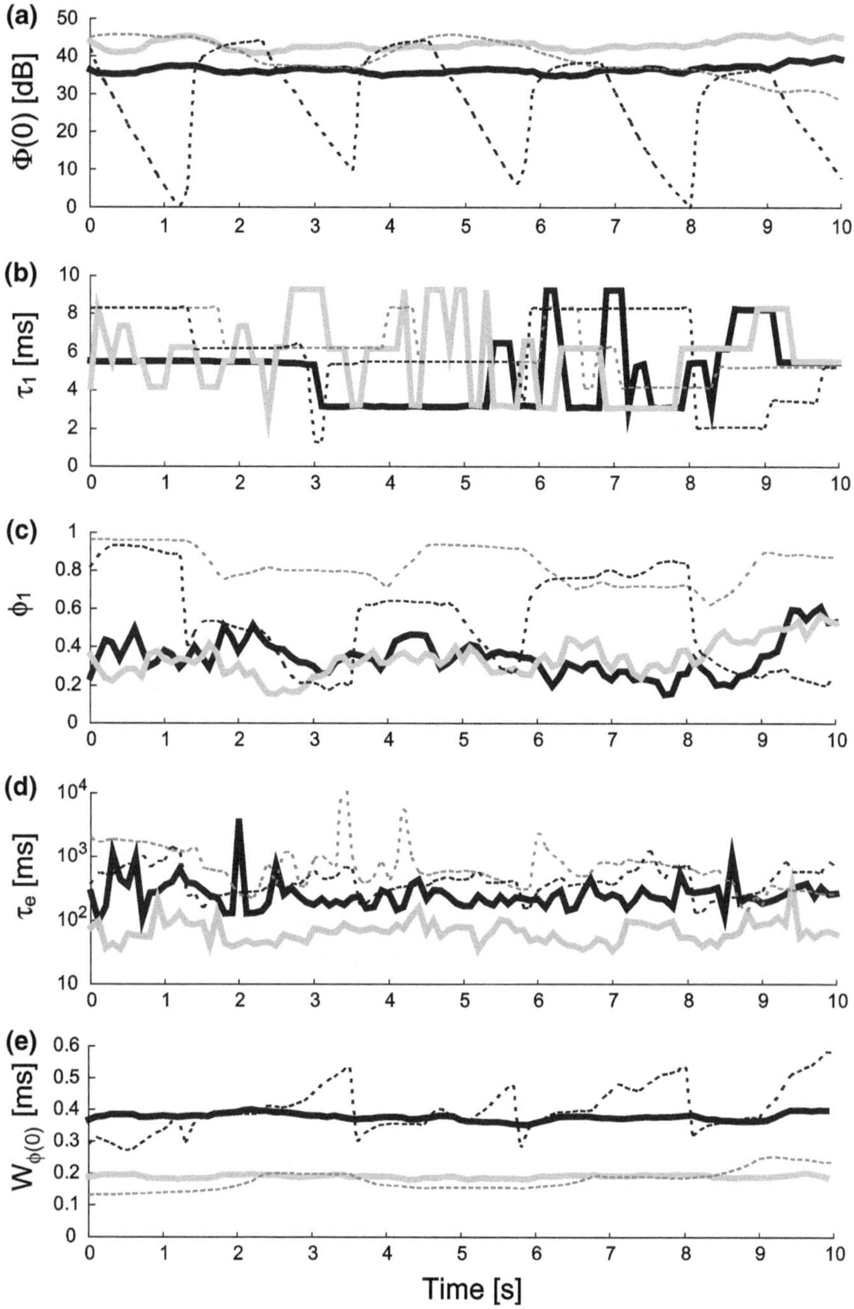

Fig. 10.2 ACF factors as a function of time for Melody A by piano (—), Melody A by trumpet (—*gray color*), Melody B by piano (—), and Melody B by organ (—*gray color*) (Shimokura et al. 2011)

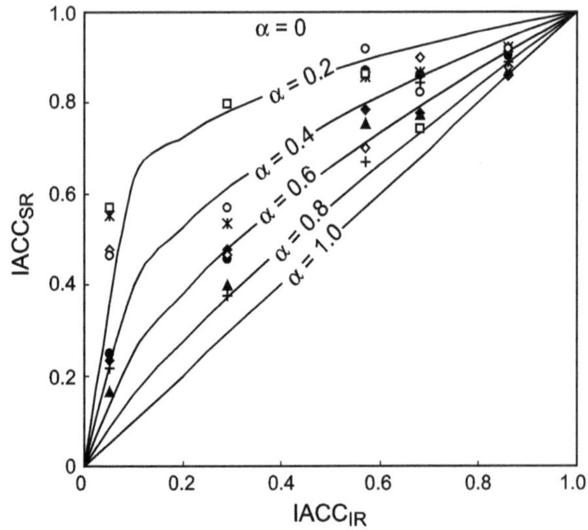

Fig. 10.3 Relationship between $IACC_{IR}$ and $IACC_{SR}$ for Melody A by glockenspiel (*division times*), harpsichord (*filled diamond*), piano (*filled circle*), and trumpet (*filled triangle*); and Melody B by harpsichord (*open diamond*), organ (*open square*), piano (*open circle*), and strings ensemble (*plus*) (Shimokura et al. 2011)

The 40 echoic music signals were obtained by the convolutions between eight anechoic music signals and five BIRs. Since the echoic music signals lasted for 10 s, the running IACF was calculated in the same manner as the running ACF. Figure 10.3 shows the relationship between $IACC_{IR}$ and $IACC_{SR}$. The $IACC_{SR}$ increased with increasing $IACC_{IR}$. The relationship can be expressed by a power function as

$$IACF_{SR} = IACC_{IR}^{\alpha}. \tag{10.7}$$

Almost all music signals fitted Eq. (10.7) with high correlations, and the α values were different according to the music signals. Figure 10.4 shows the α as a function of τ_e of the anechoic music signals. The relationship between α and τ_e is determined by an asymptotic curve regression and expressed by

$$\alpha = \frac{286}{\tau_e + 286}, \tag{10.8}$$

with high correlation ($r = 0.95$, $p < 0.01$). For white noise, τ_e is theoretically 0, and $IACC_{SR} = IACC_{IR}$ ($\alpha = 1$). For pure tones, τ_e is ∞, and $IACC_{SR} = 1$ ($\alpha = 0$). When the τ_e of a musical motif is short, the value of α approaches 1. On the other hand, when the τ_e of a musical motif is long, the value of α approaches 0. The correlation between the $IACC_{SR}$ obtained from the convolved music with the BIR and estimated $IACC_{SR}$ by Eqs. (10.7) and (10.8) was high ($r = 0.95$, $p < 0.01$).

The $IACC_{SR}$ can be estimated from the $IACC_{IR}$ and τ_e of the source signal with high accuracy. The $IACF_{SR}$ can be also estimated from the sum of $IACF_{SR}$ of band-pass noises with the frequency weighting of the source signal (Nakajima et al. 1993). Putting a loudspeaker in any direction, the $IACF_{SR}$ of band-pass noises is

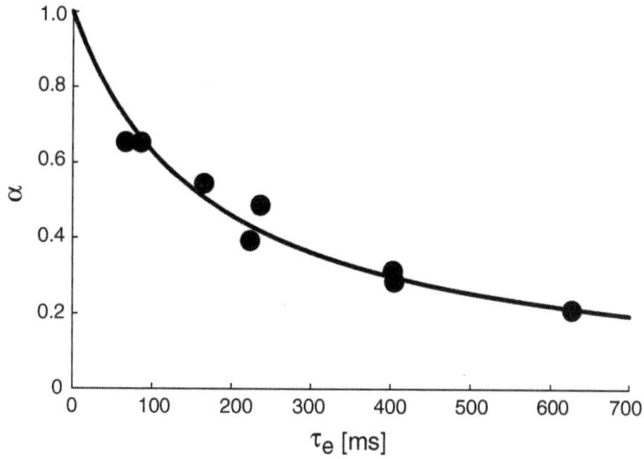

Fig. 10.4 Relationship between the exponential coefficient α and τ_e (Shimokura et al. 2011)

previously measured by a dummy head. However, this method needs to measure the horizontal angle of each reflection, so it is difficult to apply the method for the measured BIR of a hall. From the measured BIR, several kinds of IACC were introduced to evaluate the spatial impression of a hall (Hidaka et al. 1995; Hidaka and Beranek 2000). For example, the $IACC_E$ is calculated from only the early part (within 80 ms) of the BIR, and the $IACC_{E3}$ is calculated by averaging the values of the $IACC_E$ in the three octave bands (0.5, 1, and 2 kHz). The $IACC_{E3}$ is particularly correlated with the subjective rank-orderings of halls (Hidaka et al. 1995; Hidaka and Beranek 2000). Although the $IACC_{E3}$ is useful to compare the acoustical qualities of concert halls in which symphonic music is performed mainly, the $IACC_{SR}$ estimated from the $IACC_{IR}$ and τ_e of the sound source can compare the sound images of different performances in a same hall (e.g., singing and symphonic music in an opera house, or speech and organ music in a church).

10.1.2 PsychoAcoustic Experiment

The stimuli were four music signals (Melody A played by piano or trumpet, and Melody B played by piano or organ) convolved with three BIRs (BIRs 1–3). A total of 12 stimuli were obtained. Two loudspeakers were located in the anechoic chamber (3.2 m in width, 3.2 m in length, and 2.6 m in height), and the BIRs measured in the different halls were presented using the stereo dipole method. To create the parallel presentation from the left (right) loudspeaker to the left (right) ear, cross-talk canceling (CTC) filters were convolved before the output (Bauer 1961; Schroeder 1970; Kirkeby et al. 1998). The CTC filter is generated by converting a BIR measured in the anechoic chamber (Kirkeby et al. 1998) and cancels only the cross talk pathway sounds using the sound interference effect. The sound pressure level (SPL) at the listener's position was 70 dBA.

Nine listeners (A–I), 21–38 years of age with normal hearing, participated in the experiment. The scale values (SV)s of subjective diffuseness were calculated according to Case V of Thurstone's theory (Thurstone 1927). The model of Case V for all data was confirmed by the goodness-of-fit test (Mosteller 1951). The higher SV indicates that stimulus is perceived to be more diffused.

Figure 10.5 shows the SV of subjective diffuseness as a function of the $IACC_{SR}$ for the 12 stimuli. The SV had a negative correlation with the $IACC_{SR}$. The correlation coefficients were significantly high for listeners A and E, while they were

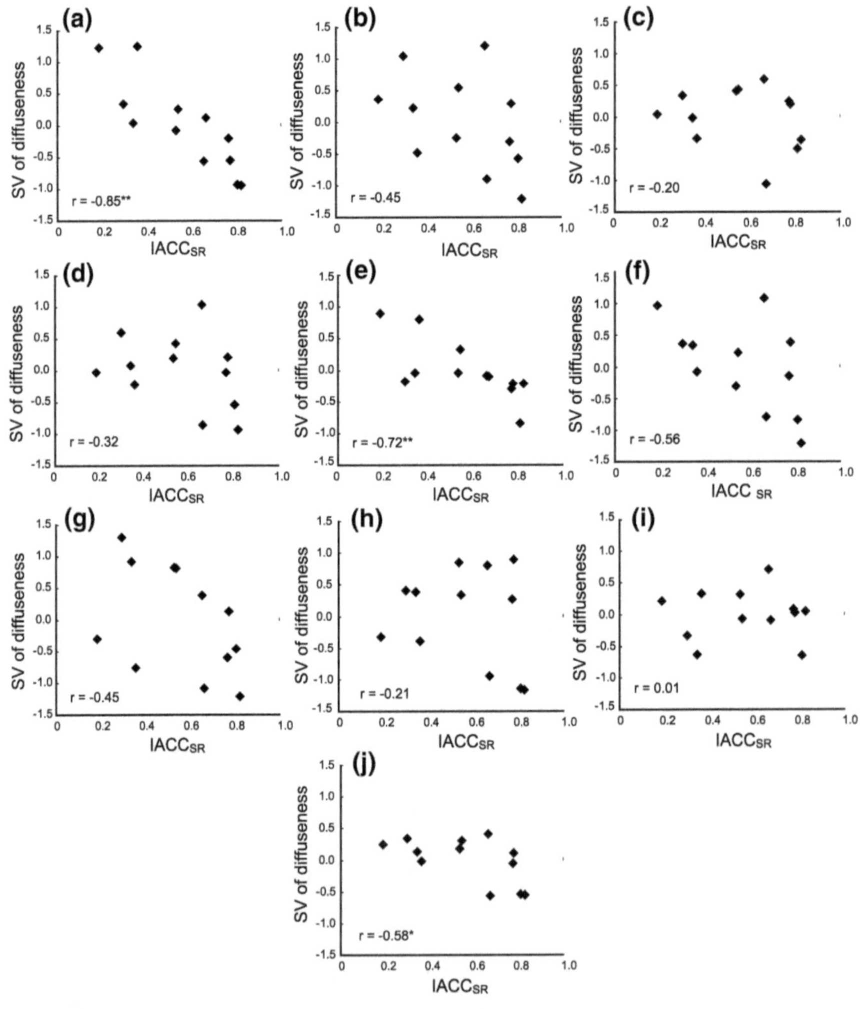

Fig. 10.5 Individual scale values of subjective diffuseness as a function of $IACC_{SR}$ of the stimuli for listeners **a** A, **b** B, **c** C, **d** D, **e** E, **f** F, **g** G, **h** H, **i** I, and **j** Total. * 5 %, ** 1 % significant levels (Shimokura et al. 2011)

low for the other listeners. The results indicated that listeners perceive the sound image of music in a hall to be more diffused as the $IACC_{SR}$ decreases. This result is consistent with the results of previous experiments for band-pass noise sources (Ando and Kurihara 1986; Singh et al. 1994). Unlike the band-pass noise sources, the correlation between $IACC_{SR}$ and subjective diffuseness is not high for the music source, suggesting that the subjective diffuseness perceived by seven listeners is not explained by the $IACC_{SR}$. The $(W_{IACC})_{SR}$ and $(\tau_{IACC})_{SR}$ were not correlated with the subjective diffuseness.

To examine which factor of music signal affects the subjective diffuseness, the contributions of the ACF factors calculated from the music signals were investigated using multiple regression analysis. The explanatory variables were $IACC_{IR}$, τ_e, and $W_{\phi(0)}$ as follows:

$$SV = a\mathrm{IACC}_{IR} + b\tau_e + cW_{\phi(0)} + d. \quad (10.9)$$

Because the $(W_{IACC})_{IR}$, $(\tau_{IACC})_{IR}$, and $\Phi(0)$ were fixed as the experimental condition, they were excluded from the explanatory variables. Because τ_1 ($r = -0.71$, $p < 0.05$) and ϕ_1 ($r = 0.93$, $p < 0.01$) were highly correlated with τ_e, they were excluded. The standardized partial regression coefficients and the correlation coefficient between the calculated and measured SVs are listed in Table 10.1. The contribution of $IACC_{IR}$ was significant for the eight listeners, and the regression coefficient a was negative for all listeners. In the factors of music signal, the contributions of τ_e and $W_{\phi(0)}$ were significant for four and two listeners, respectively. However, the regression coefficients b and c were different individually.

The individual standardized regression coefficients for the ACF factor suggest that the subjective diffuseness perceived by listeners A, C, D, and H was influenced by the τ_e. However, the contributions were not uniform among the listeners. Listener A perceived the music with shorter τ_e to be more diffused, while listeners C, D, and H perceived the music with longer τ_e to be more diffused. The judgments of listener A can be explained by the $IACC_{SR}$, as shown in Fig. 10.5a. The music

Table 10.1 Individual and total results of standardized regression coefficients for each ACF or IACF factor

Listener	$a(IACC_{IR})$	$b(\tau_e)$	$c(W_{\phi(0)})$	r
A	−0.49*	−0.66**	−0.35	0.87*
B	−0.88**	0.27	−0.05	0.92**
C	−0.69**	0.64**	0.15	0.94**
D	−0.81**	0.46*	−0.02	0.93**
E	−0.55*	−0.46	−0.38	0.79
F	−0.85**	0.14	−0.35*	0.94**
G	−0.71**	0.24	0.45*	0.87*
H	−0.60*	0.58*	−0.04	0.84*
I	−0.30	0.28	−0.53	0.69
Total	−0.90**	0.27	−0.05	0.94**

*5 %, **1 % significant levels

with shorter τ_e makes the $IACC_{SR}$ smaller, and the listener A perceives it to be more diffused. On the other hand, the judgments of listeners C, D, and H can be explained by the τ_e of sound source, which indicates its degree of reverberation included in the source. For example, when legato chordal music such as Melody B is played by an organ, the sound reverberates well, and it seems that the listeners C, D, and H perceive it to be more diffused. Since the stereo dipole presentation cannot control the SPL of lateral reflections, the LG and G_{LL}, which are related to the judgment of LEV (Bradley 1994: Bradley et al. 2000), were the same in all the simulated sound fields. However, listeners C, D, and H may have perceived to be surrounded due to the reverberation in the source signal.

The individual standardized regression coefficients for the ACF factor also suggest an effect of $W_{\phi(0)}$ on the subjective diffuseness perceived by listeners F and G. The $W_{\phi(0)}$ is related to the spectral centroid of a signal. The broad and narrow values of $W_{\phi(0)}$ mean that the low and high spectral components are strong, respectively. The low spectral components of the source signal are considered to be important in evaluating the ASW (Okano et al. 1998). The W_{IACC} of a band-pass noise in the sound field becomes broader as the low spectral components of the noise increase, and the ASW can be explained by the W_{IACC} (Sato and Ando 2002). For the music, the $W_{\phi(0)}$ was highly correlated with the $(W_{IACC})_{SR}$ (r = 0.85, $p < 0.01$). Listener G perceived the stimulus with the broader $W_{\phi(0)}$ and $(W_{IACC})_{SR}$ to be more diffused, so his perceived subjective diffuseness might be based on the ASW. However, listener F's judgment, that the music with narrow $W_{\phi(0)}$ was diffused, does not correspond to this explanation. Regarding the judgments of the other three listeners (B, E and I), the factors of ACF (i.e., τ_e and $W_{\phi(0)}$) are insufficient to explain the effect of the sound source on subjective diffuseness.

Taken together, the results indicate the following: (1) The $IACC_{SR}$ can be estimated from the $IACC_{IR}$ of BIR and the τ_e of music signal with high accuracy. (2) When the subjective diffuseness is evaluated based on BIR and music signal separately, the contributions of $IACC_{IR}$ from the BIR and τ_e or $W_{\phi(0)}$ from the music signal were significant for eight and for six in nine listeners, respectively. (3) The contributions of τ_e or $W_{\phi(0)}$ show the individual differences.

The aim of this study was to clarify whether the subjective diffuseness in a hall depends on the music signal. In this study, the $(W_{IACC})_{IR}$, $(\tau_{IACC})_{IR}$, and reverberation time were constant in the all simulated sound fields, while the music signals with different ACF factors were used to find which ACF factors influence the subjective diffuseness. Although the experimental result indicated that the ACF factors, τ_e and $W_{\phi(0)}$, were influential on the subjective diffuseness, the contributions were individual among the listeners. This suggests that individual diffuseness is optimized for each listener based on music signal that is characterized by ACF, τ_e, and $W_{\phi(0)}$.

10.2 Listening Level of Music Through Headphones in Train Car Noise Environments

Although European countries share a common recognition regarding the risk of high exposure SPL through headphones (e.g., Khatter 2011), very limited academic work has been conducted in terms of environmental noise and music listening levels. In the automobile industry, automatic gain control (AGC) devices have been developed and incorporated into car audio systems, and researchers have investigated algorithms to maintain the loudness of music within the environment of car noise (Heyde 2008; Behler et al. 2010). The listening levels of music played through a car audio system were estimated in the presence of different car noises (Suzuki et al. 1982). Although the spectral characteristics of noise inside a car do not change according to the state of movement of the car, both the spectral and temporal characteristics of noise inside a train car change dynamically according to the state of movement of the train car (Remington 1976). For example, the squeal noise produced while turning on a curve has a distinct peak at a particular frequency, and the impact noise of the train car hitting the joints of the rails on the train track can be heard periodically.

Music listening levels may be affected by the characteristics of not only the surrounding noise, but also the music itself. Studies have reported that the listening level of rock music is higher than that of pop music (Miyake and Kumashiro 1986; Dance and Wash 2008). The reason for this difference, however, has not yet been clarified, because music genres cannot be evaluated objectively. Optimum listening levels through stereo loudspeakers have been examined by the semantic differential method (Iwamiya and Suzuki 1988), with results showing that powerful and sharp music was listened to at higher volumes. However, while the relationship between the subjective impressions of music and optimum listening levels has been clarified, it is still not clear which physical factors create such impressions of the music.

In this study, optimum listening levels through headphones are examined based on physical factors extracted from an ACF of the music (Shimokura and Soeta 2012). Dynamic changes in SPLs can be observed by the series of $\Phi(0)$ within a particular moment. Music contains complex sounds. The perceived pitch and pitch strength of complex sounds can be expressed by the variables τ_1 and ϕ_1, respectively (Yost 1996; Ando et al. 1999). Variable τ_e indicates the degree of reverberation included in the source and is dependent on the type of musical instrument, the tempo of the melody, and the pattern of playing (Ando and Cariani 2009). Variable $W_{\phi(0)}$ is related to the spectral centroid of a signal (Soeta and Shimokura 2009). Broad and narrow values of $W_{\phi(0)}$ mean that the low and high spectral components are strong, respectively.

10.2.1 Experimental Methods

The train noise changed dynamically according to the state of movement of the train car. Four train noises (rolling, squeal, impact, and resonance) were used as the noise environments of the experiment in Experiment I. Only the rolling noise was used as a noise source in Experiment II. Rolling noise, caused by the surface asperities of the wheels and rails, could be heard when the train was moving. Squeal noise was generated when the train was turning on a curve, and the spectral energies increased at 300 and 800 Hz. Impact noise was generated when the wheels passed rail joints of uneven heights. Resonance noise, caused by reflected sounds in tunnels, had a spectral peak at 300 Hz (Shimokura and Soeta 2012).

Train noise was measured in a train car running in Japan (Shimokura and Soeta 2011). A surround microphone (MKV, SoundField) was located at a large area for wheelchair user. The surround microphone was composed of four closely spaced subcardioid microphone capsules. The height of the microphone was 1.3 m from the floor of the car. The noise environments in the train car were simulated in an anechoic chamber using an ambisonic method (Gerzon 1973). Using a surround microphone processor, each train noise recording made by the four microphone capsules of the surround microphone was separated into either three dimensions (*X*: longitudinal, *Y*: lateral, and *Z*: vertical axis) or omnidirectional (*W*), that is, originating from the same central point (Elen 2001). By mixing these four-dimensional train noises, the noise environments were reproduced using eight loudspeakers in a cubic configuration. The noise levels, L_{train}, during this duration were set at 64, 68, 72, 76, and 80 dBA. The variable L_{train} refers to the A-weighted equivalent continuous SPL, L_{Aeq}, of the train noise measured by the dummy head (KU100, Neumann) in the anechoic chamber.

The Japanese pop song "Flavor of Life" by Hikaru Utada was used as a music stimulus in Experiment I. In the song, the singer begins singing a hook-line softly, with a synth pad in the background. The song has strong spectral power in a range of 300–800 Hz (Gray line in Fig. 10.6). Five vocal (Music *a*–*e*) and five

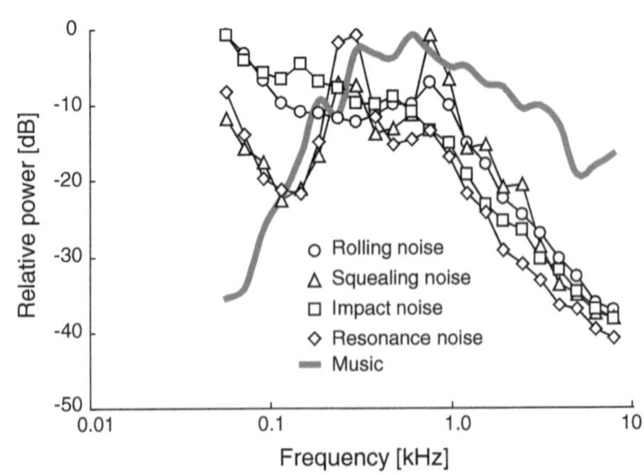

Fig. 10.6 Spectral shapes of the train noises and music used in Experiment I (Shimokura and Soeta 2012)

Table 10.2 List of music, dynamic range (DR), and median values of ACF factors

Music	Composer or player	Title	Musical ensemble	*DR* [dB]	τ_1 [ms]	ϕ_1	τ_e [ms]	$W_{\phi(0)}$ [ms]
a	K. Hirai	"Omoi ga kasanaru sono maeni"	Vocal solo	24.95	2.86	0.82	31.71	0.41
b	A. Hirahara	"Ashita"	Vocal with harp	16.95	3.93	0.68	122.48	0.99
c	T. Yamashita	"Fragil"	Vocal with chorus	19.38	3.79	0.47	49.01	0.44
d	H. Utada	"Flavor of life"	Vocal with band	8.38	3.22	0.31	34.67	0.24
e	K. Inaba (B'z)	"Sayonara nanka iwasenai"	Vocal with band	10.22	3.42	0.22	37.60	0.18
f	J. S. Bach	BWV988 "Goldberg variations"	Cembalo solo	26.26	4.81	0.59	354.25	0.34
g	F. F. Chopin	As-dur op.53 "Polonaise"	Piano solo	15.03	5.56	0.56	253.36	0.40
h	Sonny Rollins	"More than you know"	Saxophone with band	40.16	3.51	0.64	105.82	0.42
i	Charles Parker	"Confirmation"	Saxophone with band	16.02	2.31	0.45	34.33	0.15
j	L. Beethoven	Symphony no. 7	Orchestra	31.40	2.53	0.45	51.10	0.23

instrumental (Music *f*–*j*) music samples were used in Experiment II (Table 10.2). These vocal and instrumental music samples were selected according to the number of instruments involved in performing each piece. For example, Music *a* is sung solo, while Music *e* is sung accompanied by a lead guitar, bass guitar, keyboard, and drums. Music *d* is the same piece of music as that used in Experiment I, although a different rendition, accompanied by a lead guitar, bass guitar, keyboard, and drums, was used in Experiment II. Music *f* and *g* are played using a cembalo and piano solo, respectively, while Music *j* is played by an orchestra. Figure 10.7 shows the spectral shapes of each of the music samples.

The music was presented binaurally through an intra-concha headphone (MDR-E10LP, Sony), because intra-concha headphones are typically the default headphone type sold with portable audio players (Hodgetts et al. 2007). The duration of music was 18 s. Eight listeners (A–H), all of whom reported not having any hearing disorders, adjusted their listening level of the music to their individual optimum level (L_{music}) under the given conditions. The variable L_{music} refers to the SPL of music measured by an ear simulator, including an IEC 60711 coupler (Type 4157, BK). First, the listeners adjusted L_{music} in a silent condition; the default listening level was 0 dB. Second, the listeners adjusted L_{music} in a noise condition; the default listening level was L_{music} in silence. The number of replications for each noise condition was set at five.

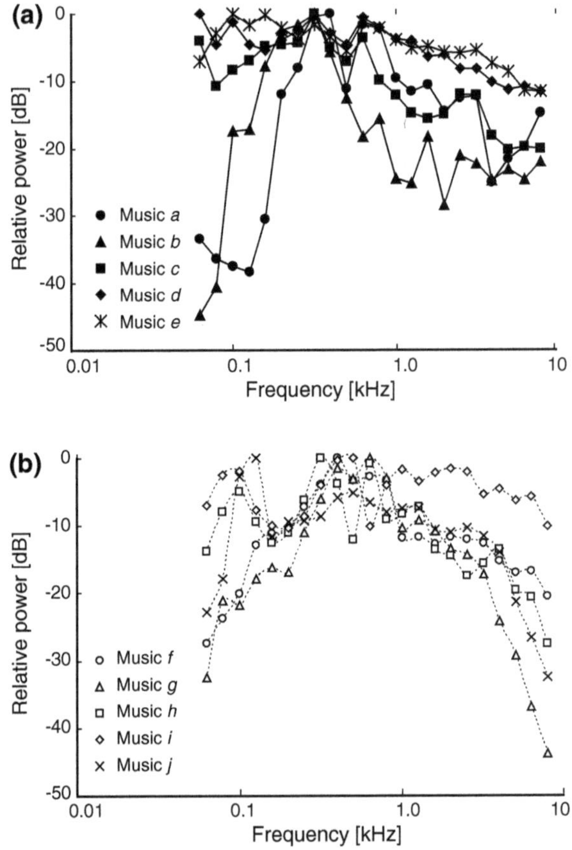

Fig. 10.7 Spectral shapes of **a** vocal and **b** instrumental music samples used in Experiment II (Shimokura and Soeta 2012)

Because ACF factors in music vary as a function of time, they were calculated in the integral interval (2*T*) that moves along with the duration of music. In this study, the running ACF using a 2*T* of 0.5 s with 0.1 s sliding steps was calculated. As a consequence of the ACF calculation sliding along the music (18 s) by the 0.1-s step, 180 factors could be obtained. The 180 factors were converted into a histogram; Figs. 10.8 and 10.9 show the cumulative frequency curves based on the histogram. For example, Music *d* and *e* (the vocal music with band music) had narrow dynamic ranges, while Music *h* (slow-paced saxophone with band music) had a broad dynamic range. The distribution of τ_1 in Music *a* was concentrated to 2.9 ms, which corresponds to the formant frequency of the voice (345 Hz), while the τ_1 of Music *b*, *c*, *d*, and *e* were distributed widely within 10 ms because of the background instrumental music (Fig. 10.9b). For the vocal music samples, ϕ_1 changed greatly according to the number of instruments involved. The large ϕ_1 of Music *a* (vocal solo) was realized by the clear harmonic components of the voice. When the instrumental sound was added to the voice, ϕ_1 became smaller. For the instrumental music, τ_e changed greatly depending on the number of instruments (Fig. 10.9d). In

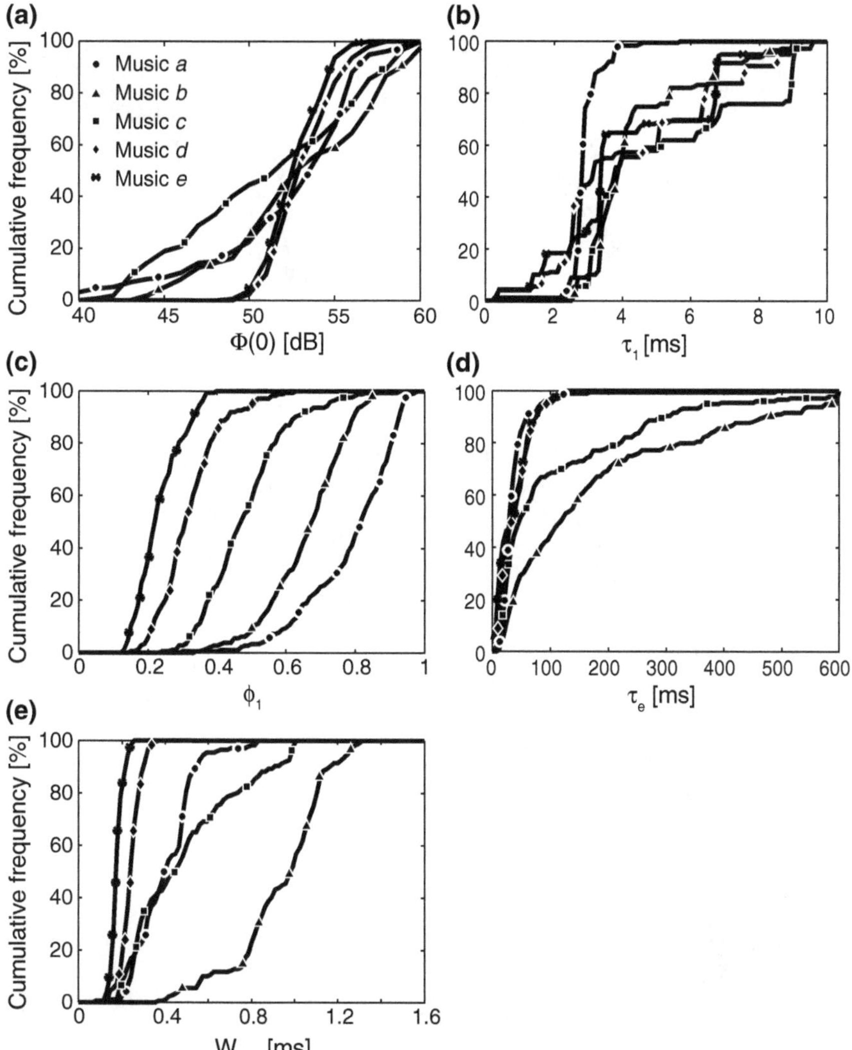

Fig. 10.8 Cumulative frequencies as a function of **a** $\Phi(0)$, **b** τ_1, **c** ϕ_1, **d** τ_e, and **e** $W_{\phi(0)}$ from the vocal music (Shimokura and Soeta 2012)

Music *f* or *g*, for example, the music is played by one musical instrument (cembalo or piano). In this case, one instrumental sound continues for a long duration in the legato or adagio passages, causing τ_e to be long. On the other hand, in Music *h*, *i*, or *j*, the music is played by a few musical instruments, and thus, τ_e becomes shorter. The $W_{\phi(0)}$ of Music *b* was larger than that of Music *e*, because Music *b* has a shortage of high-frequency components compared with Music *e*.

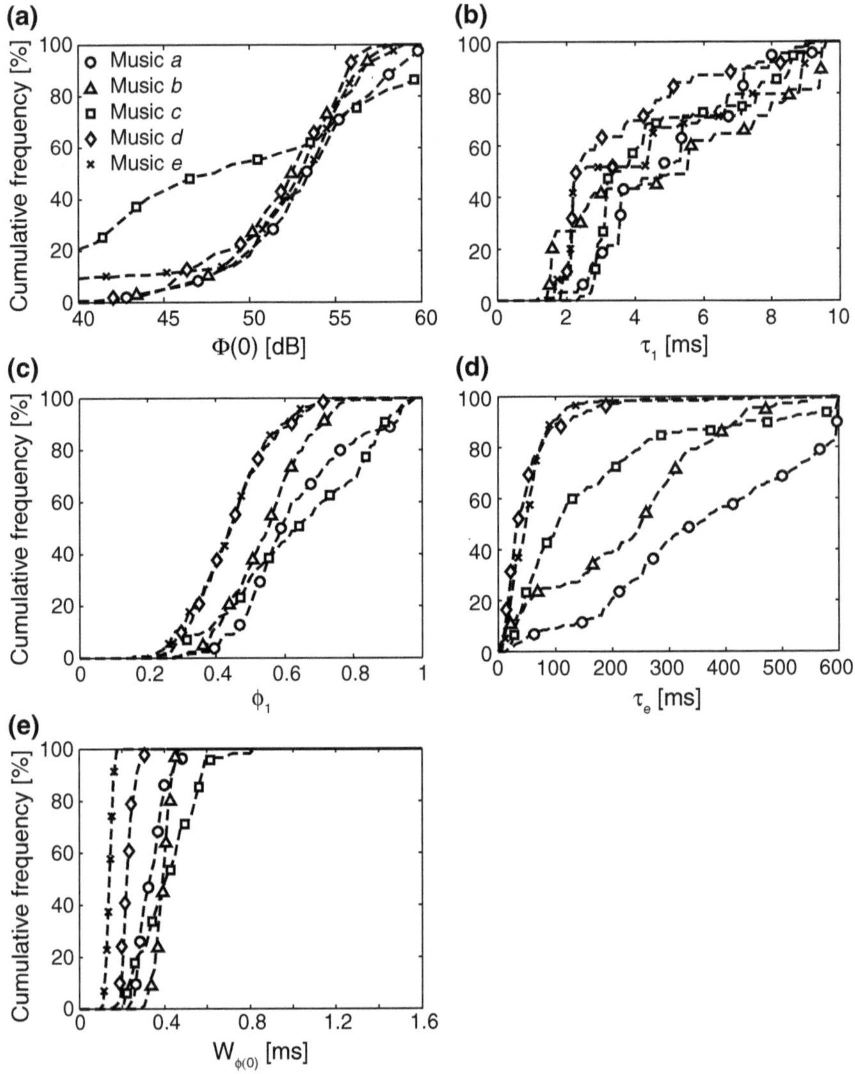

Fig. 10.9 Cumulative frequencies as a function of **a** $\Phi(0)$, **b** τ_1, **c** ϕ_1, **d** τ_e, and **e** $W_{\phi(0)}$ from the instrumental music samples (Shimokura and Soeta 2012)

10.2.2 Effects of Noise Source

Figure 10.10 shows the relationship between L_{train} and L_{music} as a parameter of train noise in Experiment 1. L_{music} increased along with an increase in L_{train}. The differences of the averaged L_{music} for the rolling, squeal, impact, and resonance noises

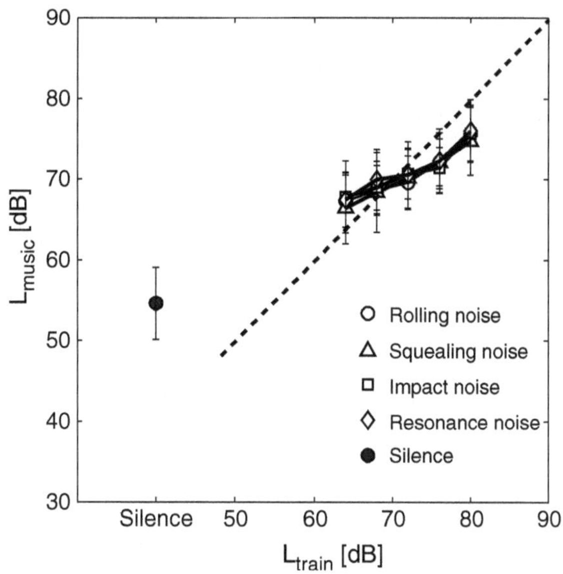

Fig. 10.10 L_{music} in the train noises and silence as a function of L_{train}

were within 0.9 dB. Because the effect of train noise type was small, L_{music} was estimated by the explanatory variable of L_{train} as follows:

$$L_{music} = 0.33 * L_{train} + 47.16. \quad (10.10)$$

These values include L_{music} in silence, in which L_{train} was 17 dBA, which corresponded to the background noise level in the anechoic chamber. Equation (10.10) estimated L_{music} with high accuracy. The correlation coefficient between the measured and calculated L_{music} was 0.96 ($p < 0.01$), and their averaged residual was 1.27 dB.

Regardless of their different spectral and temporal characteristics, the L_{music} values of the rolling, squeal, impact, and resonance noises were almost the same. This finding may be explained by the loudness as evaluated by the τ_e of the noise. The perceived loudness of band-pass, complex, and iterated rippled noises is influenced by the τ_e of those noises even when the SPL is identical (Sato et al. 2001, 2002; Soeta et al. 2004, 2007; Soeta and Nakagawa 2008). When τ_e is between 10 and 100 ms, the loudness increases as τ_e becomes longer. When τ_e is below 10 ms or over 100 ms, the loudness is not influenced by τ_e (Soeta and Nakagawa 2008). The τ_e of the train noises was too short to show any clear effect on L_{music}. The differences among the averaged L_{music} values were within the just-noticeable difference of SPL added by environmental noise (ISO 3382-1:2009).

Although L_{music} was uniquely determined by L_{train}, L_{train} through headphones changes depending on the sound insulation properties of the headphones (Hirahara et al. 2010). The insulation property of the intra-concha headphones used in this study was measured using a method similar to that used in previous research

(Hirahara et al. 2010). The insulation level was 3 dB in a range of 1–2 kHz, and −10 dB in a range of 3–5 kHz, which agrees with the findings of previous research (Hirahara et al. 2010). The SPLs of train noise at the ear canal (L_{train}') can be obtained by calculating the train noise and the insulation properties of the headphones. The insulation levels ($L_{train} - L_{train}'$) were 0.98, 0.80, 0.98, and 0.93 dB for the rolling, squeal, impact, and resonance noises, respectively. Because the insulation properties of other types of headphones (e.g., inner- and over-the-ear headphones) can be obtained using the same method, Eq. (10.10) can be generalized by the SPL at the ear canal (L_{train}') as follows:

$$L_{music} = a_1 * L_{train}' + b_1. \tag{10.11}$$

For the intra-concha headphones used in these experiments, a_1 and b_1 were 0.33 and 47.11, respectively, when the regression analysis was recalculated for the L_{music}. The correlation between the measured and calculated L_{music} for all listeners was also high ($r = 0.94$, $p < 0.01$), and their averaged residual was 1.46 dB.

10.2.3 Effects of Music Source

Figure 10.11 shows the relationship between L_{train} and L_{music} as a parameter of the music in Experiment II. L_{music} had a tendency to increase along with higher L_{train}, similar to the results of Experiment I. While L_{music} was not influenced by train noise type in Experiment I, it was influenced by music type. The L_{music} values of Music *e* and Music *j* were the largest, while the L_{music} values of Music *b* and Music *h* were the smallest for the vocal and instrumental music samples, respectively. The differences of the averaged L_{music} for the 10 music were within 11.6 dB.

To examine which ACF factor of music affects L_{music}, the correlation coefficients between the averaged L_{music} in each music sample and the ACF factors of the music were investigated. Because ACF factors vary as a function of time and are not normally distributed, the represented value for each factor was determined by the dynamic range (DR = $\Phi(0)_{max} - \Phi(0)_{min}$) and the median values of τ_1, ϕ_1, τ_e, and $W_{\phi(0)}$. Results showed that the correlation coefficients between the averaged L_{music} and ACF factors were higher for ϕ_1 ($r = -0.76$, $p < 0.05$) and $W_{\phi(0)}$ ($r = -0.77$, $p < 0.01$), while there were no correlations for DR ($r = -0.38$), τ_1 ($r = -0.50$), and τ_e ($r = -0.49$). There was also no correlation between ϕ_1 and $W_{\phi(0)}$ ($r = 0.58$).

The contributions of the ACF factors to L_{music} calculated from the music signals were investigated using multiple regression analysis. The explanatory variables were L_{train}, ϕ_1, and $W_{\phi(0)}$ as follows:

$$L_{music} = 0.29 * L_{train} - 11.9 * \phi_1 - 9.55 * W_{\phi(0)} + 59.46. \tag{10.12}$$

As in Experiment I, the values include L_{music} in silence. The correlation coefficient between the measured and calculated L_{music} was 0.94 ($p < 0.01$). Their averaged

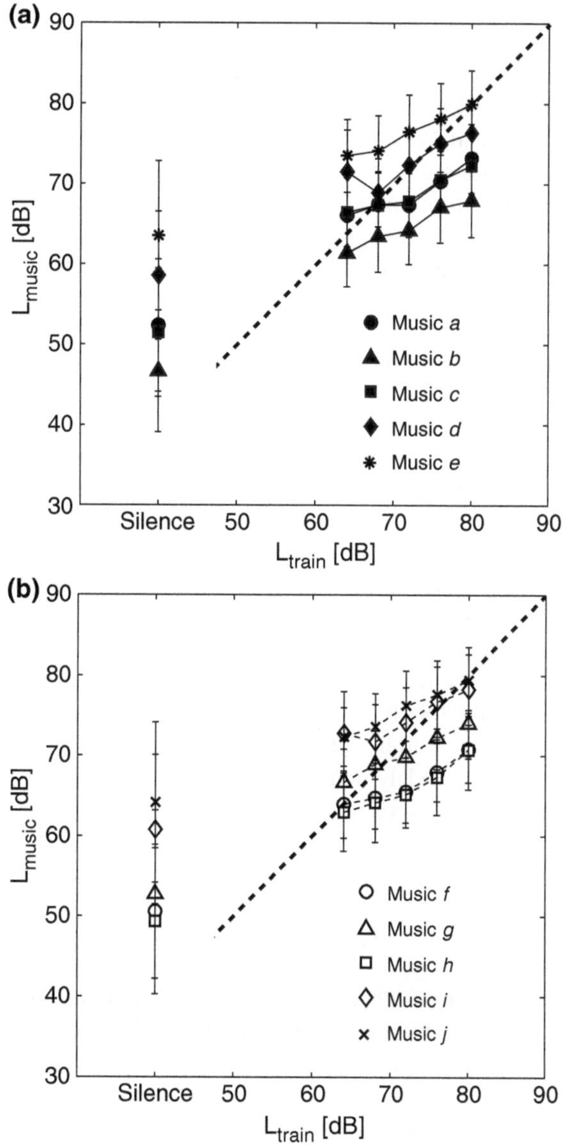

Fig. 10.11 L_{music} of **a** vocal and **b** instrumental music samples as a function of L_{train}

residual was 2.39 dB. The standardized regression coefficients for L_{train}, ϕ_1, and $W_{\phi(0)}$ were 0.81, −0.26, and −0.28, respectively.

The negative regression coefficient for ϕ_1 means participants listened to music at higher listening levels when ϕ_1 calculated from the music samples was smaller. A smaller ϕ_1 means that the perceived pitch was weak. The ϕ_1 of the music performed by a solo vocalist or a single instrument is large, but it becomes smaller when background music is added. In both vocal music and instrumental music, melody lines are assigned to the sounds of the voice and lead instrument, respectively. Therefore, it can be concluded that listeners turned up the volume when the pitch of the melody was difficult to catch. Some studies have reported the listening levels of rock music to be higher than those of pop music (Miyake and Kumashiro 1986; Dance and Wash 2008). The background music of rock usually contains strong drum beats and intensive electric guitar sounds. A listener of rock music may therefore turn up the volume to catch the vocals among such background music.

The negative regression coefficient for $W_{\phi(0)}$ means participants listened to the music at higher listening levels when $W_{\phi(0)}$ calculated from the music samples was narrower. A narrower $W_{\phi(0)}$ means that high spectral components were strong. The optimum listening level is high for music with strong high-frequency components (Iwamiya and Suzuki 1988). As noted above, it is possible that listeners will adjust listening levels to listen to the melody of music. In such a case, the sound is amplified at not only the desired but also all frequency ranges, and therefore, L_{music} increases. In contrast to ϕ_1 and $W_{\phi(0)}$, the τ_e of the music did not influence L_{music} significantly in this study. Indeed, loudness has been noted to increase as τ_e becomes longer and remains high for τ_e longer than 100 ms (Sato et al. 2001, 2002; Soeta et al. 2004, 2007; Soeta and Nakagawa 2008). In this study, the τ_e of Music *b*, *f*, *g*, and *h* was more than 100 ms (music with long τ_e), and τ_e of Music *a*, *c*, *d*, *e*, *i*, and *j* concentrated around 40 ms (music with short τ_e). A one-way analysis of variance revealed that the averaged L_{music} was significantly lower for music with long τ_e ($p < 0.05$). The averaged L_{music} values in music with long and short τ_e were 66.8 and 72.8 dB, respectively. Therefore, music with a longer τ_e, which has higher loudness, could be sufficiently heard at a low L_{music}. Thus, if music signals have τ_e that varies widely in a range between 10 and 100 ms, then an effect of τ_e on L_{music} may be observed.

Considering the insulation properties of headphones, such as those expressed in Eqs. (10.11) and (10.12) can be generalized as follows:

$$L_{\text{music}} = a_2 * L_{\text{train}}' + b_2 * \phi_1 + c_2 * W_{\phi(0)} + d_2. \qquad (10.13)$$

For the intra-concha headphones used in this study, a_2, b_2, c_2, and d_2 were 0.29, −11.90, −9.55, and 59.38, respectively. The correlation between the measured and calculated L_{music} for all listeners was also high ($r = 0.94$, $p < 0.01$), and their averaged residual was 2.39 dB. Inserting the L_{train}' and the ACF music factors ($\phi_1 = 0.43$ and $W_{\phi(0)} = 0.28$) from Experiment I, Eq. (10.13) could also estimate L_{music} with a high correlation ($r = 0.88$, $p < 0.01$) and low average residual (2.29 dB).

Although L_{train} was limited to 80 dBA in these experiments, the noise levels in actual train cars reach more than 80 dBA (Yoshihisa et al. 1968). In the case of these experiments, the L_{music} levels were 68–71 dB for Music *b* and 81–84 dB for Music *e*, according to Eq. (10.13). According to the WHO, the allowable exposure duration to a noise level of 85 dBA is 45 min over a 24-hr period (Berglund et al. 1999). This statistic suggests that listeners whose commute time is longer than 45 min should choose music with higher ϕ_1 and broader $W_{\phi(0)}$ to avoid the risk of hearing loss. If portable music devices included an installed microphone that measured environmental noise and a computer processor to calculate the ACF factors of the selected music, then a system that automatically adjusts the optimum listening level according to L_{train}', ϕ_1, and $W_{\phi(0)}$ values could be realized and help reduce the risk of hearing loss. Although the definite target L_{music} can be determined by these factors, it is difficult to adjust the L_{music} smoothly according to the precise moments of noise and music situations. Immediate changes of L_{music} can annoy the listener. Optimum smoothing is also an important task for AGCs incorporated into car audio systems (Behler et al. 2010). To develop AGC-incorporated portable music devices, optimum smoothing algorithms and the subjective evaluations of listeners should be studied.

The present study examined the listening levels of music (L_{music}) presented through headphones in the noisy environment of a train car for various types of train noise and various types of music. Furthermore, L_{music} was estimated according to physical factors extracted from the train noise and music. An overall finding of this study is that the listeners increase L_{music} as the train car noise level (L_{train}) rises. When the types of train noise (rolling, squeal, impact, and resonance) were varied, the effect was not significant. Therefore, L_{music} can be estimated by L_{train} with high accuracy. When the types of music (five vocal and five instrumental samples) were changed, the contributing ratio of music type (71.6 %) to L_{music} was larger than that of L_{train} (26.4 %). In this case, L_{music} can be estimated by L_{train} and the ACF factors ϕ_1 and $W_{\phi(0)}$, which correspond to the pitch strength and spectral centroid of the music, respectively. Hence, L_{music} depends on the music type and not the train noise type.

10.3 Subject Preference for Birdsongs

De Coensel et al. (2011) have found that introducing bird sounds could increase the pleasantness of a soundscape. Bird and water sounds have been usually evaluated as the most effective and favorable sounds to improve urban sound environments among the various natural sounds (Guastavino 2006; Jeon et al. 2010; Hong and Jeon 2013). But it has not clarified that which birdsongs are appropriate for a soundscape or person. To find the parameters that affect preference for birdsongs, subjective preference was determined and predicted by the ACF factors.

Eight stimuli were selected from 37 birdsongs. They are black paradise fly-catcher (BP), cuckoo (CC), Japanese grosbeak (JG), jay (JY), jungle nightjar (JN), little cuckoo (LC), oriental scops owl (OS), and ruddy kingfisher (RK). The stimuli were presented binaurally through headphones (HD650, Sennheiser). Twelve listeners aged from 20 to 39 year with no history of hearing disorders participated in the experiment. They sat in a comfortable thermal environment in a soundproof room and heard the auditory stimuli. Paired-comparison tests were performed for all combinations of pairs (i.e., 28 pairs (N(N − 1)/2, N = 8)) of stimuli, interchanging the order that the stimuli in each pair were presented in each session and presenting the pairs in random order. Six sessions were conducted for each participant. All stimuli were presented at the same SPL (60 dB). The duration of the stimuli was between 0.5 and 1.0 s, and the silent interval between stimuli was 1.0 s. After the presentation of the two stimuli, listeners were asked to judge which of the two sound signals were more preferred. The SVs of the preference were calculated according to Case V of Thurstone's theory (case V; Thurstone 1927).

The SVs of the preference for the eight birdsongs are presented in Fig. 10.12. The relationships between the SV of preference and each ACF factor are presented in Fig. 10.13. SV of preference was increased with increasing ϕ_1.

To calculate the effects of each objective factor on averaged SVs of the preference, multiple regression analyses were conducted using a linear combination of ACF factors and their variations by stepwise procedures. The final model indicated ϕ_1 and τ_1 were significant factors:

$$\mathrm{SV}_{\mathrm{preference}} \approx \mathrm{a}_1 * \phi_1 + \mathrm{a}_2 * \tau_1 + \mathrm{c}. \tag{10.14}$$

The relationships between the measured and calculated preference for different ACF factors are shown in Fig. 10.14. The correlation coefficient was 0.98 ($p < 0.01$). Standardized partial regression coefficients of each variable, a_1 and a_2 in Eq. (10.14), were 0.90 and 0.35, respectively. ϕ_1 has the greatest effect on subjective preference. Perceived preference increased with increasing ϕ_1 and τ_1, suggesting that birdsongs with higher regularity and lower pitch cause higher preference.

In a similar way, SVs of preference for each listener were predicted by multiple regression analyses using a linear combination of ACF factors and their variations by stepwise procedure. The significant factors, the standardized regression coefficients, and the correlation coefficients between measure and calculated SV of preference are listed in Table 10.3. Subjective preference for birdsongs of most listeners can be predicted by ϕ_1. The positive standardized regression coefficients suggest most listeners to prefer songbirds with higher regularity. Therefore, ACF factor, especially ϕ_1, may be helpful for selecting sounds in a soundscape.

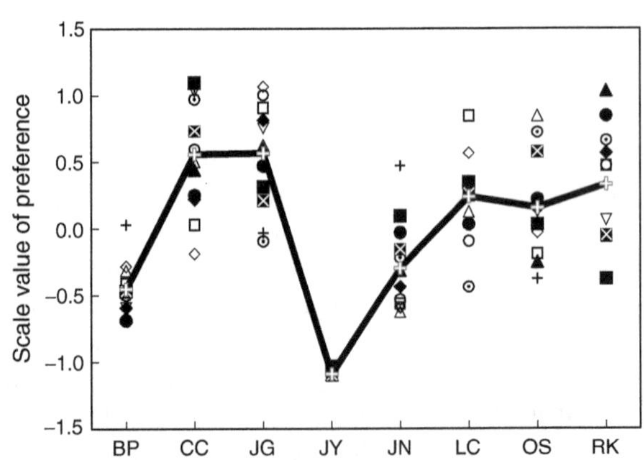

Fig. 10.12 The scale values of the preference for the eight birdsongs. Each symbol indicates each listener

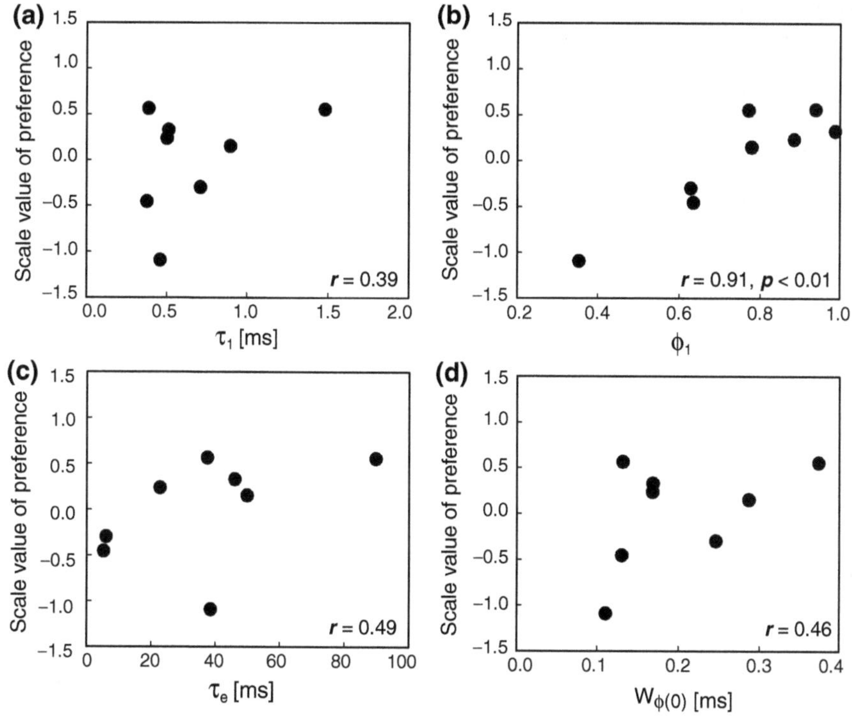

Fig. 10.13 The relationships between the scale value of preference and each ACF factor

Fig. 10.14 The relationships between the measured and calculated preference for ACF factor

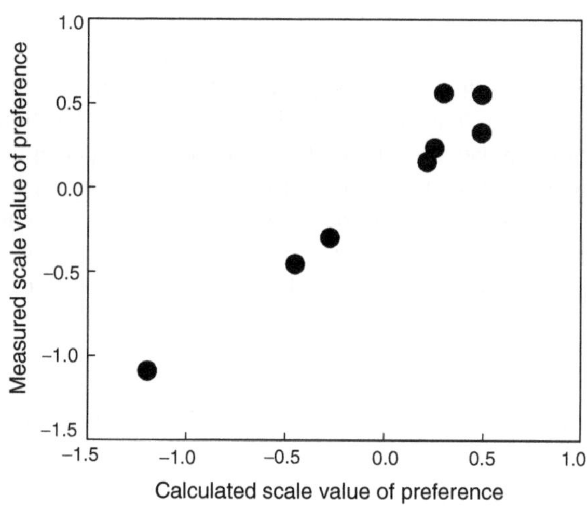

Table 10.3 Individual and total results for significant factors, the standardized regression coefficients, and the correlation coefficients between measure and calculated scale value of preference for birdsongs

Listener	Factor 1	Factor 2	r
A	ϕ_1 0.92		0.92
B			
C	ϕ_1 0.94		0.94
D	ϕ_1 0.97		0.97
E			
F	ϕ_1 0.74		0.74
G	ϕ_1 0.93		0.93
H	ϕ_1 0.94		0.94
I	ϕ_1 0.87		0.87
J	ϕ_1 0.98	VAR_τ_1 0.59	0.94
K			
L	ϕ_1 0.68	τ_1 0.51	0.94
Total	ϕ_1 0.90	τ_1 0.35	0.98

10.4 Urban Soundscape Design

It is notable that the definition of the term soundscape is itself not yet standardized. Based on the definition by the world soundscape project (Truax 2000), the working definition "the totality of all sounds within a location with emphasis on the relationship between individual's or society's perception of understanding of and interaction with the sonic environment" has been proposed (Davies et al. 2013).

Some researchers tried to find objective assessments for soundscape. Ratings of pleasantness and vibrancy for soundscape were not predicted well by any of the usual psychoacoustic variables, such as loudness, roughness, fluctuation strength, and sharpness, or variables based on averaged spectral shape (Hall et al. 2013). Selected 73 soundscape recordings were successfully separated to several groups with different factors extracted from the ACFs, fluctuation spectra, and so on (Chen and Zhao 2014). It suggests that ACF and IACF factors can assess soundscape by introducing other factors or ambisonic recordings for sound source separation (Elen 2001).

References

Ando Y (1977) Subjective preference in relation to objective parameters of music sound fields with a single echo. J Acoust Soc Am 62:1439–1441

Ando Y (1998) Architectural acoustics: blending sound sources, sound fields, and listeners. AIP Press Springer, New York

Ando Y (2002) Correlation factors describing primary and spatial sensations of sound fields. J Sound Vib 258:405–417

Ando Y, Alrutz H (1982) Perception of coloration in sound fields in relation to the autocorrelation function. J Acoust Soc Am 71:616–618

Ando Y, Cariani P (2009) Auditory and visual sensations. Springer, New York

Ando Y, Kurihara Y (1986) Nonlinear response in evaluating the subjective diffuseness of sound fields. J Acoust Soc Am 80:833–836

Ando Y, Sato S, Sakai H (1999) Fundamental subjective attributes of sound fields based on the model of auditory-brain system. In: Sendra JJ (ed) Computational acoustics in architecture. Southampton, WIT, pp 63–99

Barron M (1971) The subjective effects of first reflections in concert halls: the need for lateral reflections. J Sound Vib 15:475–494

Barron M, Marshall AH (1981) Spatial impression due to early lateral reflections in concert halls: the derivation of a physical measure. J Sound Vib 77:211–232

Bauer BB (1961) Stereophonic earphones and binaural loudspeakers. J Audio Eng Soc 9:148–151

Behler G, Guski M, Vorlander M (2010) Automatic loudness control. In: Proceedings of the 20th international congress on acoustics

Beranek LL (1996) Concert and opera halls: how they sound. Acoustical Society of America, New York

Berglund B, Lindvall T, Schwela DH (1999) Guidelines for community noise. World Health Organisation, Geneva

Bradley JS (1994) Comparison of concert hall measurements of spatial impression. J Acoust Soc Am 96:3525–3535

Bradley JS, Soulodre GA (1995) The influence of late arriving energy on spatial impression. J Acoust Soc Am 97:2263–2271

Bradley JS, Reich RD, Norcross SG (2000) On the combined effects of early- and late-arriving sound on spatial impression in concert halls. J Acoust Soc Am 108:651–661

Chen G, Zhao Y (2014) A new method for soundscape categorization on the basis of temporal parameters. In: Proceedings of the 21st international congress on sound and vibration, China National Convention Center, Beijing, 13–17 July 2014

Damaske P, Ando Y (1972) Interaural cross-correlation for multichannel loudspeaker reproduction. Acustica 27:232–238

Dance S, Wash P (2008) IPods listening levels on London underground. Proceedings of the Acoustics 08:3491–3495

Davies WJ, Adams MD, Bruce NS, Cain R, Carlyle A, Cusack P, Hall DA, Hume KI, Irwin A, Jennings P, Marselle M, Plack CJ, Poxon J (2013) Perception of soundscapes: an interdisciplinary approach. Appl Acoust 74:224–231

De Coensel B, Vanwetswinkel S, Botteldooren D (2011) Effects of natural sounds on the perception of road traffic noise. J Acoust Soc Am 129:EL148–EL153

Elen R (2001) Ambisonics: the surround alternative. In: Proceedings of the third annual surround conference and technology showcase, pp 1–4

Gerzon MA (1973) Periphony: with-height sound reproduction. J Audio Eng Soc 21:2–10

Guastavino C (2006) The ideal urban soundscape: investigating the sound quality of French cities. Acta Acust Acust 92:945–951

Hall DA, Irwin A, Edmondson-Jones M, Phillips S, Poxon JEW (2013) An exploratory evaluation of perceptual, psychoacoustic and acoustical properties of urban soundscapes. Appl Acoust 74:248–254

Heyde N (2008) Loudness control of level in automotive infotainment system. Diploma Thesis Institute of Technical Acoustics, RWTH Aachen University
Hidaka T, Beranek LL (2000) Objective and subjective evaluations of twenty-three opera houses in Europe, Japan, and the Americans. J Acoust Soc Am 107:368–383
Hidaka T, Beranek LL, Okano T (1995) Interaural cross-correlation, lateral fraction, and low- and high-frequency sound levels as measures of acoustical quality in concert halls. J Acoust Soc Am 98:988–1007
Hirahara T, Aoyama H, Otani M (2010) Acoustic characteristics of earphones and issues of IEC60711 coupler. J Acoust Soc Jpn 66:45–65 (in Japanese)
Hodgetts WE, Rieger JM, Szarko RA (2007) The effects of listening environment and earphone style on preferred listening levels of normal hearing adults using an MP3 player. Ear Hear 28:290–297
Hong JY, Jeon JY (2013) Designing sound and visual components for en-hancement of urban soundscapes. J Acoust Soc Am 134:2026–2036
International Organization for Standardization ISO 3382-1:2009 Acoustics-measurement of room acoustics parameters part 1: performance spaces
Iwamiya S, Suzuki T (1988) Optimum listening level of music. Ann Physiol Anthropol 7:175–182
Jeon JY, Lee PJ, You J, Kang J (2010) Perceptual assessment of quality of urban soundscapes with combined noise sources and water sounds. J Acoust Soc Am 127:1357–1366
Khatter K (2011) Personal music players and hearing loss: are we deaf to the risks? Open Med 5: e137–e138
Kirkeby O, Nelson PA, Hamada H (1998) The "stereo dipole"? A virtual source imaging system using two closely spaced loudspeakers. J Audio Eng Soc 46:387–395
Miyake S, Kumashiro M (1986) Effects of listening to music with headphones on hearing? Especially under noisy conditions? J UOEH 8:391–404
Mosteller F (1951) Remarks on the method of paired comparisons III. Psychometrika 16:207–218
Nakajima T, Yoshida J, Ando Y (1993) A simple method of calculating the interaural cross-correlation function for a sound field. J Acoust Soc Am 93:885–891
Okano T, Beranek LL, Hidaka T (1998) Relations among interaural cross-correlation coefficient (IACCE), lateral fraction (LFE), and apparent source width (ASW) in concert halls. J Acoust Soc Am 104:255–265
Remington PJ (1976) Wheel/rail noise? Part I: characterization of the wheel/rail dynamic system. J Sound Vib 46:359–379
Sato S, Ando Y (2002) Apparent source width (ASW) of complex noises in relation to the interaural cross-correlation function. J Temporal Des Arch. Environ 2:29–32
Sato S, Kitamura T, Sakai H, Ando Y (2001) The loudness of "complex noise" in relation to the factors extracted from the autocorrelation function. J Sound Vib 241:97–103
Sato S, Kitamura T, Ando Y (2002) Loudness of sharply (2068 dB/Octave) filtered noises in relation to the factors extracted from the autocorrelation function. J Sound Vib 250:47–52
Schroeder MR (1970) Digital simulation of sound transmission in reverberant space. J Acoust Soc Am 47:424–431
Shimokura R, Soeta Y (2011) Characteristics of train noise in ground and underground stations with side and island platforms. J Sound Vib 330:1621–1633
Shimokura R, Soeta Y (2012) Listening level of music through headphones in train car noise environments. J Acoust Soc Am 132:1407–1416
Shimokura R, Tronchin L, Cocchi A, Soeta Y (2011) Subjective diffuseness of music signals convolved with binaural impulse responses. J Sound Vib 330:3526–3537
Singh PK, Ando Y, Kurihara Y (1994) Individual subjective diffuseness responses of filtered noise sound fields. Acustica 80:471–477
Soeta Y, Nakagawa S (2008) Effect of the repetitive components of a noise on loudness. J Temporal Des Arch Environ 8:1–7
Soeta Y, Shimokura R (2009) Comparison of noise characteristics in airplanes and high-speed trains. J Temporal Des Arch Environ 9:22–25

Soeta Y, Maruo T, Ando Y (2004) Annoyance of bandpass filtered noises in relation to the factor extracted from autocorrelation function. J Acoust Soc Am 116:3275–3278
Soeta Y, Yanai K, Nakagawa S, Kotani K, Horii K (2007) Loudness in relation to iterated rippled noise. J Sound Vib 304:415–419
Suzuki Y, Sone T, Kanasashi H, Mafune Y, Soma J (1982) The optimum level of music listened to in the presence of noise. J Acoust Soc Jpn 3:55–65
Thurstone LL (1927) A law of comparative judgement. Psychol Rev 34:273–289
Truax B (2000) Acoustic communication, 2nd edn. Greenwood Publishing Group, Santa Barbara
Yoshihisa N, Fukai M, Yamaguchi T (1968) Noise levels inside subway cars in Western Europe, U.S.A. and Japan. J Acoust Soc Jpn 24:69–75 (in Japanese)
Yost WA (1996) Pitch strength of iterated rippled noise. J Acoust Soc Am 100:3329–3335

Index

Y. Soeta and Y. Ando, *Neurally Based Measurement and Evaluation of Environmental Noise*, Mathematics for Industry 20,
DOI 10.1007/978-4-431-55432-5

MIX
Papier aus verantwortungsvollen Quellen
Paper from responsible sources
FSC® C105338

If you have any concerns about our products,
you can contact us on
ProductSafety@springernature.com

In case Publisher is established outside the EU,
the EU authorized representative is:
Springer Nature Customer Service Center GmbH
Europaplatz 3, 69115 Heidelberg, Germany

Printed by Libri Plureos GmbH
in Hamburg, Germany